GALOIS THEORIES

T0276144

Books in this series

Galois Theories

Francis Borceux
Université Catholique de Louvain

George Janelidze
Georgian Academy of Sciences, Tbilisi

CAMBRIDGE
UNIVERSITY PRESS

CAMBRIDGE UNIVERSITY PRESS
Cambridge, New York, Melbourne, Madrid, Cape Town, Singapore, São Paulo

Cambridge University Press
The Edinburgh Building, Cambridge CB2 8RU, UK

Published in the United States of America by Cambridge University Press, New York

www.cambridge.org
Information on this title: www.cambridge.org/9780521803090

First published 2001
This digitally printed version 2008

A catalogue record for this publication is available from the British Library

Library of Congress Cataloguing in Publication data
Borceux, Francis, 1948–
Galois theories / F. Borceux, G. Janelidze.
p.   cm. – (Cambridge studies in advanced mathematics; 72)
Includes bibliographical references and index.
ISBN 0 521 80309 8
1. Galois theory.   I. Janelidze, G. (George), 1952–
II. Title.   III. Series.
QA214.B67   2001
512′.3–dc21   00-04863   CIP

ISBN 978-0-521-80309-0 hardback
ISBN 978-0-521-07041-6 paperback

Contents

Preface

E. Galois (1811–1832) would certainly be surprised to see how often his name is mentioned in the mathematical books and articles of the twentieth century, in topics which are so far from his original work.

Since antiquity, mathematicians have been able to solve polynomial equations of degree 1 or 2. Formulæ for solving the equations of degree 3 or 4 were found during the sixteenth century. But it was only during the nineteenth century that the problem of equations of higher degree reached a final answer: the impossibility of solving by radicals a general equation of degree at least 5, and some methods for finding some solutions by radicals when these exist. Galois and Abel certainly played a key role in the development of this theory. All these results can be found in almost every book on Galois theory ... and that's the reason why we considered it useless to present them once more in the present book.

A strong peculiarity of those developments about solving equations is that the methods used to reach the final goal proved to be much more interesting than the problem to be solved. Nobody uses the formulæ for solving cubic or quartic equations ... but their consideration forced the discovery of complex numbers. And the impossibility proof for equations of higher degree led to specifying the notion of group, on the interest of which it is unnecessary to comment.

Let us now sketch, in modern language, the central result used by Galois to prove his celebrated theorem. A field extension $K \subseteq L$ is a Galois extension when every element of L is the root of a polynomial $p(X) \in K[X]$ which factors in $L[X]$ into linear factors and all of whose roots are simple. The Galois group $\mathsf{Gal}\,[L : K]$ of this extension is the group of all field automorphisms of L which fix all the elements of K. The classical Galois theorem asserts that when $K \subseteq L$ is a finite dimensional Galois extension, the subgroups $G \subseteq \mathsf{Gal}\,[L : K]$ of the Galois group classify exactly the intermediate field extensions $K \subseteq M \subseteq L$. An elementary presentation of this classical Galois theory is given in chapter 1.

The environment for the theory named after Galois has been extended and changed many times, as displayed in the rough scheme of key words on page ix. Moreover, even this scheme would be incomplete if one tried to include the Galois descent and cohomology, Tannaka duality in connection with the Grothendieck motivic Galois theory, and some other closely related topics.

The solid arrows in the scheme represent generalizations of various constructions and results, and the dotted arrows represent "inspirations". Probably each of those arrows would be a good subject for a whole book, and so no reasonable description of them in a few pages can be produced.

The first step of generalization of the classical Galois theory is to replace the intermediate field extensions $K \subseteq M \subseteq L$ by more general commutative algebras over the field K. Given a field extension $K \subseteq L$, a K-algebra A is split by L when each element of A is the root of a polynomial $p(X) \in K[X]$ which factors in $L[X]$ into distinct linear factors. Of course, when $K \subseteq L$ is a Galois extension, every intermediate field extension $K \subseteq M \subseteq L$ is a K-algebra split by L. Conversely every intermediate algebra $K \subseteq A \subseteq L$ is necessarily a field.

Chapter 2, inspired by the work of Grothendieck, proves that the classical Galois theorem of chapter 1 is a local segment of a more general equivalence of categories. Given a finite dimensional Galois field extension $K \subseteq L$, the Galois theorem asserts now that the category of finite dimensional K-algebras split by L is equivalent to the category of finite $\mathsf{Gal}\,[L : K]$-sets, that is, finite sets provided with an action of the Galois group. The classical Galois theorem of chapter 1 is recaptured by observing that the subgroups $G \subseteq \mathsf{Gal}\,[L : K]$ correspond bijectively with the quotients of the Galois group in the category of $\mathsf{Gal}\,[L : K]$-sets; via the equivalence of categories, these are in bijection with the split algebras $K \subseteq A \subseteq L$, which turn out to be the intermediate fields.

In chapter 3 we handle the case of an arbitrary Galois extension of fields $K \subseteq L$, not necessarily finite dimensional. In that case, the Galois group $\mathsf{Gal}\,[L : K]$ comes naturally equiped with a profinite topology, that is, the structure of a compact Hausdorff space whose topology admits a base of closed–open subsets. The classical version of the Galois theorem asserts now that the closed subgroups of the profinite Galois group

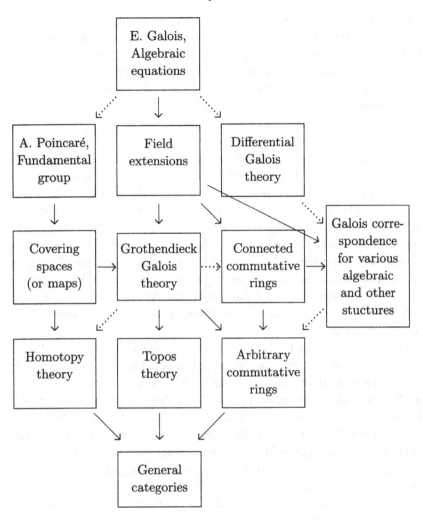

Scheme 1: The contexts of Galois theories

classify the intermediate extensions $K \subseteq M \subseteq L$. The Galois theorem of Grothendieck extends in an analogous way, yielding now an equivalence between the category of K-algebras split by L and the category of profinite $\mathsf{Gal}\,[L : K]$-spaces, with continuous action of the Galois group.

The next and very crucial generalization is to replace the ground field K by a commutative ring R and thus develop the Galois theory for rings. The notion of Galois extension of commutative rings was first defined (independently of A. Grothendieck) by M. Auslander and O. Goldman (see [3]), and the Galois theory of those extensions was developed by S.U. Chase, D.K. Harrison, and A. Rosenberg (see [21]) and G.J. Janusz (see [54]), and many others (see the references in [24]). A.R. Magid in [67] develops the Grothendieck Galois theory of commutative rings in full generality.

Passing from fields to rings requires introducing a new ingredient: the spectrum of the ring. The idempotent elements $e = e^2$ of a commutative ring S constitute a boolean algebra, for the operations $e \wedge e' = ee'$ and $e \vee e' = e + e' - ee'$. The spectrum of this boolean algebra, constituted of the set of its ultrafilters with the Stone topology, is called the *Pierce spectrum* of the ring S. It is a profinite space, that is, a compact Hausdorff space whose topology is generated by its closed-open subsets. Of course since a field K admits 0 and 1 as its only idempotents, its Pierce spectrum is a singleton, which explains why the spectrum of a field never appeared up to now. Nevertheless, given a Galois field extension $K \subseteq L$, the K-algebra $L \otimes_K L$ has non-trivial idempotents; the ring case will show evidence that these idempotents determine the Galois group of the extension.

Every morphism of fields is injective, thus mentioning a field extension is the same as mentioning a morphism of fields. The Galois theory of commutative rings will be developed for Galois morphisms of rings $\sigma \colon R \longrightarrow S$, also called morphisms of Galois descent. Here *descent* refers to σ being an effective descent morphism in the dual of the category of commutative rings, that is, pulling back along this morphism in the dual of the category of rings yields a monadic functor between the corresponding slice categories. In other words, tensoring along this morphism in the category of rings yields a comonadic situation. The definition of σ being of *Galois* descent, and also the definition of being an R-algebra split by σ, are given in terms of the Pierce spectrum functor and its adjoint, which maps a profinite space X onto the ring $\mathcal{C}(X, R)$ of continuous functions from the space X to the ring R provided with the discrete topology. In fact the adjunction formed by Spec and $\mathcal{C}(-, R)$ localizes as an adjunction between the categories of S-algebras and that of profinite spaces over Spec(S); an R-algebra A is split by σ when the counit of this adjunction is an isomorphism at the S-algebra $S \otimes_R A$. And σ is of *Galois* descent when for each profinite space over Spec(S), the corresponding S-algebra given by adjunction, seen as an R-algebra, is split by σ. These categorical definitions generalize the classical corresponding notions in the case of fields and constitute the core of a categorical approach to Galois theory. The interested reader will find in section A.1 a historical discussion throwing light on the evolution of the notions, from the classical algebraic case to the categorical one.

But the major difference with the case of fields is the replacement of the Galois group by a Galois groupoid. The groupoids appear in Galois theory of commutative rings first in the papers of O.E. Villamayor and D. Zelinsky (see [73] and [74]). A.R. Magid uses what he calls Galois groupoids instead of Galois groups, and those groupoids have a profinite topology which makes them not equivalent (as topological groupoids) to any kind of a topological family of groups – unlike the ordinary groupoids (although R. Brown has good reasons to say that even ordinary groupoids should never be replaced by families of groups!).

Let us now explain how groupoids enter the story. Given a Galois descent morphism $\sigma \colon R \longrightarrow S$, the objects of the Galois groupoid are the elements of the spectrum of S, while the arrows are the elements of the spectrum of the cokernel pair of σ. The construction of the spectrum is functorial and contravariant; the two canonical morphisms from S to the cokernel pair of σ define the "domain" and "codomain" operations of the Galois groupoid. Observe that by definition, this groupoid lives in the category of profinite spaces. Again, in the case of fields, the spectrum of S is reduced to a singleton, thus the groupoid has a single object and therefore is a group: it is exactly the classical Galois group.

There is a last piece of the puzzle to generalize: the profinite spaces on which the topological Galois group acts. A set on which a group acts is exactly a presheaf on that group considered as a one object category. Thus the category to consider here will be that of internal presheaves on the internal Galois groupoid, in the category of profinite spaces. The Galois theorem for rings asserts that this category of internal presheaves on the Galois groupoid is equivalent to the category of R-algebras split by σ. Chapter 4 is devoted to developing this Galois theory of rings.

Chapter 5 is the core of this book. Taking our inspiration from the situation for rings, we first formalize the categorical context in which a general Galois theorem holds, and then give some applications. The Pierce spectrum functor between the category of rings and that of profinite spaces is replaced by an arbitrary functor between arbitrary categories. We exhibit the assumptions required to infer a Galois theorem proving an equivalence of categories between "split algebras" for this adjunction and the internal presheaves on some internal Galois groupoid. The main sources of inspiration for this are [36], [37], [39], [41].

The central extensions of groups are not usually considered as a part of

Galois theory, and therefore they are not included in the scheme above. However, they turn out to be precisely the objects split over extensions in a certain "non-Grothendieck" special case of categorical Galois theory, as explained in section 5.2.

Next we devote attention to another particularization of the general Galois theory of section 5.1: the case of semi-left-exact reflections, of which the situation of chapter 4 is a special case. Such an adjunction is given by a full reflective subcategory $r \dashv i \colon \mathcal{R} \xrightarrow{\longleftarrow} \mathcal{C}$ such that for each object $C \in \mathcal{C}$, one keeps a full reflective subcategory $r_C \dashv i_C \colon \mathcal{R}/r(C) \xrightarrow{\longleftarrow} \mathcal{C}/C$ by localizing the situation over the object C. This property holds in particular when the reflection r is left exact, from which we borrowed the terminology. We further particularize this study in a topological context. Eilenberg and Whyburn have studied the monotone–light factorization of a continuous map. Our Galois theory allows an elegant treatment of various aspects of this theory in the general context of compact Hausdorff spaces. We exhibit in particular Galois descent morphisms related to the Stone–Čech compactification.

Chapter 6 focuses on the notion of covering map of topological spaces, defined classically as a continuous map $f \colon A \longrightarrow B$ such that every point in B has an open neighbourhood U whose inverse image is a disjoint union of open subsets, each of which is mapped homeomorphically onto U by f. One usually says that $A = (A, f)$ is a covering space over B (or of B) when $f \colon A \longrightarrow B$ is a covering map.

If B is connected and locally connected, and has a universal (i.e. the "largest" connected) covering (E, p), then all connected coverings of B are quotients of (E, p) and there is a bijection between (the isomorphic classes of) them and the subgroups of the automorphism group $\mathsf{Aut}(E, p)$. That bijection is constructed precisely as the standard Galois correspondence for separable field extensions, but in the dual category $(\mathsf{Top}/B)^{\mathsf{op}}$ of bundles over B.

Actually this result appears in most books on algebraic topology only in the special case when the Chevalley fundamental group $\mathsf{Aut}(E, p)$ is isomorphic to the usual Poincaré fundamental group of B. The general case was first studied by C. Chevalley (see [23], where however the Galois correspondence is not explicitly mentioned) who actually called $\mathsf{Aut}(E, p)$ the Poincaré group.

Finally, in chapter 7, we show first that it is possible to get a Galois theorem in the general context of descent theory, without necessarily a Galois assumption. This yields in particular a Galois theorem for every field extension $K \subseteq L$, without any further assumption. The price to pay is that the Galois group or the Galois groupoid must now be replaced by the more general notion of precategory. But in some cases of interest, even without any Galois assumption, the Galois precategory turns out to be again a groupoid. For example, the Galois theorem for toposes, due to Joyal and Tierney, enters the context of our generalized theory, without any Galois assumption. It asserts that every topos is equivalent to a category of étale presheaves on an open étale groupoid.

This book originated in a French manuscript that the first author wrote for his students. The second author convinced him to transform these notes into an actual book in English, and offered later to write an additional chapter, which became chapter 6 of the present book. Of course, most of the material of chapters 4, 5 and 7 is originally also due to the second author and his coauthors, and certainly the role of these coauthors in the genesis of the results should be emphasized here.

We tried to make this book accessible to a wide audience. First courses in algebra and general topology, together with some familiarity with the categorical notions of limit and adjoint functors, are sufficient to read it. The first chapters require even fewer prerequisites.

We thank all those who helped us or supported us in the preparation of this book, even if we cannot cite all of them. We cite first the participants in the Louvain-la-Neuve category seminar, with whom the original French version of this work was first discussed. Among them, Gilberte van den Bossche is worth a special mention, and the first author wants to dedicate this book to her memory. We thank also R. Brown, A. Carboni, G.M. Kelly, A.R. Magid, R. Paré, D. Schumacher, R.H. Street, W. Tholen, and all others whose ideas and proofs have provided a big part of the material in this book. The second author wants to add that most of his joint work with the persons just cited was carried out during his various visits to them and supported by their universities and national councils, especially of Australia, Belgium, Canada, Italy and the UK.

Finally we express our special thanks to René Lavendhomme and Hvedri Inassaridze who were our teachers a long time ago, and to Saunders Mac Lane, since like most category-theorists in the world, we feel we are his students too – but also for his important particular advice concerning categorical Galois theory.

1

Classical Galois theory

Convention *In this chapter, all fields we consider are commutative.*

This chapter develops the basic aspects of Galois theory for fields. We study first the notions of algebraic, separable, normal and Galois field extensions, before exhibiting the Galois correspondence between Galois extensions and field automorphisms.

1.1 Algebraic extensions

Given a field extension $K \subseteq L$, the composite

$$K \times L \rightarrowtail L \times L \xrightarrow{\times} L$$

provides L with the structure of a vector space on K, thus also of a K-algebra. We write $[L : K]$ or $\dim [L : K]$ for the dimension of L as a K-vector space.

Let us recall that every proper ideal $I \subsetneq L$ of a field L is necessarily trivial. Indeed, if I contains a non zero element i, then $l = ii^{-1}l \in I$ for every $l \in L$, thus $I = L$. Since the kernel of a field homomorphism is an ideal not containing 1, this ideal must be zero, proving that every field homorphism is injective.

We write $K[X]$ for the ring of polynomials with coefficients in the field K.

Definition 1.1.1 Let $K \subseteq L$ be a field extension. An element $l \in L$ is algebraic over K when there exists a non-zero polynomial $p(X) \in K[X]$ such that $p(l) = 0$. The extension $K \subseteq L$ is algebraic when all elements of L are algebraic over K.

Proposition 1.1.2 *Every finite dimensional field extension is algebraic.*

Proof   Given $l \in L$, the sequence of elements $1, l, l^2, l^3, \ldots, l^n, \ldots$ necessarily yields a dependence relation

$$a_n l^n + a_{n-1} l^{n-1} + \cdots + a_2 l^2 + a_1 l + a_0 = 0, \quad a_i \in K,$$

since $[L : K]$ is finite. Putting

$$p(X) = a_n X^n + a_{n-1} X^{n-1} + \cdots + a_2 X^2 + a_1 X + a_0$$

yields $p(l) = 0$.   □

Proposition 1.1.3 *Let $K \subseteq L$ be a field extension and $l \in L$ an element which is algebraic over K. There exists a unique polynomial $p(X) \in K[X]$ such that*

(i) *the leading coefficient of $p(X)$ is 1,*
(ii) *$p(l) = 0$,*
(iii) *the degree of $p(X)$ is minimal among the polynomials $q(X) \in K[X]$ satisfying $q(l) = 0$.*

This polynomial $p(X)$ is irreducible and is called the minimal polynomial of l. When $q(X) \in K[X]$ and $q(l) = 0$, $p(X)$ divides $q(X)$.

Proof   By 1.1.1, we choose a polynomial $p(X) \in K[X]$ of minimal degree among those satisfying $p(l) = 0$; there is of course no restriction in assuming that its leading coefficient is 1. If $p(X)$ could be decomposed as the product of two polynomials in $K[X]$, then l would be a root of one of these; the minimality of the degree of $p(X)$ implies therefore that $p(X)$ is irreducible in $K[X]$.

If $q(X)$ is a polynomial as in the statement, we consider the euclidean division of $q(X)$ by $p(X)$, yielding

$$q(X) = p(X)\alpha(X) + r(X), \quad \text{degree of } r(X) < \text{degree of } p(X).$$

Since $q(l) = p(l) = 0$, it follows that $r(l) = 0$ and by minimality of the degree of $p(X)$, we get $r(X) = 0$.

The uniqueness of $p(X)$ satisfying those conditions follows at once from the last statement and condition (i).   □

Proposition 1.1.4 *In the conditions of proposition 1.1.3, the smallest subfield $K(l)$ of L containing K and l is isomorphic to the quotient $K[X]/\langle p(X) \rangle$, where $\langle p(X) \rangle \subseteq K[X]$ is the principal ideal generated by $p(X)$. Moreover, the dimension $[K(l) : K]$ of the extension $K(l)$ equals the degree of the minimal polynomial $p(X)$ of l.*

Proof Writing

$$p(X) = X^n + a_{n-1}X^{n-1} + \cdots + a_1 X + a_o$$

it follows at once that we have an isomorphism

$$\frac{K[X]}{\langle p(X) \rangle} \cong \left\{ k_{n-1}X^{n-1} + \cdots + k_1 X + k_0 \middle| k_i \in K \right\}$$

where the operations on the right hand side are defined modulo the relation

$$X^n = -a_{n-1}X^{n-1} - \cdots - a_1 X - a_0.$$

Observe that the dimension of $K[X]$ over K is indeed n, the degree of $p(X)$.

The subalgebra $K(l) \subseteq L$ generated by K and l is clearly given by

$$K(l) = \left\{ q(l) \middle| q(X) \in K[X] \right\}.$$

Using the minimal polynomial of l, every occurrence of l^n can be replaced by terms of lower degree, thus

$$K(l) = \left\{ k_{n-1}l^{n-1} + \cdots + k_0 \middle| k_i \in K \right\}.$$

This is a K-vector space of dimension at most n. Multiplying by $a \in K(l) \subseteq L$, $a \neq 0$, is a K-linear endomorphism of $K(l)$, which is injective since L is a field. By finiteness of the dimension of $K(l)$, this endomorphism is a linear isomorphism, proving that a is invertible in $K(l)$. Thus the subalgebra $K(l) \subseteq L$ is in fact the subfield $K(l) \subseteq L$ generated by K and l.

Next we consider the ring homomorphism

$$\gamma \colon \frac{K[X]}{\langle p(X) \rangle} \longrightarrow K(l), \quad q(X) \mapsto q(l), \quad \text{degree } q(X) < n.$$

This K-linear map is injective since $q(l) = 0$ implies $q(X) = 0$, because the degree of $q(X)$ is stricly less than the degree of the minimal polynomial $p(X)$ (see 1.1.3). Therefore it is bijective, because the first space has dimension n and the second has dimension at most n. Thus γ is an isomorphism. $\square$

Definition 1.1.5 Let $K \subseteq L$ be a field extension. Two elements $l_1, l_2 \in L$ are conjugate over K when they are algebraic over K and have the same minimal polynomial.

A link with the usual notion of conjugate complex numbers should certainly be exhibited. This is the object of the following example.

Example 1.1.6 Consider the field extension $\mathbb{R} \subseteq \mathbb{C}$. For every complex number $l = a + b\,i$ which is not already a real, that is $b \neq 0$, one gets at once

$$p(X) = \big(X - (a + b\,i)\big)\big(X - (a - b\,i)\big) = X^2 - 2aX + (a^2 + b^2) \in \mathbb{R}[X]$$

which is an irreducible polynomial over $\mathbb{R}$, since it is a product of linear polynomials not in $\mathbb{R}[X]$. The degree 2 of $p(X)$ is certainly minimal in $\mathbb{R}[X]$ for allowing $p(a + b\,i) = 0$, since $a + b\,i$ is not a real. Therefore $p(X)$ is the minimal polynomial of $a + b\,i$ (see 1.1.3) and $a + b\,i$, $a - b\,i$ are conjugate complex numbers over the reals.

Definition 1.1.7 Let $K \subseteq L$ be a field extension. A field homomorphism $f \colon L \longrightarrow L$ is called a K-homomorphism when it fixes all elements of K, that is, $f(k) = k$ for every element $k \in K$.

Proposition 1.1.8 *Let $K \subseteq L$ be an algebraic field extension. Then every K-endomorphism of L is necessarily an automorphism. We shall write $\mathrm{Aut}_K(L)$ for the group of K-automorphisms of L.*

Proof Consider $l \in L$ with minimal polynomial $p(X)$ over K. For every K-endomorphism f of L, one has

$$p\big(f(l)\big) = f\big(p(l)\big) = f(0) = 0$$

which proves that $f(l)$ is conjugate to l over K. Therefore f induces a map

$$f_l \colon \{l' \in L | p(l') = 0\} \longrightarrow \{l' \in L | p(l') = 0\}, \quad l' \mapsto f(l').$$

The set on which f_l acts is finite, as the set of roots of $p(X)$ in L. Since f is injective as a field homomorphism, f_l is injective as well and thus surjective, because it acts on a finite set. In particular $l = f_l(l') = f(l')$ for some conjugate l' of l, which proves the surjectivity of f. $\qquad\square$

1.2 Separable extensions

Let us recall that the derivative of a polynomial

$$p(X) = a_n X^n + a_{n-1} X^{n-1} + \cdots + a_2 X^2 + a_1 X + a_0$$

in $K[X]$ is the polynomial

$$p'(X) = na_n X^{n-1} + (n-1)a_{n-1}X^{n-2} + \cdots + 2a_2 X + a_1.$$

The classical formulæ for the derivative of a sum and a product are valid for polynomials over an arbitrary field.

Remark 1.2.1 With the previous notation, when $a_n \neq 0$, $p(X)$ has degree n. The corresponding coefficient na_n of the derivative vanishes when $n = 0$ in K, that is, when the characteristic of the field K divides n. Thus $p'(X)$ has degree $n - 1$ if and only if the characteristic of K does not divide n. This is the place to recall that the exponents of a polynomial in $K[X]$ are natural numbers, not elements of K; when computing a derivative, the exponent n which is a natural number enters the coefficients of the derivative in the form $1 + \cdots + 1$ (n times), $1 \in K$, which is an element of K, possibly 0.

Proposition 1.2.2 *Consider a field K, an element $a \in K$ and a polynomial $p(X) \in K[X]$. The following conditions are equivalent:*

(i) *a is a multiple root of $p(X)$;*
(ii) *$p(a) = 0$ and $p'(a) = 0$.*

Proof Assuming (i), we can write $p(X) = (X-a)^k q(X)$ with $k \geq 2$. Therefore

$$p'(X) = k(X-a)^{k-1}q(X) + (X-a)^k q'(X).$$

Since $k - 1 \geq 1$, this implies $p'(a) = 0$.

Conversely, $p(X) = (X-a)q(X)$ since a is a root of $p(X)$, thus

$$p'(X) = q(X) + (X-a)q'(X).$$

Putting $X = a$, there remains $q(a) = 0$ since $p'(a) = 0$. This implies $q(X) = (X-a)s(X)$ and thus $p(X) = (X-a)^2 s(X)$. $\qquad\square$

Definition 1.2.3 A field extension $K \subseteq L$ is separable when

(i) the extension is algebraic,
(ii) the roots of the minimal polynomial of every $l \in L$ are all simple.

Proposition 1.2.4 *Let $K \subseteq L$ be a field extension in characteristic zero. If $l \in L$ is algebraic over K, all roots of the minimal polynomial of l over K are simple.*

Proof Write $p(X)$ for the minimal polynomial of l; it suffices to verify
that l is a simple root of $p(X)$. If l is a multiple root of $p(X)$, then $p(X)$
has degree at least 2 and $p'(l) = 0$ (see 1.2.2), with $p'(X)$ a polynomial
of degree at least 1 (see remark 1.2.1). This contradicts the minimality
condition in 1.1.3.   □

Corollary 1.2.5 *In characteristic zero, every algebraic extension is separable.*   □

Proposition 1.2.6 *Let $K \subseteq M \subseteq L$ be field extensions. If $K \subseteq L$ is
separable, then $M \subseteq L$ is separable as well.*

Proof Every $l \in L$ has a minimal polynomial $p(X) \in K[X]$. Since
$K[X] \subseteq M[X]$, the extension $M \subseteq L$ is algebraic. The minimal polynomial $q(X) \in M[X]$ of l is a factor of $p(X)$ in $M[X]$, thus all its roots in
L are distinct.   □

1.3 Normal extensions

Definition 1.3.1 A field extension $K \subseteq L$ is normal when

(i) the extension is algebraic,
(ii) for every element $l \in L$, the minimal polynomial of l over K
     factors entirely in $L[X]$ in polynomials of degree 1.

An algebraic field extension $K \subseteq L$ with L algebraically closed is thus
necessarily normal.

Proposition 1.3.2 *Let $K \subseteq M \subseteq L$ be field extensions. If $K \subseteq L$ is
normal, then $M \subseteq L$ is normal as well.*

Proof Every $l \in L$ has a minimal polynomial $p(X) \in K[X]$. Since
$K[X] \subseteq M[X]$, the extension $M \subseteq L$ is algebraic. The minimal polynomial $q(X) \in M[X]$ of l divides $p(X)$ in $M[X]$, thus factors in $L[X]$ into
polynomials of degree 1, since so does $p(X)$.   □

Proposition 1.3.3 *Let $K \subseteq L$ be a normal, finite dimensional field
extension. For every intermediate field extension $K \subseteq M \subseteq L$, every
K-homomorphism $M \longrightarrow L$ extends to a K-automorphism of L.*

Proof Consider the situation

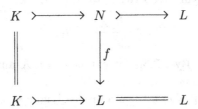

where f is a K-homomorphism between arbitrary field extensions; we want to prove that f can be extended to $N(l)$, for an arbitrary element $l \in L$. If $N = L$, we are done; otherwise choose $l \in L$, $l \notin N$, with minimal polynomial $p(X) \in K[X]$. Since $[L : N]$ is finite, the extension $N \subseteq L$ is algebraic (see 1.1.2). Let $q(X) \in N(X)$ be the minimal polynomial of l over N. It follows by 1.1.3 that $q(X)$ divides $p(X)$ in $N[X]$ and thus, since $p(X)$ decomposes in $L[X]$ into linear factors, so does $q(X)$.

On the other hand f induces a ring homomorphism

$$\overline{f} \colon N[X] \longrightarrow L[X]$$

obtained by applying f to the coefficients of every polynomial. Since $q(X)$ divides $p(X)$ in $N[X]$ and $\overline{f}$ is a ring homomorphism, $\overline{f}(q(X))$ divides $\overline{f}(p(X))$ in $L[X]$. But $\overline{f}(p(X)) = p(X)$ since f is a K-homomorphism. Again since $p(X)$ factors in L into polynomials of degree 1, the same conclusion applies to $\overline{f}(q(X))$. We choose a root l' of $\overline{f}(q(X))$ in L. Thus $\overline{f}(q(X))$ is in the ideal of $L[X]$ generated by $X - l'$. Therefore, combining with 1.1.4, we get a K-homomorphism

$$N(l) \cong \frac{N[X]}{\langle q(X) \rangle} \xrightarrow{\ \tilde{f}\ } \frac{L[X]}{\langle X - l' \rangle} \cong L(l') \cong L$$

extending f.

Coming back to the situation in the statement, we put first $N = M$, from which we get a K-extension $M(l) \longrightarrow L$. One repeats the process with $N = M(l)$, and so on. We must reach $N = L$ after finitely many steps, because $[L : K]$ is finite. □

Proposition 1.3.4 *Let $K \subseteq L$ be a finite dimensional normal extension. The following conditions are equivalent:*

   (i) *the elements l_1, l_2 of L are conjugate over K;*
   (ii) *there exists a K-automorphism $f \colon L \longrightarrow L$ such that $f(l_1) = l_2$.*

Proof   Assuming condition (i), we consider the minimal polynomial
$p(X)$ of l_1, l_2. Applying 1.1.4 twice, we get a K-isomorphism

$$K(l_1) \cong \frac{K[X]}{\langle p(X) \rangle} \cong K(l_2),$$

mapping l_1 onto l_2. By 1.3.3, this extends to a K-automorphism of L
still mapping l_1 onto l_2.

Conversely, write $p(X)$ for the minimal polynomial of l_1 over K. Since
f is a homomorphism, one gets at once

$$p(l_2) = p\big(f(l_1)\big) = f\big(p(l_1)\big) = f(0) = 0.$$

Since $p(X)$ is irreducible, it is also the minimal polynomial of l_2 and
therefore l_1, l_2 are conjugate (see 1.1.5). $\square$

1.4 Galois extensions

Definition 1.4.1 A field extension $K \subseteq L$ is Galois when it is normal
and separable. The group $\mathsf{Aut}_K(L)$ of K-automorphisms of L is called
the Galois group of this extension and is denoted by $\mathsf{Gal}\,[L:K]$.

Proposition 1.4.2 *Let $K \subseteq M \subseteq L$ be field extensions. If $K \subseteq L$ is a
Galois extension, then $M \subseteq L$ is a Galois extension as well.*

Proof   By propositions 1.2.6 and 1.3.2. $\square$

Let us now introduce some notation and constructions. Let $K \subseteq L$
be a Galois field extension.

- Given an intermediate field extension $K \subseteq M \subseteq L$, via proposition
  1.4.2 we consider the Galois group $\mathsf{Gal}\,[L:M] = \mathsf{Aut}_M(L)$ of those
  automorphisms of L which fix M.
- Given a subgroup $G \subseteq \mathsf{Gal}\,[L:K]$, we write

$$\mathsf{Fix}\,(G) = \{l \in L \mid \forall g \in G \ \ g(l) = l\}.$$

  $\mathsf{Fix}\,(G)$ is clearly a subfield of L since each $g \in G$ is a field automorphism, and it contains K since each $g \in G$ is a K-automorphism:

$$K \subseteq \mathsf{Fix}\,(G) \subseteq L.$$

- $\#X$ indicates as usual the number of elements (the *cardinal*) of
  the set X.

We now reduce our attention to Galois extensions even if some results, like proposition 1.4.4, are valid more generally.

Definition 1.4.3 A Galois connection between two posets A, B consists in two order reversing maps

$$f \colon A \longrightarrow B, \quad g \colon B \longrightarrow A$$

with the property

$$a \le gf(a), \quad b \le fg(b)$$

for all elements $a \in A$ and $b \in B$.

The reader familiar with category theory will observe that viewing A and B as categories and f, g as contravariant functors, this is just the usual definition of two adjoint functors. Indeed, viewing f, g as covariant functors between A and the dual of B, the conditions for a Galois connection present f as left adjoint to g. The situation is reversed if one works with B and the dual of A.

Proposition 1.4.4 *Let $K \subseteq L$ be a Galois field extension. The maps*

$$\{M \mid K \subseteq M \subseteq L\} \underset{\mathsf{Fix}}{\overset{\mathsf{Gal}}{\rightleftarrows}} \{G \mid G \subseteq \mathsf{Gal}\,[L:M]\}$$

defined as above constitute a Galois connection.

Proof   Gal and Fix are contravariant functors between posets, so the announced adjunction property reduces to the trivial relations

$$\mathsf{Fix}\left(\mathsf{Gal}\,(M)\right) = M \subseteq \mathsf{Fix}\left(\mathsf{Gal}\,[L:M]\right), \quad G \subseteq \mathsf{Gal}\left(\mathsf{Fix}\,(G)\right). \quad \square$$

Theorem 1.4.5 (Galois theorem)   *Let $K \subseteq L$ be a finite dimensional Galois extension of fields. In this case, the adjunction in 1.4.4 is a contravariant isomorphism. Moreover, for every intermediate field extension $K \subseteq M \subseteq L$*

$$\dim\,[L:M] = \#\mathsf{Gal}\,[L:M].$$

Proof   Let us first prove the last statement, by induction on $\dim\,[L:M]$.

When $\dim\,[L:M] = 1$, $M = L$ and the unique automorphism of L fixing L is the identity.

Let us now assume the result for all extensions $M' \subseteq L'$ of dimension $k < n$ and let us consider $M \subseteq L$ of dimension n. If $l \in L \setminus M$ has minimal polynomial $p(X) \in M[X]$ of degree r, let us consider $M \subseteq$

$M(l) \subseteq L$. Since $l \notin M$, then $M \neq M(l)$ and thus $\dim [L : M(l)] < n$. But

$$\dim [M(l) : M] \cdot \dim [L : M(l)] = \dim [L : M]$$

and thus $\dim [L : M(l)] = \frac{n}{r}$. By inductive assumption, there are exactly $\frac{n}{r}$ $M(l)$-automorphisms

$$f_1, \ldots, f_{\frac{n}{r}} : L \longrightarrow L.$$

On the other hand, writing $l_1, \ldots, l_r$ for the roots of $p(X)$ in L, we get for each index j an M-automorphism $g_j : L \longrightarrow L$ such that $g_j(l) = l_j$ (see 1.3.4). Let us define $h_{ij} : L \longrightarrow L$ by $h_{ij} = g_j \circ f_i$. This yields n automorphisms h_{ij} of L. We shall prove that they are exactly all the elements of $\mathrm{Gal}\,[L : M]$.

First of all, these h_{ij} are M-automorphisms, thus elements of $\mathrm{Gal}\,[L : M]$. Moreover these elements are distinct, because

$$
\begin{aligned}
h_{ij} = h_{i'j'} \;\; &\Rightarrow \;\; g_j f_i(l) = g_{j'} f_{i'}(l) \\
&\Rightarrow \;\; g_j(l) = g_{j'}(l) \qquad \text{since } f_i,\, f_j \text{ fix } M(l) \\
&\Rightarrow \;\; l_j = l_{j'} \qquad\quad\; \text{by definition of } g_j,\, g_{j'} \\
&\Rightarrow \;\; j = j' \qquad\qquad \text{by separability of } M \subseteq L \\
&\qquad\qquad\qquad\qquad\; \text{(see 1.2.6).}
\end{aligned}
$$

Thus $g_j = g_{j'}$ and since this is a monomorphism, $f_i = f_{i'}$.

Finally observe that when an automorphism $f : L \longrightarrow L$ fixes M, from $p(f(l)) = f(p(l)) = f(0) = 0$, we deduce $f(l) = l_j$ for some j. Therefore $(g_j^{-1} \circ f)(l) = g_j^{-1}(l_j) = l$ and $g_j^{-1} \circ f$ is an M-automorphism fixing l; it is thus an $M(l)$-automorphism, meaning that $g_j^{-1} \circ f = f_i$ for some i. Therefore $f = g_j \circ f_i = h_{ij}$.

Next let us prove the formula $\dim [L : \mathrm{Fix}\,(G)] = \#G$ for every subgroup $G \subseteq \mathrm{Gal}\,[L : K]$. It suffices to prove $\dim [L : \mathrm{Fix}\,(G)] \leq \#G$ since, by the first part of the proof and proposition 1.4.4, this will imply

$$\#G \leq \#\mathrm{Gal}\,[L : \mathrm{Fix}\,(G)] = \dim [L : \mathrm{Fix}\,(G)] \leq \#G.$$

On the other hand observe that, again by the first part of the proof,

$$\dim [L : K] = \#\mathrm{Gal}\,[L : K]$$

which shows that $\mathrm{Gal}\,[L : K]$ and thus G are finite. Let us say that $\#G = n$; we must prove that $\dim [L : \mathrm{Fix}\,(G)] \leq n$.

We develop the proof by reduction *ad absurdum*. We thus choose $l_1, \ldots, l_{n+1}$ in L, linearly independent over $\mathrm{Fix}\,(G)$. Let us also write

$g_1, \ldots, g_n$ for the n elements of G. Let us consider the homogeneous system of equations

$$\left.\begin{array}{c} g_1(l_1)X_1 + \cdots + g_1(l_{n+1})X_{n+1} = 0, \\ \vdots \\ g_n(l_1)X_1 + \cdots + g_n(l_{n+1})X_{n+1} = 0. \end{array}\right\} \quad (1)$$

Since there are more unknowns than equations, there exists a non-zero solution. Let us choose such a non-zero solution which possesses the minimal number of non zero components. There is no restriction in assuming that this solution has the form $(\alpha_0, \ldots, \alpha_r, 0, \ldots, 0)$ with each α_i non-zero. This yields a system

$$\left.\begin{array}{c} g_1(l_1)X_1 + \cdots + g_1(l_r)X_r = 0, \\ \vdots \\ g_n(l_1)X_1 + \cdots + g_n(l_r)X_r = 0 \end{array}\right\} \quad (2)$$

admitting a solution all of whose components are non-zero. Let us then fix $g \in G$ and let us apply this g to all equations, evaluated in $\alpha_1, \ldots, \alpha_r$:

$$\left.\begin{array}{c} gg_1(l_1)g(\alpha_1) + \cdots + gg_1(l_r)g(\alpha_r) = 0, \\ \vdots \\ gg_n(l_1)g(\alpha_1) + \cdots + gg_n(l_r)g(\alpha_r) = 0. \end{array}\right\} \quad (3)$$

But the elements gg_i are just a permutation of the elements of G, thus the system (3) can be rewritten, up to a permutation of the equations,

$$\left.\begin{array}{c} g_1(l_1)g(\alpha_1) + \cdots + g_1(l_r)g(\alpha_r) = 0, \\ \vdots \\ g_n(l_1)g(\alpha_1) + \cdots + g_n(l_r)g(\alpha_r) = 0. \end{array}\right\} \quad (4)$$

Let us multiply system (4) by α_r and system (2), evaluated at the α_i, by $g(\alpha_r)$; let us substract the results. This yields

$$\left.\begin{array}{l} g_1(l_1)\big(\alpha_r g(\alpha_1) - \alpha_1 g(\alpha_r)\big) + \cdots \\ \qquad + g_1(l_{r-1})\big(\alpha_r g(\alpha_{r-1}) - \alpha_{r-1}g(\alpha_r)\big) = 0, \\ \vdots \\ g_n(l_1)\big(\alpha_r g(\alpha_1) - \alpha_1 g(\alpha_r)\big) + \cdots \\ \qquad + g_n(l_{r-1})\big(\alpha_r g(\alpha_{r-1}) - \alpha_{r-1}g(\alpha_r)\big) = 0 \end{array}\right\} \quad (5)$$

which yields a solution of system (1) with an additional zero component. By choice of the original solution, this one is the zero solution, yielding

$\alpha_r g(\alpha_i) = \alpha_i g(\alpha_r)$ for all $i \leq r - 1$. This can be rewritten as

$$\alpha_i \alpha_r^{-1} = g(\alpha_i)g(\alpha_r)^{-1} = g(\alpha_i)g(\alpha_r^{-1}) = g(\alpha_i \alpha_r^{-1}).$$

Since $g \in G$ is arbitrary, we get $\alpha_i \alpha_r^{-1} \in \mathsf{Fix}\,(G)$; let us put $m_i = \alpha_i \alpha_r^{-1} \in \mathsf{Fix}\,(G)$. We thus have $\alpha_i = m_i \alpha_r$ with $i < r$ and $m_i \in \mathsf{Fix}\,(G)$; putting $m_r = 1 \in \mathsf{Fix}\,(G)$, we get $\alpha_i = m_i \alpha_r$ for all $i \leq r$. The first equation of system (2) evaluated at the α_i yields

$$
\begin{aligned}
0 &= g_1(l_1)\alpha_1 + \cdots + g_1(l_r)\alpha_r \\
&= g_1(l_1)m_1\alpha_1 + \cdots + g_1(l_r)m_r\alpha_r \\
&= \alpha_r\big(g_1(l_1)m_1 + \cdots + g_1(l_r)m_r\big) \\
&= \alpha_r\big(g_1(l_1)g_1(m_1) + \cdots + g_1(l_r)g_1(m_r)\big) \quad \text{since } m_i \in \mathsf{Fix}\,(G) \\
&= \alpha_r g_1(l_1 m_1 + \cdots + l_r m_r).
\end{aligned}
$$

We know that $\alpha_r \neq 0$ and g_1 is injective, from which $l_1 m_1 + \cdots + l_r m_r = 0$, which contradicts the linear independance of the l_i over $\mathsf{Fix}\,(G)$. This concludes the proof that $\dim\big[L : \mathsf{Fix}\,(G)\big] = \#G$.

The rest is easy. Starting from $K \subseteq M \subseteq L$, we get $M \subseteq \mathsf{Fix}\,\big(\mathsf{Gal}\,[L : M]\big)$ by proposition 1.4.4. It remains to see that given $l \in L \setminus M$, there exists $f: L \longrightarrow L$ fixing M but not l. If $p(X)$ is the minimal polynomial of l over M, then $p(X)$ has not degree 1 since $l \notin M$; thus $p(X)$ has at least two distinct roots in L, since $M \subseteq L$ is a Galois extension by proposition 1.4.2. Let $l' \neq l$ be such a root of $p(X)$; by proposition 1.3.4, there exists an M-automorphism f of L such that $f(l) = l'$.

Choose now $G \subseteq \mathsf{Gal}\,[L : K]$. By proposition 1.4.4 we have $G \subseteq \mathsf{Gal}\,\big[L : \mathsf{Fix}\,(G)\big]$. What we have already proved of the present theorem then yields

$$\#G \leq \#\mathsf{Gal}\,\big[L : \mathsf{Fix}\,(G)\big] = \dim\big[L : \mathsf{Fix}\,(G)\big] = \#G.$$

Therefore, since the cardinals are finite, $G = \mathsf{Gal}\,\big[L : \mathsf{Fix}\,(G)\big]$. □

The following lemma will help completing the statement of theorem 1.4.5.

Lemma 1.4.6 *Let $K \subseteq M \subseteq L$ be finite dimensional field extensions, with $K \subseteq L$ a Galois extension. For every $f \in \mathsf{Gal}\,[L : K]$,*

$$f \cdot \mathsf{Gal}\,[L : M] \cdot f^{-1} = \mathsf{Gal}\,\big[L : f(M)\big].$$

Proof Given $g \in \mathsf{Gal}\,[L : M]$ and $m \in M$,

$$(f \circ g \circ f^{-1})(f(m)) = (f \circ g)(m) = f(m)$$

since g fixes M. Thus $f \circ g \circ f^{-1}$ fixes $f(M)$, meaning $f \circ g \circ f^{-1} \in$ Gal $[L : f(M)]$. This already proves

$$f \cdot \text{Gal}\,[L : M] \cdot f^{-1} \subseteq \text{Gal}\,[L : f(M)].$$

Putting $g = f^{-1}$ and $N = f(M)$, we get in an analogous way

$$f^{-1} \cdot \text{Gal}[L : f(M)] \cdot f = g \cdot \text{Gal}[L : N] \cdot g^{-1} \subseteq \text{Gal}[L : g(N)] = \text{Gal}\,[L : M],$$

that is, multiplying on the left by f and on the right by f^{-1},

$$\text{Gal}\,[L : f(M)] \subseteq f \cdot \text{Gal}\,[L : M] \cdot f^{-1}. \qquad \square$$

Theorem 1.4.7 *In the conditions of theorem 1.4.5, the extensions $K \subseteq M \subseteq L$, with $K \subseteq M$ a normal extension, correspond via the bijection with the normal subgroups $G \subseteq \text{Gal}\,[L : K]$. Moreover,*

$$\text{Gal}\,[M : K] \cong \frac{\text{Gal}\,[L : K]}{\text{Gal}\,[L : M]},$$

again when $K \subseteq M$ is a normal extension.

Proof Let us begin with $K \subseteq M \subseteq L$ with $K \subseteq M$ normal and prove that Gal $[L : M]$ is normal in Gal $[L : K]$. Given $f \in \text{Gal}\,[L : K]$, let us prove first that $f(M) \subseteq M$. Indeed, if $m \in M$ admits $p(X) \in K[X]$ as minimal polynomial, then $p(f(m)) = f(p(m)) = f(0) = 0$, thus $f(m)$ is one of the roots of $p(X)$. Since $K \subseteq M$ is normal, $f(m) \in M$. Moreover since f is injective and dimensions are finite, we have $f(M) = M$. Applying lemma 1.4.6, we then find

$$f \cdot \text{Gal}\,[L : M] \cdot f^{-1} = \text{Gal}\,[L : f(M)] = \text{Gal}\,[L : M].$$

Let us start now from $K \subseteq M \subseteq L$ with Gal $[L : M] \subseteq \text{Gal}\,[L : K]$ a normal subgroup. Choose $m \in M$ with minimal polynomial $p(X) \in K[X]$; this polynomial $p(X)$ factors in $L[X]$ into distinct factors of degree 1, and we must prove these belong to $M[X]$. If $m \neq l \in L$ is another root of $p(X)$, we get by proposition 1.3.4 the existence of a K-automorphism f of L such that $f(m) = l$. By normality of Gal $[L : M]$,

$$f \cdot \text{Gal}\,[L : M] \cdot f^{-1} = \text{Gal}\,[L : M]$$

while by lemma 1.4.6

$$f \cdot \text{Gal}\,[L : M] \cdot f^{-1} = \text{Gal}\,[L : f(M)].$$

This yields Gal $[L : M] = \text{Gal}\,[L : f(M)]$ and thus $M = f(M)$, by the isomorphism of theorem 1.4.5. This proves $l = f(m) \in M$.

It remains to prove the last formula. There is a group homomorphism obtained by taking the restriction to M:

$$\theta \colon \mathsf{Gal}\,[L : K] \longrightarrow \mathsf{Gal}\,[M : K], \quad f \mapsto f\,|_M;$$

indeed, as just observed, the normality of M implies $f(M) = M$. By proposition 1.3.3, this homomorphism θ is surjective. Thus

$$\mathsf{Gal}\,[M : K] \cong \frac{\mathsf{Gal}\,[L : K]}{\mathsf{Ker}\,\theta}$$

and it remains to prove $\mathsf{Ker}\,\theta = \mathsf{Gal}\,[L : M]$. And indeed $f \in \mathsf{Gal}\,[L : K]$ is in $\mathsf{Ker}\,\theta$ when its restriction to M is the identity on M, that is, when $f \in \mathsf{Gal}\,[L : M]$. $\qquad\qquad\square$

2

Galois theory of Grothendieck

This chapter does not present the Galois theory of Grothendieck in its full generality, that is in the context of schemes: this would require a long technical introduction. But the spirit of Grothendieck's approach is applied to the context of Galois theory for fields.

Convention *In this chapter, all fields, rings and algebras we consider are commutative and have a unit.*

2.1 Algebras on a field

Let us recall that an algebra A on a field K is a vector space A on K provided with a multiplication which makes it a ring and which satisfies $k(aa') = (ka)a'$ for all elements $k \in K$ and $a, a' \in A$. Let us emphasize the fact that all rings we consider are commutative and have a unit.

A typical example of a K-algebra is the ring $K[X]$ of polynomials with coefficients in K. Observe that given such a polynomial

$$p(X) = k_n X^n + \cdots + k_0$$

and a K-algebra A, we get at once a polynomial function

$$p\colon A \longrightarrow A, \quad a \mapsto p(a) = k_n a^n + \cdots + k_0 \cdot \mathbf{1}$$

where $\mathbf{1}$ is the unit of A. It is well known that distinct polynomials can give rise to the same polynomial function, for example the polynomials X and X^2 on the field of integers modulo 2.

Another immediate example of a K-algebra is the power K^n, where all operations are defined componentwise.

An ideal $I \subseteq A$ of a K-algebra A is an ideal I of the ring A; it is in particular a vector subspace of A. The quotient A/I is then again a

K-algebra. On the other hand the kernel of a K-algebra homomorphism $f \colon A \longrightarrow B$ is an ideal of the K-algebra A.

Proposition 2.1.1 *Let K be a field and A a K-algebra. The following conditions are equivalent:*

(i) *A is a field;*
(ii) *A has only trivial ideals.*

Proof (i) $\Rightarrow$ (ii) was already observed at the beginning of section 1.1. Conversely, if $0 \neq a \in A$, then aA is an ideal of A as K-algebra. Since $0 \neq a = a \cdot 1 \in aA$, we get $aA = A$, from which we get $a' \in A$ with $aa' = 1$. $\qquad\qquad\square$

Corollary 2.1.2 *Let K be a field and $f \colon A \longrightarrow B$ a surjective homomorphism of K-algebras. If B is a field, then the kernel of f is a maximal ideal of A.*

Proof We have $B \cong A/\mathsf{Ker}\, f$. If $\mathsf{Ker}\, f \subseteq I$, with I an ideal, we get a corresponding quotient $q \colon A/\mathsf{Ker}\, f \longrightarrow A/I$. The kernel of q is then an ideal of the field $B = A/\mathsf{Ker}\, f$, thus is trivial. If $\mathsf{Ker}\, q = (0)$, then $\mathsf{Ker}\, f = I$. If $\mathsf{Ker}\, q = A/\mathsf{Ker}\, f$, then $A/I = (0)$ by surjectivity of q and $A = I$. $\qquad\qquad\square$

Proposition 2.1.3 *Let K be a field. Every ideal of the K-algebra $K[X]$ is principal.*

Proof Let $I \subseteq K[X]$ be a non zero ideal and $p(X)$ a non zero polynomial in I, whose degree is minimal in I. For every polynomial $q(X) \in K[X]$ let us perform the euclidean division

$$q(X) = p(X) \cdot \alpha(X) + \beta(X), \quad \text{degree } \beta(X) < \text{degree } p(X).$$

Since $p(X)$ and $q(X)$ are in the ideal I, we get $\beta(X) \in I$ and by minimality of the degree of $p(X)$, it follows that $\beta(X) = 0$. Therefore $p(X)$ divides $q(X)$ and I is the principal ideal generated by $p(X)$. $\qquad\square$

Proposition 2.1.4 *Let K be a field and $p(X) \in K[X]$. The following conditions are equivalent:*

(i) *the polynomial $p(X)$ is irreducible;*
(ii) *the ideal $\langle p(X) \rangle$ generated by $p(X)$ is maximal;*
(iii) *the K-algebra $K[X]/\langle p(X) \rangle$ is a field.*

Proof (i) $\Rightarrow$ (ii) Let $\langle p(X) \rangle \subseteq I$ with I an ideal; by 2.1.3 $I = \langle s(X) \rangle$ for some polynomial $s(X)$. Thus $p(X) \in \langle s(X) \rangle$, from which follows the existence of $r(X)$ such that $p(X) = r(X) \cdot s(X)$. If $s(X)$ is a non zero constant, then $\langle s(X) \rangle = K[X]$. Otherwise, $r(X)$ is a non zero constant since $p(X)$ is irreducible, from which $\langle p(X) \rangle = \langle s(X) \rangle$.

(ii) $\Rightarrow$ (iii) Consider the quotient $q \colon K[X] \longrightarrow K[X]/\langle p(X) \rangle$. Every ideal $I \subseteq K[X]/\langle p(X) \rangle$ induces an ideal $q^{-1}(I) \supseteq \langle p(X) \rangle$. Since $\langle p(X) \rangle$ is maximal, $q^{-1}(I) = \langle p(X) \rangle$ or $q^{-1}(I) = K[X]$, that is, $I = qq^{-1}(I) = (0)$ or $I = qq^{-1}(I) = K[X]/\langle p(X) \rangle$. Thus the ideals of $K[X]/\langle p(X) \rangle$ are trivial and it is a field by proposition 2.1.1.

(iii) $\Rightarrow$ (i) Let $p(X) = s(X) \cdot r(X)$ be a factorization of $p(X)$. It follows that $\langle p(X) \rangle \subseteq \langle s(X) \rangle$, from which $\langle s(X) \rangle / \langle p(X) \rangle$ is an ideal of $K[X]/\langle p(X) \rangle$. If this ideal is (0), then $\langle s(X) \rangle = \langle p(X) \rangle$ and $p(X)$ divides $s(X)$, thus $r(X)$ is a constant. If this ideal is $K[X]/\langle p(X) \rangle$, then the constant polynomial 1 is in $\langle s(X) \rangle$ up to a polynomial in $\langle p(x) \rangle$, that is

$$1 = u(X) \cdot s(X) + v(X) \cdot p(X) = s(X) \cdot \big(u(X) + v(X) \cdot r(X) \big).$$

This implies that $s(X)$ is a constant. $\qquad\square$

Let us now study the algebraic elements of a K-algebra and their minimal polynomials.

Definition 2.1.5 Let K be a field and A a K-algebra. An element $a \in A$ is algebraic when there exists a polynomial $p(X) \in K[X]$ with $p(a) = 0$. The K-algebra A itself is called "algebraic" when all its elements are algebraic.

Proposition 2.1.6 *Let K be a field. Every finite dimensional K-algebra is algebraic.*

Proof As for proposition 1.1.2. $\qquad\square$

Proposition 2.1.7 *Let K be a field, A a K-algebra and $0 \neq a \in A$ an algebraic element. There exists a unique polynomial $p(X) \in K[X]$ such that*

(i) *the leading coefficient of $p(X)$ is 1,*
(ii) *$p(a) = 0$,*
(iii) *if $q(X) \in K[X]$ with $q(a) = 0$, then $p(X)$ divides $q(X)$.*

This polynomial $p(X)$ is called the minimal polynomial of a.

Proof As for proposition 1.1.3. Observe that the irreducibility of $p(X)$ is not asserted (see example 2.1.9). And of course there is a trivial reason for this, since $A = K[X]/\langle p(X) \rangle$, with $p(X)$ any reducible polynomial, is a possible situation to which this proposition applies. □

Corollary 2.1.8 *Let K be a field, A a K-algebra and $0 \neq a \in A$ an algebraic element. When the algebra A is an integral domain, the minimal polynomial of a is irreducible.*

Proof Write $p(X)$ for the minimal polynomial of a. If $p(X) = r(X) \times s(X)$, then $0 = s(a) \times r(a)$, from which $s(a) = 0$ or $r(a) = 0$ since A is an integral domain. By minimality of the degree of $p(X)$, it follows that $r(X)$ or $s(X)$ is constant. □

Example 2.1.9 (A reducible minimal polynomial) Let us consider a field K and the K-algebra K^2, of dimension 2 over K. Given $k \in K$, the only root of the first degree polynomial $X - k$ in K^2 is $k \cdot \mathbf{1}$, where $\mathbf{1} = (1,1)$ is the unit of K^2. In particular the minimal polynomial of the element $(1,0) \in K^2$ has not degree 1. But since $(1,0)^2 = (1,0)$, this element is a root in K^2 of $X^2 - X$, which is thus its minimal polynomial. Observe that this polynomial is reducible: $X^2 - X = X(X - 1)$.

Proposition 2.1.10 *Let K be a field, A a K-algebra and $0 \neq a \in A$ an algebraic element with minimal polynomial $p(X)$ of degree n. The K-subalgebra $K(a) \subseteq A$ generated by a is isomorphic to*

$$K(a) \cong \frac{K[X]}{\langle p(X) \rangle} \cong \{k_0 + k_1 X + \cdots + k_{n-1} X^{n-1} \mid k_i \in K\}$$

where, in this last expression, the operations are defined modulo $p(X)$.

Proof One has trivially

$$K(a) \cong \{q(a) \mid q(X) \in K[X]\}.$$

The map

$$\{k_0 + k_1 X + \cdots + k_{n-1} X^{n-1} \mid k_i \in K\} \longrightarrow K(a), \quad r(X) \mapsto r(a)$$

is surjective. Indeed, every polynomial $q(X)$ can be written as $q(X) = p(X)s(X) + r(X)$ where $r(X)$ has degree at most $n - 1$ and, of course, $q(a) = r(a)$. This map is also injective. Indeed, given $r(X)$ and $s(X)$ of degrees at most $n-1$, $r(a) = s(a)$ implies $(r-s)(a) = 0$ with $(r-s)(X)$ of

degree at most $n-1$; by minimality of the degree of $p(X)$, $(r-s)(X) = 0$
and $r(X) = s(X)$. □

Corollary 2.1.11 *Let K be a field and A a K-algebra. If A is an
integral domain, every non-zero algebraic element of A is invertible.*

Proof By corollary 2.1.8 and propositions 2.1.10 and 2.1.4. □

Let us conclude this section with two general results on K-algebras.

Proposition 2.1.12 (Chinese lemma) *Let K be a field and*

$$(f_i \colon A \longrightarrow\!\!\!\!\!\rightarrow B_i)_{1 \leq i \leq n}$$

a finite family of surjective homomorphisms of K-algebras. If

$$i \neq j \implies \mathsf{Ker}\, f_i + \mathsf{Ker}\, f_j = A,$$

*then the corresponding factorization $f \colon A \longrightarrow \prod_{1 \leq i \leq n} B_i$ is surjective
as well.*

Proof For every pair $i \neq j$ of indices, let us choose $\alpha_{ij} \in \mathsf{Ker}\, f_i$ and
$\beta_{ij} \in \mathsf{Ker}\, f_j$ such that $\alpha_{ij} + \beta_{ij} = 1$. It follows that

$$f_j(\alpha_{ij}) = f_j(\alpha_{ij}) + f_j(\beta_{ij}) = f_j(\alpha_{ij} + \beta_{ij}) = f_j(1) = 1.$$

We then put

$$\alpha_j = \prod_{i \neq j} \alpha_{ij}$$

and observe immediately that

$$\begin{cases} f_j(\alpha_j) &=\quad 1, \\ f_k(\alpha_j) &=\quad 0 \quad \text{if } k \neq j. \end{cases}$$

Then fix $b = (b_i)_{1 \leq i \leq n} \in \prod_{1 \leq i \leq n} B_i$ and for each index i, choose $a_i \in A$
such that $f_i(a_i) = b_i$. The element

$$a = \sum_{1 \leq k \leq n} \alpha_k a_k$$

is such that

$$f_j(a) = \sum_{1 \leq k \leq n} f_j(\alpha_k) f_j(a_k) = f_j(a_j) = b_j,$$

from which

$$f(a) = \big(f_j(a)\big)_{1 \leq j \leq n} = (b_j)_{1 \leq j \leq n} = b. \qquad \square$$

Proposition 2.1.13 *Let K be a field and $n \in \mathbb{N}$ an integer. Every ideal I of the K-algebra K^n has the form*

$$I = \{(k_i)_{1 \leq i \leq n} | \forall i \in J \; k_i = 0\}$$

where $J \subseteq \{1, \dots, n\}$ is an arbitrary subset of indices.

Proof It is obvious that the subsets I as described are ideals of K^n. Conversely, if $I \subseteq K^n$ is an ideal, put

$$J = \{j | 1 \leq j \leq n, \; \forall (a_i)_{1 \leq i \leq n} \in I \; a_j = 0\}$$

and write

$$I_J = \{(a_i)_{1 \leq i \leq n} | \forall i \in J \; a_i = 0\}.$$

We have $I \subseteq I_J$ by definition of J and it remains to prove the converse inclusion.

For each index $j \notin J$, there exists thus an element $k^j = (k_i^j)_{1 \leq i \leq n} \in I$ with $k_j^j \neq 0$. Let us write $e_i \in K^n$ for the element whose ith component is 1, while the other components are 0. One observes at once that when $j \notin J$, $k^j e_j = k_j^j e_j$. Therefore if $x = (x_i)_{1 \leq i \leq n} \in I_J$,

$$x = \sum_{1 \leq i \leq n} x_i e_i = \sum_{i \notin J} x_i e_i = \sum_{i \notin J} \frac{x_i e_i}{k_i^i} k_i^i = \sum_{i \notin J} \frac{x_i e_i}{k_i^i} k^i,$$

which is an element of I since each k^i is in I. $\square$

2.2 Extension of scalars

In this section, we study the properties of K-algebras in relation with the consideration of a field extension $K \subseteq L$. In section 1.1 we observed that L is a K-vector space; thus it is itself a K-algebra.

Proposition 2.2.1 *Let $K \subseteq L$ be a finite dimensional field extension and $K \subseteq A \subseteq L$ an intermediate K-algebra. In these conditions, the algebra A is itself a field.*

Proof The algebra A is an integral domain since so is L; it is algebraic by finite dimensionality (see proposition 2.1.6). One concludes the proof by corollary 2.1.11. $\square$

Proposition 2.2.2 *Let $K \subseteq L$ be a field extension. Every L-algebra B is trivially a K-algebra, by restriction of the scalar multiplication to the* *elements of K. On the other hand every K-algebra A yields an L-algebra $L \otimes_K A$, where the multiplication of this algebra is determined by*

$$(l \otimes a)(l' \otimes a') = (ll') \otimes (aa')$$

and the scalar multiplication by

$$l(l' \otimes a) = (ll') \otimes a,$$

for $l, l' \in L$ and $a, a' \in A$. These constructions extend to functors

$$\text{L-Alg} \longrightarrow K\text{-Alg}, \quad B \mapsto B, \quad K\text{-Alg} \longrightarrow L\text{-Alg}, \quad A \mapsto L \otimes_K A,$$

the second functor being left adjoint to the first one.

Proof Only the adjunction requires a comment. With the previous notation, we must exhibit natural isomorphisms

$$\text{Hom}_L(L \otimes_K A, B) \cong \text{Hom}_K(A, B).$$

Given $f \colon L \otimes_K A \longrightarrow B$, one considers

$$f' \colon A \longrightarrow B, \quad a \mapsto f(1 \otimes a)$$

and given $g \colon A \longrightarrow B$, one constructs

$$g' \colon L \otimes_K A \longrightarrow B, \quad l \otimes a \mapsto lg(a).$$

Notice that $l\big(f(1 \otimes a)\big) = f(l \otimes a)$ by L-linearity of f, from which the result follows at once. $\qquad\square$

Corollary 2.2.3 *Let $K \subseteq L$ be a field extension and A a K-algebra. The following isomorphism holds:*

$$\text{Hom}_K(A, L) \cong \text{Hom}_L(L \otimes_K A, L). \qquad\square$$

Proposition 2.2.4 *Let $K \subseteq L$ be a field extension and $p(X) \in K[X]$ a polynomial. The following isomorphism holds:*

$$L \otimes_K \frac{K[X]}{\langle p(X) \rangle} \cong \frac{L[X]}{\langle p(X) \rangle},$$

where in the right hand side, $p(X)$ is viewed as a polynomial with coefficients in L.

Proof It is straightforward to observe that the maps

$$L \otimes_K \frac{K[X]}{\langle p(X) \rangle} \longrightarrow \frac{L[X]}{\langle p(X) \rangle}, \quad \sum_{i=1}^n l_i \otimes [q_i(X)] \mapsto [\sum_{i=1}^n l_i q_i(X)],$$

$$\frac{L[X]}{\langle p(X) \rangle} \longrightarrow L \otimes_K \frac{K[X]}{\langle p(X) \rangle}, \quad [\sum_{i=1}^n l_i X^i] \mapsto \sum_{i=1}^n l_i \otimes [X^i],$$

are correctly defined and describe the required isomorphism.   □

Proposition 2.2.5 *Let $K \subseteq L$ be a field extension and $p(X) \in K[X]$ a polynomial. There exists a bijection between*

(i) *the roots of $p(X)$ in L,*

(ii) *the homomorphisms of K-algebras $\dfrac{K[X]}{\langle p(X) \rangle} \longrightarrow L$.*

Proof   An element $l \in L$ such that $p(l) = 0$ yields a corresponding well-defined evaluation morphism of K-algebras

$$\mathrm{ev}_l \colon \frac{K[X]}{\langle p(X) \rangle} \longrightarrow L, \quad [q(X)] \mapsto q(l).$$

Conversely, given a morphism $f \colon \dfrac{K[X]}{\langle p(X) \rangle} \longrightarrow L$ of K-algebras, let us put $l = f([X])$, where $[X]$ denotes the equivalence class of the polynomial $X \in K[X]$. It follows at once that l is a root of $p(X)$ since

$$p(l) = p\Big(f([X])\Big) = f\Big(p([X])\Big) = f\Big([p(X)]\Big) = f(0) = 0$$

since f, as a homomorphism of K-algebras, fixes the elements of K, thus the coefficients of $p(X)$.

Starting with a root l of $p(X)$, it is immediate that $\mathrm{ev}_l([X]) = l$. Next, beginning with f as above, for every polynomial $q(X) \in K[X]$,

$$\mathrm{ev}_{f([X])}\Big([q(X)]\Big) = q\Big(f([X])\Big) = f\Big(q([X])\Big) = f\Big([q(X)]\Big),$$

again since f fixes K, thus the coefficients of $q(X)$.   □

Theorem 2.2.6 *Let $K \subset L$ be a field extension and A a K-algebra. The homomorphisms of K-algebras $A \longrightarrow L$ are linearly independent over K, in the vector space of K-linear maps $A \longrightarrow L$.*

Proof By corollary 2.2.3,

$$\mathrm{Hom}_K(A, L) \cong \mathrm{Hom}_L(L \otimes_K A, L),$$

from which it suffices to prove that for every L-algebra B, the homomorphisms of L-algebras $B \longrightarrow L$ are linearly independent over L, from which, a fortiori, the linear independence over K follows.

Every homomorphism of L-algebras $f\colon B \longrightarrow L$ is surjective, because given $l \in L$, one has $l = l \cdot 1 = l \cdot f(1) = f(l \cdot 1)$. As a consequence, $L \cong B/\mathsf{Ker}\, f$ where, by corollary 2.1.2, $\mathsf{Ker}\, f$ is a maximal ideal of B. If $f, g\colon B \longrightarrow\!\!\!\!\!\to L$ are distinct homomorphisms of B-algebras, they are thus quotient maps and therefore their kernels must be distinct; by maximality of these kernels, $\mathsf{Ker}\, f + \mathsf{Ker}\, g = B$. Let us then consider a finite family $f_i\colon B_i \longrightarrow\!\!\!\!\!\to L$ of distinct homomorphisms of K-algebras, such that $\sum_{1 \leq i \leq n} l_i f_i = 0$, for some $l_i \in L$. The Chinese lemma applies (see 2.1.12), thus the map

$$B \longrightarrow L^n, \quad b \mapsto \big(f_i(b_i)\big)_{1 \leq i \leq n}$$

is surjective. If at least one l_i is non zero, the equation $\sum_{1 \leq i \leq n} l_i X_i = 0$ is that of a proper linear subspace of L^n and, since $\sum_{1 \leq i \leq n} l_i f_i = 0$, this proper subspace contains the image of the previous map, which contradicts its surjectivity. Therefore all l_i are zero. $\qquad\square$

2.3 Split algebras

In the first chapter, we were interested in Galois extensions of fields, that is, algebraic field extensions $K \subseteq L$ such that the minimal polynomial $p(X) \in K[X]$ of each element $l \in L$ factors in $L[X]$ into factors of degree 1 with distinct roots. This recollection indicates at once the spirit of the next definition.

Definition 2.3.1 Let $K \subseteq L$ be a field extension and A a K-algebra. The extension L splits the K-algebra A when

   (i) the algebra A is algebraic over K,

   (ii) the minimal polynomial $p(X) \in K[X]$ of every element of A factors in $L[X]$ into factors of degree 1 with distinct roots.

The K-algebra A is an étale K-algebra when it is split by the algebraic closure of K.

Proposition 2.3.2 *Let $K \subseteq L$ be a field extension. The following conditions are equivalent:*

   (i) $K \subseteq L$ *is a Galois extension;*

   (ii) *the extension L splits the K-algebra L.* $\qquad\square$

Theorem 2.3.3 *Let $K \subseteq L$ be a field extension of finite dimension m and A a K-algebra of finite dimension n. The following conditions are equivalent:*

(i) *the extension L splits the K-algebra A;*

(ii) *the following map, called the "Gelfand transformation", is an isomorphism of L-algebras –*

$$\mathsf{Gel}\colon L \otimes_K A \longrightarrow L^{\mathsf{Hom}_L(L \otimes_K A, L)},$$
$$l \otimes a \mapsto \big(f(l \otimes a)\big)_{f \in \mathsf{Hom}_L(L \otimes_K A, L)};$$

(iii) *the following map is an isomorphism of L-algebras –*

$$L \otimes_K A \longrightarrow L^{\mathsf{Hom}_K(A, L)}, \quad l \otimes a \mapsto \big(lg(a)\big)_{g \in \mathsf{Hom}_K(A, L)};$$

(iv) $\#\mathsf{Hom}_L(L \otimes_K A, L) = n$;

(v) $\#\mathsf{Hom}_K(A, L) = n$;

(vi) $L \otimes_K A$ *is isomorphic to L^n as an L-algebra;*

(vii) $\forall x \in L \otimes_K A, \ x \neq 0, \ \exists f \in \mathsf{Hom}(L \otimes_K A, L)$ *such that $f(x) \neq 0$;*

where $\#$ is the cardinality symbol.

Proof The vector space of K-linear maps $L \otimes_K A \longrightarrow L$ has dimension $(mn)m$ over K, thus by 2.2.6 $\mathsf{Hom}_L(L \otimes_K L, L)$ is finite. By proposition 2.1.6, the algebra A is algebraic over K. Putting $B = L \otimes_K A$ in the proof of theorem 2.2.6 yields at once the surjectivity of the Gelfand transformation. For the sake of clarity, we split the proof into three lemmas.

Lemma 2.3.4 *Conditions* (ii) *to* (vii) *of theorem 2.3.3 are equivalent.*

Proof The equivalence of (ii), (iii) and the equivalence of (iv), (v) follow at once from corollary 2.2.3.

(ii) $\Rightarrow$ (iv) The K-vector space $L \otimes_K A$ has dimension mn and the K-vector space $L^{\mathsf{Hom}(L \otimes_K A, L)}$ has dimension $m \cdot \#\mathsf{Hom}(L \otimes_K A, L)$. If the Gelfand transformation is an isomorphism, the equality of these dimensions yields $n = \#\mathsf{Hom}(L \otimes_K A, L)$. This also proves (ii) $\Rightarrow$ (vi).

(iv) $\Rightarrow$ (ii) We know already that the Gelfand transformation is surjective; if its domain and codomain have the same finite dimension, it is an isomorphism.

(vi) $\Rightarrow$ (iv) We must prove that $\#\mathsf{Hom}_L(L^n, L) = n$. Observe that the projections $p_i\colon L^n \longrightarrow L$ constitute n distinct homomorphisms of L-algebras, linearly independent over L by theorem 2.2.6. Since the space $\mathsf{Lin}_L(L^n, L)$ of all L-linear maps has dimension n, the p_i are all the morphisms of L-algebras, again by theorem 2.2.6.

(ii) $\Leftrightarrow$ (vii) Condition (vii) expresses precisely the injectivity of the Gelfand transformation, which is already known to be surjective. $\qquad\square$

Lemma 2.3.5 *In the conditions of theorem 2.3.3, the class of those K-algebras satisfying the equivalent conditions* (ii) *to* (vii) *is stable under subobjects, quotients, finite products and tensor products. Moreover if a K-algebra A admits two subalgebras A_1, A_2 satisfying conditions* (ii) *to* (vii), *the same holds for the subalgebra of A generated by the elements of A_1 and A_2.*

Proof Condition (vii) is trivially stable under subobjects.

Consider now a quotient $A \relbar\joinrel\twoheadrightarrow Q$ of a K-algebra A of dimension n, which satisfies conditions (ii) to (vii) of theorem 2.3.3. Since tensoring with L has a right adjoint by proposition 2.2.2, it preserves quotients, from which we obtain a quotient of L-algebras

$$L^n \cong L \otimes_K A \relbar\joinrel\twoheadrightarrow L \otimes_K Q.$$

By proposition 2.1.13, the kernel of this second quotient is an ideal $J \subseteq L^n$ of the form

$$J = \{(l_i)_{1 \le i \le n} | \forall i \in X \ l_i = 0\}, \quad X \subseteq \{1, \dots, n\}.$$

Putting $x = \#X$, we get $L \otimes_K Q \cong L^n/J \cong L^{n-x}$ and by condition (vi) of theorem 2.3.3, it remains to prove that Q has dimension $n - x$ over K. Since L has dimension m over K and L^n/J has dimension $n - x$ over L, it follows that L/J has dimension $m(n-x)$ over K. On the other hand $L \otimes_K Q$ has dimension $m \cdot \dim_K Q$ over K, from which $\dim_K Q = n - x$ since $L \otimes_K Q \cong L^n/J$.

To treat the case of finite products, observe first that tensoring with L is an additive, thus finite product preserving, functor

$$L \otimes_K - : \mathsf{Vect}_K \longrightarrow \mathsf{Vect}_L$$

between categories of vector spaces. Therefore if A, A' are K-algebras of respective dimensions n, n' and satisfying conditions (ii) to (vii) of theorem 2.3.3, then $A \times A'$ has dimension $n + n'$ and

$$L \otimes_K (A \times A') \cong (L \otimes_K A) \times (L \otimes_K A') \cong L^n \times L^{n'} \cong L^{n+n'}.$$

Thus condition (vi) is satisfied by $A \times A'$.

For the tensor product, with the same notation and applying once more the fact that tensoring with L commutes with finite products, $A \otimes_K A'$ has dimension nn' and

$$L \otimes_K (A \otimes_K A') \cong (L \otimes_K A) \otimes_K A' \cong L^n \otimes_K A'$$

$$\cong (L \otimes_K A')^n \cong \left(L^{n'} \right)^n \cong L^{nn'}.$$

Again condition (vi) of theorem 2.3.3 is satisfied by $A \otimes_K A'$.

To prove the last assertion, observe that the subalgebra generated by A_1 and A_2 is precisely

$$A_1 \cdot A_2 = \left\{ \sum_{i=1}^{k} a_i^1 a_i^2 \,\middle|\, a_i^1 \in A_1, \ a_i^2 \in A_2 \right\}.$$

This algebra can be presented as a quotient

$$A_1 \otimes_K A_2 \longrightarrow\!\!\!\!\!\rightarrow A_1 \cdot A_2, \quad a_1 \otimes a_2 \mapsto a_1 a_2,$$

from which the result follows by the previous parts of the proof. $\square$

Lemma 2.3.6 *In the conditions of theorem 2.3.3, L splits the K-algebra A if and only if its Gelfand transformation is an isomorphism. That is, conditions* (i) *and* (ii) *of theorem 2.3.3 are equivalent.*

Proof   (i) $\Rightarrow$ (ii) Let $a \in A$ have minimal polynomial $p(X) \in K[X]$ of degree n. In L, $p(X)$ admits n distinct roots, thus $\#\mathsf{Hom}_K(K(a), L) = n$. Applying propositions 2.1.10 and 2.2.5, we deduce that $K(a) \cong \frac{K[X]}{\langle p(X) \rangle}$ satisfies condition (v) of theorem 2.3.3, thus condition (ii) by lemma 2.3.4. Since A is finite dimensional over K, it is the K-algebra generated by finitely many such subalgebras $K(a_i)$, from which one deduces the conclusion by the last assertion in lemma 2.3.5, iterated finitely many times.

(ii) $\Rightarrow$ (i)   Write $p(X) \in K[X]$ for the minimal polynomial, with degree n, of a given element $a \in A$. One has $\frac{K[X]}{\langle p(X) \rangle} \cong K(a)$ by proposition 2.1.10 and $K(a) \subseteq A$ satisfies condition (ii) of theorem 2.3.3 by assumption on A and lemma 2.3.5 (stability under subobjects). By lemma 2.3.4, it follows that $\#\mathsf{Hom}_K\left(\frac{K[X]}{\langle p(X) \rangle}, L \right) = n$ which implies, by proposition 2.2.5, that $p(X)$ has n distinct roots in L. $\square$

2.4 The Galois equivalence

Let us recall that given a group G, whose composition law is written multiplicatively, a left G-set is a set X provided with a left action of G

$$G \times X \longrightarrow X, \quad (g, x) \mapsto gx$$

which satisfies the axioms $1x = x$ and $g(g'x) = (gg')x$, for all elements $x \in X$ and $g, g' \in G$. A morphism $f \colon X \longrightarrow Y$ of left G-sets is a map $f \colon X \longrightarrow Y$ which respects the action of G, that is, $f(gx) = gf(x)$ for all $x \in X$ and $g \in G$. Let us observe that G itself becomes a G-set with the multiplication of G as action. For a group G, we write G-Set for the category of left G-sets and their morphisms, and G-Set$_f$ for the full subcategory of finite left G-sets. From now on, we shall refer to left G-sets just as G-sets.

Proposition 2.4.1 *Let G be a group. There exists a bijection between*

- *the subgroups of G,*
- *the quotients of the G-set G.*

This bijection puts a subgroup H in correspondence with the quotient G-set G/H.

Proof With every subgroup $H \subseteq G$ is associated the equivalence relation

$$x \equiv y \quad \text{iff} \quad x^{-1}y \in H;$$

the quotient set is written G/H. This quotient set can easily be provided with the structure of a G-set, by putting $g[x] = [gx]$, for all elements $g, x \in G$. This definition is easily seen to be independent of the choice of x in the equivalence class $[x]$: indeed, if $x \equiv y$, then

$$(gx)^{-1}(gy) = x^{-1}g^{-1}gy = x^{-1}y \in H$$

from which $[gx] = [gy]$. By construction, the projection $G \longrightarrow\!\!\!\!\!\rightarrow G/H$ is a morphism of G-sets.

Observe that $1 \equiv x$ precisely when $x \in H$, that is, $H = [1]$.

Conversely, starting with a quotient $p \colon G \longrightarrow\!\!\!\!\!\rightarrow Q$ of G-sets, let us consider the equivalence class $[1]$ of the unit of G. It follows at once that $[1]$ is a subgroup of G. Indeed if $x, y \in [1]$,

$$[x^{-1}] = [x^{-1}1] = x^{-1}[1] = x^{-1}[x] = [x^{-1}x] = [1],$$
$$[xy] = x[y] = x[1] = [x1] = [x] = [1].$$

Observe finally that the quotient Q of G is that induced by the subgroup $[1]$. Indeed, if $[x] = [y]$,

$$[x^{-1}y] = x^{-1}[y] = x^{-1}[x] = [x^{-1}x] = [1]$$

and on the other hand if $x^{-1}y \in [1]$,

$$[y] = [xx^{-1}y] = x[x^{-1}y] = x[1] = [x1] = [x]. \qquad \square$$

Proposition 2.4.2 *Let G be a group. Every G-set X is a sum of quotients of the G-set G. When X is finite, this sum is finite.*

Proof A sum of G-sets is their disjoint union, with the original action of G on each piece of the disjoint union. Given $x \in X$, it follows at once that

$$Gx = \{gx | g \in G\}$$

is a sub-G-set of X isomorphic to a quotient of the G-set G, that is,

$$G \longrightarrow\!\!\!\!\!\rightarrow Gx, \quad g \mapsto gx.$$

If $y \in X \setminus Gx$, observe that Gx and Gy are disjoint. Indeed, $gx = g'y$ would imply $y = (g')^{-1}gx \in Gx$. The result follows at once. $\qquad \square$

Theorem 2.4.3 (Galois theorem) *Let $K \subseteq L$ be a finite dimensional Galois extension of fields. Let us write $\mathsf{Gal}\,[L : K]$ for the group of K-automorphisms of L and $\mathsf{Gal}\,[L : K]\text{-Set}_f$ for the category of finite $\mathsf{Gal}\,[L : K]$-sets. Let us also write $\mathsf{Split}_K(L)_f$ for the category of those finite dimensional K-algebras which are split by L. The functor on $\mathsf{Split}_K(L)_f$, represented by L, factors through the category $\mathsf{Gal}\,[L : K]\text{-Set}_f$:*

$$\mathsf{Hom}_K(-, L)\colon \mathsf{Split}_K(L)_f \longrightarrow \mathsf{Gal}\,[L : K]\text{-Set}_f, \quad A \mapsto \mathsf{Hom}_K(A, L)$$

with $\mathsf{Gal}\,[L : K]$ acting by composition on $\mathsf{Hom}_K(L)$. This factorization functor is a contravariant equivalence of categories.

Proof The action of $\mathsf{Gal}\,[L : K]$ is thus given by

$$\mathsf{Gal}\,[L : K] \times \mathsf{Hom}_K(A, L) \longrightarrow \mathsf{Hom}_K(A, L), \quad (g, f) \mapsto g \circ f.$$

For the sake of clarity, we split the proof into five lemmas. Lemmas 2.4.6, 2.4.7 and 2.4.8 imply the result at once. $\qquad \square$

Lemma 2.4.4 *In the conditions of theorem 2.4.3, for every algebra $A \in \mathsf{Split}_K(L)_f$, we get the structure of a $\mathsf{Gal}\,[L : K]$-set on $L \otimes_K A$ by putting*

$$\mathsf{Gal}\,[L : K] \times (L \otimes_K A) \longrightarrow L \otimes_K A, \quad (g, l \otimes a) \mapsto g(l) \otimes a.$$

Via the Gelfand isomorphism of theorem 2.3.3, this action becomes

$$\mathsf{Gal}\,[L : K] \times L^{\mathsf{Hom}_K(A,L)} \longrightarrow L^{\mathsf{Hom}_K(A,L)},$$

$$(g, \varphi) \mapsto \left[f \mapsto g\big(\varphi(g^{-1} \circ f)\big) \right]$$

where $\varphi \colon \mathsf{Hom}_K(A, L) \longrightarrow L$ and $f \in \mathsf{Hom}_K(A, L)$.

Proof Let us fix an element $g \in \mathsf{Gal}\,[L : K]$ and consider the morphism

$$\gamma \colon L^{\mathsf{Hom}_K(A,L)} \longrightarrow L^{\mathsf{Hom}_K(A,L)}, \quad \big(\gamma(\varphi)\big)(f) = g\big(\varphi(g^{-1} \circ f)\big).$$

One computes at once that

$$
\begin{aligned}
\Big((\gamma \circ \mathsf{Gel})(l \otimes a)\Big)(f) &= \Big(\gamma\big(\mathsf{Gel}(l \otimes a)\big)\Big)(f) \\
&= g\Big(\mathsf{Gel}(l \otimes a)(g^{-1} \circ f)\Big) \\
&= g\Big(l\big((g^{-1} \circ f)(a)\big)\Big) \\
&= g\Big(lg^{-1}\big(f(a)\big)\Big) \\
&= g(l)gg^{-1}\big(f(a)\big) \\
&= g(l)f(a) \\
&= \mathsf{Gel}\Big((g \otimes \mathsf{id})(l \otimes a)\Big)(f) \\
&= \Big((\mathsf{Gel} \circ (g \otimes \mathsf{id}))(l \otimes a)\Big)(f).
\end{aligned}
$$

This proves the commutativity of the diagram

$$
\begin{array}{ccc}
L \otimes_K A & \xrightarrow{\;\;\mathsf{Gel}\;\cong\;} & L^{\mathsf{Hom}_K(A,L)} \\
{\scriptstyle g \otimes \mathsf{id}}\big\downarrow & & \big\downarrow{\scriptstyle \gamma} \\
L \otimes_K A & \xrightarrow[\;\mathsf{Gel}\;]{\cong} & L^{\mathsf{Hom}_K(A,L)}
\end{array}
$$

and this expresses precisely the equivalence between the two formulations of the statement.

The fact of having the structure of a $\mathsf{Gal}\,[L : K]$-set is obvious.   □

Lemma 2.4.5 *In the conditions of theorem 2.4.3, for every algebra* $A \in \mathsf{Split}_K(L)_f$, *one has*

$$A \cong \mathsf{Fix}_{\mathsf{Gal}\,[L:K]}(L \otimes_K A)$$
$$= \{x \in L \otimes_K A \,|\, \forall g \in \mathsf{Gal}\,[L:K]\ (g \otimes \mathsf{id})(x) = x\}.$$

Proof First of all let us identify A with a subset of $L \otimes_K A$ via the inclusion

$$A \cong K \otimes_K A \rightarrowtail L \otimes_K A, \quad a \mapsto 1 \otimes a.$$

(Every vector space is flat: tensoring with a vector space preserves monomorphisms.) For every $g \in \mathsf{Gal}\,[L : K]$, one obviously gets $(g \otimes \mathsf{id})(1 \otimes a) = 1 \otimes a$, which proves already that $A \subseteq \mathsf{Fix}_{\mathsf{Gal}\,[L:K]}(L \otimes_K A)$.

To prove the equality, let us recall that the K-algebra A is finite dimensional over K, thus, as a K-vector space, is isomorphic to K^n for some $n \in \mathbb{N}$. Let us consider the commutative diagram

$$
\begin{array}{ccc}
L \otimes_K K^n & \xrightarrow{\ g \otimes \mathsf{id}\ } & L \otimes_K K^n \\
\Big\downarrow{\scriptstyle\cong} & & \Big\downarrow{\scriptstyle\cong} \\
L^n & \xrightarrow[\ g^n\ ]{} & L^n
\end{array}
$$

for every $g \in \mathsf{Gal}\,[L : K]$. It reduces the problem to considering those points of L^n which are fixed by all g^n. With the notation of section 1.4 and by the classical Galois theorem (see 1.4.5),

$$\mathsf{Fix}_{\mathsf{Gal}\,[L:K]}(L \otimes_K K^n) \cong \Big(\mathsf{Fix}\,(\mathsf{Gal}\,[L : K])\Big)^n \cong K^n.$$

Remembering the form of the isomorphism

$$L^n \xrightarrow{\ \cong\ } L \otimes_K K^n, \quad (l_i)_{1 \le i \le n} \mapsto \sum_{i=1}^{n} l_i \otimes e_i,$$

where e_i is the ith vector of the canonical basis of K^n, we obtain

$$
\begin{aligned}
\mathsf{Fix}_{\mathsf{Gal}\,[L:K]}(L \otimes_K K^n) \ &\cong\ \left\{ \sum_{i=1}^{n} k_i \otimes e_i \,\middle|\, k_i \in K \right\} \\
&\cong\ \left\{ 1 \otimes \left(\sum_{i=1}^{n} k_i e_i \right) \,\middle|\, k_i \in K \right\} \\
&\cong\ A
\end{aligned}
$$

which concludes the proof.

Lemma 2.4.6 *The functor described in theorem 2.4.3 is full.*

Proof Let us fix two K-algebras A and B in $\mathsf{Split}_K(L)$, and a morphism of $\mathsf{Gal}[L:K]$-sets

$$\varphi\colon \mathsf{Hom}_K(B,L) \longrightarrow \mathsf{Hom}_K(A,L).$$

This yields at once a map

$$L^\varphi\colon L^{\mathsf{Hom}_K(A,L)} \longrightarrow L^{\mathsf{Hom}_K(B,L)},$$
$$(l_f)_{f\in\mathsf{Hom}_K(A,L)} \mapsto (l_{\varphi(h)})_{h\in\mathsf{Hom}_K(B,L)}.$$

In lemma 2.4.4, we described the structures of $\mathsf{Gal}[L:K]$-sets on these powers of L; let us observe that L^φ is a morphism of $\mathsf{Gal}[L:K]$-sets for these structures. Indeed, let $g \in \mathsf{Gal}[L:K]$; to avoid ambiguity with taking the image under the map g, let us write $*$ for the action of g on the $\mathsf{Gal}[L:K]$-sets. One has

$$
\begin{aligned}
L^\varphi\big(g*(l_f)_{f\in\mathsf{Hom}_K(A,L)}\big) &= L^\varphi\big(g(l_{g^{-1}\circ f})\big)_{f\in\mathsf{Hom}_K(A,L)} \\
&= \big(g(l_{g^{-1}\circ\varphi(h)})\big)_{h\in\mathsf{Hom}_K(B,L)} \\
&= g*\big(l_{\varphi(h)}\big)_{h\in\mathsf{Hom}_K(B,L)} \\
&= g*L^\varphi\big((l_f)_{f\in\mathsf{Hom}_K(A,L)}\big).
\end{aligned}
$$

Since L^φ is a morphism of $\mathsf{Gal}[L:K]$-sets, it factors through the corresponding $\mathsf{Gal}[L:K]$-subset of those points which are fixed by the action of every element g. By lemma 2.4.5 and using the Gelfand isomorphism of theorem 2.3.3, this yields the following situation:

$$
\begin{array}{ccccc}
A & \xrightarrow{\ \cong\ } & \mathsf{Fix}_{\mathsf{Gal}[L:K]}(L\otimes_K A) & \xrightarrow{\ \cong\ } & \mathsf{Fix}_{\mathsf{Gal}[L:K]}\big(L^{\mathsf{Hom}_K(A,L)}\big) \\
&&&& \Big\downarrow{\scriptstyle L^\varphi} \\
B & \xleftarrow{\ \cong\ } & \mathsf{Fix}_{\mathsf{Gal}[L:K]}(L\otimes_K B) & \xleftarrow{\ \cong\ } & \mathsf{Fix}_{\mathsf{Gal}[L:K]}\big(L^{\mathsf{Hom}_K(B,L)}\big)
\end{array}
$$

Let us write $\psi\colon A \longrightarrow B$ for this composite; we shall prove that $\varphi = \mathsf{Hom}_K(\psi,L)$.

Given $h \in \mathsf{Hom}_K(B,L)$, consider the following diagram:

$$L \otimes_K B \xrightarrow[\cong]{\mathrm{Gel}_B} L^{\mathrm{Hom}_K(B,L)}$$

$$i_B \uparrow \qquad\qquad \downarrow p_h$$

$$B \xrightarrow{\ h\ } L$$

where $i_B(b) = 1 \otimes b$ and p_h is the projection of index h. This diagram is commutative since

$$(p_h \circ \mathrm{Gel}_B \circ i_B)(b) = (p_h \circ \mathrm{Gel}_B)(1 \otimes b) = p_h\left((h'(b))_{h' \in \mathrm{Hom}_K(B,L)}\right) = h(b).$$

Let us also write $\overline{\varphi}\colon L \otimes_K B \longrightarrow L \otimes_K B$ for the morphism corresponding to L^φ by the Gelfand isomorphism of theorem 2.3.3. The following diagram is commutative by definition of ψ and $\overline{\psi}$:

$$
\begin{array}{ccccc}
A & \xrightarrow{\ i_A\ } & L \otimes_K A & \xrightarrow[\cong]{\mathrm{Gel}_A} & L^{\mathrm{Hom}_K(A,L)} \\
\downarrow{\scriptstyle\psi} & & \downarrow{\scriptstyle\overline{\varphi}} & & \downarrow{\scriptstyle L^\varphi} \\
B & \xrightarrow[\ i_B\ ]{} & L \otimes_K B & \xrightarrow[\mathrm{Gel}_B]{\cong} & L^{\mathrm{Hom}_K(B,L)}
\end{array}
$$

It yields

$$
\begin{aligned}
\mathrm{Hom}_K(\psi, L)(h) &= h \circ \psi \\
&= p_h \circ \mathrm{Gel}_B \circ i_B \circ \psi \\
&= p_h \circ \mathrm{Gel}_B \circ \overline{\varphi} \circ i_A \\
&= p_h \circ L^\varphi \circ \mathrm{Gel}_A \circ i_A \\
&= p_{\varphi(h)} \circ \mathrm{Gel}_A \circ i_A \\
&= \varphi(h),
\end{aligned}
$$

which concludes the proof. $\square$

Lemma 2.4.7 *The functor described in theorem 2.4.3 is faithful.*

Proof With the notation of lemma 2.4.6, consider a second morphism $\psi'\colon A \longrightarrow B$ such that $\mathrm{Hom}_K(\psi', L) = \varphi$. For every $h \in \mathrm{Hom}_K(B,L)$, we get

$$
\begin{aligned}
p_h \circ \mathsf{Gel}_B \circ i_B \circ \psi' &= h \circ \psi' \\
&= \varphi(h) \\
&= p_{\varphi(h)} \circ \mathsf{Gel}_A \circ i_A \\
&= p_h \circ L^\varphi \circ \mathsf{Gel}_A \circ i_A \\
&= p_h \circ \mathsf{Gel}_B \circ i_B \circ \psi.
\end{aligned}
$$

Since this relation holds for every projection p_h, it follows that

$$
\mathsf{Gel}_B \circ i_B \circ \psi' = \mathsf{Gel}_B \circ i_B \circ \psi
$$

and since both Gel_B and i_B are injective, $\psi' = \psi$. $\qquad\square$

Lemma 2.4.8 *The functor described in theorem 2.4.3 is essentially surjective on the objects.*

Proof Consider first a subgroup $H \subseteq \mathsf{Gal}\,[L:K]$ and the corresponding $\mathsf{Gal}\,[L:K]$-quotient-set $\mathsf{Gal}\,[L:K]/H$ as in proposition 2.4.1. We shall prove that

$$
\frac{\mathsf{Gal}\,[L:K]}{H} \cong \mathsf{Hom}_K\big(\mathsf{Fix}\,(H), L\big).
$$

Considering the inclusion $\mathsf{Fix}\,(H) \subseteq L$, we get by functoriality a morphism of finite $\mathsf{Gal}\,[L:K]$-sets

$$
\rho \colon \mathsf{Gal}\,[L:K] \cong \mathsf{Hom}_K(L, L) \longrightarrow \mathsf{Hom}_K\big(\mathsf{Fix}\,(H), L\big)
$$

sending $f \colon L \longrightarrow L$ onto its restriction $f \colon \mathsf{Fix}\,(H) \longrightarrow L$; let us prove that this defines a quotient map. Considering the diagram

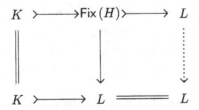

and proposition 1.3.3, we conclude already that every morphism of K-algebras $\mathsf{Fix}\,(H) \longrightarrow L$ is indeed the restriction of a morphism $L \longrightarrow L$; thus the map ρ is a quotient map.

To prove that $\mathsf{Hom}_K\big(\mathsf{Fix}\,(H), L\big)$ is precisely the expected quotient $\mathsf{Gal}\,[L:K]/H$, it remains to prove that two morphisms of K-algebras

$f, g \colon L \overset{\longrightarrow}{\longrightarrow} L$ have the same restriction to $\mathsf{Fix}\,(H)$ if and only if $f^{-1} \circ g \in H$. Indeed, f and g have the same restriction to $\mathsf{Fix}\,(H)$ precisely when $f^{-1} \circ g$ fixes the points of $\mathsf{Fix}\,(H)$, that is

$$f^{-1} \circ g \in \mathsf{Gal}\,\big[L : \mathsf{Fix}\,(H)\big] = H$$

by the classical Galois theorem (see 1.4.5) and using its notation.

Every quotient of the $\mathsf{Gal}\,[L : K]$-set $\mathsf{Gal}\,[L : K]$ is thus isomorphic to an object of the form $\mathsf{Hom}_K(A, L)$ for some $A \in \mathsf{Split}_K(L)_f$. On the other hand every finite $\mathsf{Gal}\,[L : K]$-set is, by proposition 2.4.2, a finite disjoint union of quotients of $\mathsf{Gal}\,[L : K]$. Since the category $\mathsf{Split}_K(L)_f$ has finite products by lemma 2.3.5, it suffices, for concluding the proof, to prove that the contravariant functor $\mathsf{Hom}_K(-, L)$ of theorem 2.4.3 transforms finite products into finite sums.

Consider for this two algebras $A, B \in \mathsf{Split}_K(L)_f$, of respective dimensions n and m. Composing with the projections

$$A \overset{\longleftarrow}{} A \times B \overset{\longrightarrow}{} B$$

yields maps

$$\mathsf{Hom}_K(A, L) \rightarrowtail \mathsf{Hom}_K(A \times B, L) \leftarrowtail \mathsf{Hom}_K(B, L)$$

which are injective since the projections are surjective. The corresponding subsets are disjoint, as observed by checking the actions on the elements $(1, 0)$ and $(0, 1)$ of $A \times B$. Theorem 2.3.3 and various previous arguments show that

$$
\begin{aligned}
\#\mathsf{Hom}_K(A \times B, L) &= \#\mathsf{Hom}_L\big(L \otimes_K (A \times B), L\big) \\
&= \#\mathsf{Hom}_K\big((L \otimes_K A) \times (L \otimes_K B), L\big) \\
&= \#\mathsf{Hom}_L(L^n \times L^m, L) \\
&= \#\mathsf{Hom}_L(L^{n+m}, L) \\
&= n + m \\
&= \#\mathsf{Hom}_K(A, L) + \#\mathsf{Hom}_K(B, L).
\end{aligned}
$$

This concludes the proof of this lemma, thus also the proof of theorem 2.4.3. $\qquad\qquad\square$

To conclude this chapter, it remains to observe that the Galois theorem we have just proved (theorem 2.4.3) contains the classical Galois theorem (theorem 1.4.5). Indeed, the contravariant equivalence of theorem 2.4.3 implies in particular the existence of an isomorphism between the lattice of subobjects M

$$K \rightarrowtail M \rightarrowtail L$$

in $\mathsf{Split}_K(L)_f$ and the lattice of quotients $\mathsf{Hom}_K(M, L)$,

$$\mathsf{Gal}\,[L:K] \cong \mathsf{Hom}_K(L, L) \longrightarrow\!\!\!\!\!\to \mathsf{Hom}_K(M, L) \longrightarrow\!\!\!\!\!\to \mathsf{Hom}_K(K, L) \cong \{*\},$$

in $\mathsf{Gal}\,[L:K]$-Set_f. By propositions 2.2.1 and 2.4.1, this is precisely the classical Galois isomorphism.

3

Infinitary Galois theory

Convention *In this chapter, all fields, rings and algebras we consider are commutative and have a unit.*

In this chapter, we develop Galois theory for an arbitrary Galois extension of fields $K \subseteq L$, not necessarily finite dimensional.

3.1 Finitary Galois subextensions

Given a Galois extension of fields $K \subseteq L$, we are interested in the intermediate field extensions $K \subseteq M \subseteq L$, with $K \subseteq M$ a finite dimensional Galois extension of fields. This is what we call a "finite dimensional Galois subextension" of $K \subseteq L$.

Proposition 3.1.1 *Let $K \subseteq L$ be a Galois extension of fields. Consider $l \in L$ with minimal polynomial $p(X) \in K[X]$, admitting in L the roots $l_1, \ldots, l_n$. Then $K \subseteq K(l_1, \ldots, l_n)$ is a finite dimensional Galois extension of fields.*

Proof $K(l_1, \ldots, l_n)$ is finite dimensional over K by proposition 2.1.10. It is the union, in the category of K-algebras, of the subalgebras $K(l_i)$, for all i. Each l_i admits $p(X)$ as minimal polynomial, thus by proposition 2.2.5

$$\#\mathrm{Hom}_K\big(K(l_i), K(l_1, \ldots, l_n)\big)$$

$$= \#\mathrm{Hom}_K\left(\frac{K[X]}{\langle p(X) \rangle}, K(l_1, \ldots, l_n)\right)$$

$$= \text{number of roots of } p(X) \text{ in } K(l_1, \ldots, l_n)$$

$$= \text{degree of } p(X)$$

36

$$= \dim {}_K K(l_i).$$

This proves by theorem 2.3.3 that $K(l_1, \dots, l_n)$ splits $K(l_i)$. By lemma 2.3.5, $K(l_1, \dots, l_n)$ splits $K(l_1) \cup \cdots \cup K(l_n) = K(l_1, \dots, l_n)$, which proves by proposition 2.3.2 that $K \subseteq K(l_1, \dots, l_n)$ is a Galois extension. $\qquad \square$

Proposition 3.1.2 *Let $K \subseteq L$ be a Galois extension of fields and $K \subseteq M \subseteq L$ an intermediate field extension, with M finite dimensional over K. Every K-homomorphism of fields $f \colon M \longrightarrow M$ extends to a K-homomorphism $g \colon L \longrightarrow L$.*

Proof We recall that a field L is algebraically closed when every polynomial in $L[X]$ has a root in L, thus by induction, factors in $L[X]$ as a product of polynomials of degree 1. We shall use freely the fact that every field K has an algebraic closure $\overline{K}$, which contains in particular all algebraic extensions of K. Moreover, every field homomorphism $f \colon K \longrightarrow M$ extends to a homomorphism $\overline{f} \colon \overline{K} \longrightarrow \overline{M}$ between the algebraic closures, and when f is an isomorphism, so is $\overline{f}$ (see [71]).

Using the previous notation, from the assumption we get an isomorphism $\overline{f} \colon \overline{M} \longrightarrow \overline{M}$ extending f. Since L is algebraic over K, we get $K \subseteq L \subseteq \overline{M}$ and it remains to prove that $\overline{f}(L) \subseteq L$. Given $l \in L$ with minimal polynomial $p(X) \in K[X]$, we have

$$p\bigl(\overline{f}(l)\bigr) = \overline{f}\bigl(p(l)\bigr) = \overline{f}(0) = 0$$

since f, and thus also $\overline{f}$, fix the coefficients of $p(X)$. Thus $\overline{f}(l)$ is a root of $p(X)$ and, since $K \subseteq L$ is a Galois extension, $\overline{f}(l) \in L$. $\qquad \square$

Corollary 3.1.3 *Let $K \subseteq L$ be a Galois extension of fields. One has*

$$K = \bigl\{l \in L \,\big|\, \forall f \in \mathsf{Gal}\,[L:K] \ \ f(l) = l\bigr\} = \mathsf{Fix}\,\bigl(\mathsf{Gal}\,[L:K]\bigr).$$

Proof If $l \notin K$, let $p(X) \in K[X]$ be its minimal polynomial, with roots $l_1, \dots, l_n$ in L. Moreover $K \subseteq K(l_1, \dots, l_n)$ is a Galois extension by proposition 3.1.1. Since $l \notin K$, $p(X)$ does not have degree 1 so that we can fix $l_i \neq l$. By proposition 1.3.4, there exists

$$f \colon K(l_1, \dots, l_n) \longrightarrow K(l_1, \dots, l_n)$$

such that $f(l) = l_i \neq l$. It remains to extend f to $g \colon L \longrightarrow L$ by proposition 3.1.2, which yields $g \in \mathsf{Gal}\,[L:K]$ such that $g(l) \neq l$. $\qquad \square$

Proposition 3.1.4 *Let $K \subseteq L$ be a Galois extension of fields. The field L is the set-theoretical filtered union of the subextensions $K \subseteq M \subseteq L$ where $K \subseteq M$ is a finite dimensional Galois extension.*

Proof   If $l \in L$ has minimal polynomial $p(X) \in K[X]$ with roots $l_1, \dots, l_n$ in L, then by proposition 3.1.1,

$$l \in K(l_1, \dots, l_n) \subseteq L$$

where $K \subseteq K(l_1, \dots, l_n)$ is a finite dimensional Galois extension. The field L is thus indeed the set theoretical union of the finite dimensional Galois subextensions.

It remains to prove that this union is filtered. For this choose

$$K \subseteq M_1 \subseteq L, \quad K \subseteq M_2 \subseteq L$$

with $K \subseteq M_1$ and $K \subseteq M_2$ finite dimensional Galois extensions. The K-subalgebra $M_3 \subseteq L$ generated by M_1 and M_2 remains finite dimensional over K. Since $K \subseteq M_1$ is a Galois extension, the minimal polynomial of $l \in M_1$ factors in M_1, thus also in M_3, into distinct factors of degree 1; the same argument holds for M_2. This proves that M_3 splits both M_1 and M_2, thus M_3 splits M_3 by lemma 2.3.5. By proposition 2.3.2, this proves that $K \subset M_3$ is a Galois extension. $\square$

Proposition 3.1.5 *Let $K \subseteq L$ be a Galois extension of fields. Suppose that L splits the K-algebra A. For every finite dimensional K-subalgebra $B \subseteq A$, there exists a finite dimensional Galois subextension $K \subseteq M \subseteq L$ which splits B.*

Proof   The subalgebra B is generated over K by a finite number $b_1, \dots, b_n$ of elements. Each of these elements b_i has a minimal polynomial $p_i(X) \in K[X]$, admitting in L the roots $l_1^i, \dots, l_{m_i}^i$. The extension

$$M_i = K(l_1^i, \dots, l_{m_i}^i)$$

is a finite dimensional Galois extension by proposition 3.1.1. By proposition 3.1.4, there exists a finite dimensional Galois extension M containing $M_1, \dots, M_n$. By theorem 2.3.3 this extension M splits each $K(b_i) \subseteq B$, since

$$\#\mathsf{Hom}_K\big(K(b_i), M\big) \;=\; \#\mathsf{Hom}_K\left(\frac{K[X]}{\langle p_i(X)\rangle}, M\right)$$

$$=\; \text{number of roots of } p_i(X) \text{ in } M$$

$$= \text{ degree of } p_i(X)$$
$$= \dim {}_K K(b_i)$$

by proposition 2.2.5. By lemma 2.3.5, M splits $K(b_1) \cup \cdots \cup K(b_n) = B$.

$$\square$$

3.2 Infinitary Galois groups

The Galois group of an arbitrary Galois extension $K \subseteq L$ will be a topological group, which will turn out to be discrete when the extension is finite dimensional.

Proposition 3.2.1 *Let $K \subseteq L$ be a Galois extension of fields. In the category of groups,*

$$\mathsf{Gal}\,[L : K] = \lim{}_M \mathsf{Gal}\,[M : K]$$

where M runs through the poset of finite dimensional Galois extensions $K \subseteq M \subseteq L$ and, for $M \subseteq M'$, the corresponding morphism

$$\mathsf{Gal}\,[M' : K] \longrightarrow \mathsf{Gal}\,[M : K], \quad f \mapsto f|_M$$

is the restriction.

Proof Let us recall that given a K-homomorphism $f \colon M' \longrightarrow M'$ and an element $l \in M$, for $M \subseteq M'$, this element l has a minimal polynomial $p(X) \in K[X]$ and

$$p\big(f(l)\big) = f\big(p(l)\big) = f(0) = 0$$

since f fixes the coefficients of $p(X)$. Thus $f(l)$ is a root of $p(X)$ and therefore $f(l) \in M$ since $K \subseteq M$ is a Galois extension. This proves that f indeed restricts to a K-automorphism of M. The same argument holds replacing M' by L, which provides the projections

$$p_M \colon \mathsf{Gal}\,[L : K] \longrightarrow \mathsf{Gal}\,[M : K]$$

which are thus the restrictions to M. They clearly form a cone, since all morphisms involved are restrictions.

To prove that this cone is a limit one, it remains to choose a compatible family $f_M \in \mathsf{Gal}\,[M : K]$ and show it glues uniquely as an element $f \in \mathsf{Gal}\,[L : K]$. This follows at once from proposition 3.1.4.

Another way of writing the same proof is to consider the isomorphisms

$$\mathsf{Gal}\,[L : K] \quad = \quad \mathsf{Hom}_K(L, L)$$

$$= \text{Hom}_K(\text{colim}_M M, L)$$
$$\cong \lim_M \text{Hom}_K(M, L)$$
$$\cong \lim_M \text{Hom}_K(M, M)$$
$$= \lim_M \text{Gal}\,[M : K],$$

using once more the argument that a homomorphisms $M \longrightarrow L$ factors through M, by the Galois property of the extension. □

Definition 3.2.2 Let $K \subseteq L$ be a Galois field extension. The topological Galois group of this extension is the group $\text{Gal}\,[L : K]$ provided with the initial topology for all the projections

$$\text{Gal}\,[L : K] \cong \lim_M \text{Gal}\,[M : K] \longrightarrow \text{Gal}\,[M : K], \quad f \mapsto f|_M,$$

where M runs through the finite dimensional Galois subextensions $K \subseteq M \subseteq L$ and each $\text{Gal}\,[M : K]$ is provided with the discrete topology.

By theorem 1.4.5, all the groups $\text{Gal}\,[M : K]$ in the above definition are finite; by proposition 3.1.4, the diagram constituted by these $\text{Gal}\,[M : K]$ is cofiltered. The topological Galois group is thus a cofiltered projective limit, in the category of topological groups, of a diagram constituted of discrete finite groups: such a group is called a *profinite* group. We shall first study its topology further.

Lemma 3.2.3 *Let $K \subseteq L$ be a Galois field extension. The subgroups $\text{Gal}\,[L : M] \subseteq \text{Gal}\,[L : K]$, for $K \subseteq M \subseteq L$ a finite dimensional Galois subextension, constitute a fundamental system of open and closed neighbourhoods of id_L.*

Proof A fundamental open subset of $\text{Gal}\,[L : K]$ has the form

$$U = p_{M_1}^{-1}(X_1) \cap \cdots \cap p_{M_n}^{-1}(X_n)$$

where $X_i \subseteq \text{Gal}\,[M_i : K]$ is an arbitrary (open) subset and p_{M_i} is the corresponding projection. Notice that U is both open and closed since so is each X_i. An arbitrary open subset of $\text{Gal}\,[L : K]$ is a union of such fundamental open subsets. For U being a fundamental neighbourhood of id_L, we must have $1_{M_i} = p_{M_i}(1_L) \in X_i$ for all $i = 1, \ldots, n$. In that case U contains

$$V = p_{M_1}^{-1}(\{1_{M_1}\}) \cap \cdots \cap p_{M_n}^{-1}(\{1_{M_n}\})$$

which is still an open and closed neighbourhood of id_L. Observe that

$$f \in V \quad \Leftrightarrow \quad \forall i = 1, \ldots, n \;\; f|_{M_i} = \mathrm{id}_{M_i}$$
$$\Leftrightarrow \quad f|_M = \mathrm{id}_M$$

where M is the subfield of L generated by $M_1, \ldots, M_n$. Then $K \subseteq M$ is again a finite dimensional Galois extension, as observed in the proof of proposition 3.1.4. We have thus proved that $V = \mathsf{Gal}\,[L : M]$. $\qquad \square$

Lemma 3.2.4 *Let $K \subseteq L$ be a Galois extension of fields. The topology of the Galois group $\mathsf{Gal}\,[L : K]$ is the initial topology for all the maps*

$$\mathsf{ev}_l \colon \mathsf{Gal}\,[L : K] \longrightarrow L, \quad f \mapsto f(l)$$

where l runs through L and the codomain L of ev_l is provided with the discrete topology.

Proof The topology described in this lemma is also called the topology of pointwise convergence on $\mathsf{Gal}\,[L : K]$.

Let us prove first that each map ev_l is continuous. Since L is provided with the discrete topology, it suffices to prove that the inverse image of each point is open. For this, let us fix $l, l_0 \in L$.

$$\mathsf{ev}_l^{-1}(\{l_0\}) = \{f \in \mathsf{Gal}\,[L : K] \,|\, f(l) = l_0\}.$$

If this set is empty, it is open. Otherwise, by proposition 3.1.4, we choose a finite dimensional Galois subextension $K \subseteq M \subseteq L$ containing l and l_0. Let us write $p(X) \in K[X]$ for the minimal polynomial of l. Choosing $f \in \mathsf{ev}_l^{-1}(\{l_0\})$,

$$p(l_0) = p\big(f(l)\big) = f\big(p(l)\big) = f(0) = 0$$

and thus l_0 is a root of $p(X)$. By proposition 1.3.4, there exists a K-automorphism $g \colon M \longrightarrow M$ such that $g(l) = l_0$. Therefore

$$
\begin{aligned}
\mathsf{ev}_l^{-1}(\{l_0\}) &= \{f \in \mathsf{Gal}\,[L : K] \,|\, f(l) = l_0\} \\
&= \{f \in \mathsf{Gal}\,[L : K] \,|\, f(l) = g(l)\} \\
&= \{f \in \mathsf{Gal}\,[L : K] \,|\, (g^{-1} \circ f|_M)(l) = l\}.
\end{aligned}
$$

Consider the pullback

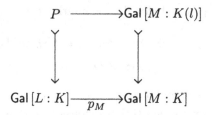

One has

$$P = \big\{ h \in \mathsf{Gal}\,[L : K] \,\big|\, h|_M(l) = l \big\}$$

and therefore

$$
\begin{aligned}
\mathrm{ev}_l^{-1}(\{l_0\}) &= \big\{ f \in \mathsf{Gal}\,[L : K] \,\big|\, g^{-1} \circ f \in P \big\} \\
&= \big\{ g \circ h \,\big|\, h \in P \big\}
\end{aligned}
$$

which is the image of the subset $P \subseteq \mathsf{Gal}\,[L : K]$ by the homeomorphism

$$\mathsf{Gal}\,[L : K] \longrightarrow \mathsf{Gal}\,[L : K], \quad h \mapsto g \circ h$$

inherited from the topological group structure. Since $\mathsf{Gal}\,[M : K]$ is provided with the discrete topology, $\mathsf{Gal}\,[M : K(l)]$ is both open and closed in $\mathsf{Gal}\,[M : K]$. Since p_M is continuous, P too is open and closed in $\mathsf{Gal}\,[L : K]$. Via the homeomorphism we have exhibited, this implies that $\mathrm{ev}_l^{-1}(\{l_0\})$ is open and closed as well.

The topology of lemma 3.2.4 is thus contained in that of lemma 3.2.3. Conversely, it suffices to prove that the fundamental open neighbourhoods $\mathsf{Gal}\,[L : M]$ of id_L in lemma 3.2.3 contain a neighbourhood of id_L for the topology of lemma 3.2.4. Indeed, M is generated by finitely many elements $l_1, \ldots, l_n$. Given $f \in \mathsf{Gal}\,[L : K]$, the condition $f \in \mathsf{Gal}\,[L : M]$ reduces to $f(l_1) = l_1, \ldots, f(l_n) = l_n$. This is equivalent to

$$f \in \mathrm{ev}_{l_1}^{-1}(l_1) \cap \cdots \cap \mathrm{ev}_{l_n}^{-1}(l_n)$$

which is a neighbourhood of id_L for the topology of lemma 3.2.4, since so is each $\mathrm{ev}_{l_i}^{-1}(l_i)$. $\square$

Corollary 3.2.5 *Let $K \subseteq L$ be a Galois extension of fields. For every $f \in \mathsf{Gal}\,[L : K]$, the subsets*

$$V_M(f) = \big\{ g \in \mathsf{Gal}\,[L : K] \,\big|\, g|_M = f|_M \big\} \subseteq \mathsf{Gal}\,[L : K]$$

for $K \subseteq M \subseteq L$ running through the arbitrary finite dimensional subextensions constitute a fundamental system of neighbourhoods of f.

Proof If M is generated by $l_1, \ldots, l_n$,

$$
\begin{aligned}
V_M(f) &= \{g \in \mathsf{Gal}\,[L:K]\,|\,g(l_1) = f(l_1), \cdots, g(l_n) = f(l_n)\} \\
&= \mathsf{ev}_{l_1}^{-1}\big(f(l_1)\big) \cap \cdots \cap \mathsf{ev}_{l_n}^{-1}\big(f(l_n)\big)
\end{aligned}
$$

which is a neighbourhod of f for the topology of lemma 3.2.4.

Conversely, every neighbourhood V of f contains by lemma 3.2.4 a neighbourhood of the form

$$
f \in \mathsf{ev}_{a_1}^{-1}(b_1) \cap \cdots \cap \mathsf{ev}_{a_m}^{-1}(b_m)
$$

where $a_i, b_i \in L$. One has $f(a_i) = b_i$ and therefore

$$
\begin{aligned}
\mathsf{ev}_{a_1}^{-1}(b_1) &\cap \cdots \cap \mathsf{ev}_{a_m}^{-1}(b_m) \\
&= \{g \in \mathsf{Gal}\,[L:K]\,|\,g(a_1) = b_1, \ldots, g(a_m) = b_m\} \\
&= \{g \in \mathsf{Gal}\,[L:K]\,|\,g(a_1) = f(a_1), \ldots, g(a_m) = f(a_m)\} \\
&= V_{K(l_1,\ldots,l_n)}(f)
\end{aligned}
$$

which concludes the proof. $\qquad\square$

Corollary 3.2.6 *Let $K \subseteq L$ be a Galois extension of fields. Given a subset $U \subseteq \mathsf{Gal}\,[L:K]$, its closure is given by*

$$
\overline{U} = \left\{ f \in \mathsf{Gal}\,[L:K] \;\middle|\;
\begin{array}{l}
\forall M \quad K \subseteq M \subseteq L \text{ with } K \subseteq M \\
\quad \text{finite dimensional Galois extension} \\
\exists g \in U \quad g|_M = f|_M
\end{array}
\right\}.
$$

Proof With shortened notation,

$$
\begin{aligned}
\overline{U} &= \{f\,|\,\forall M \ \dim_K(M) \text{ finite}, \ V_M(f) \cap U \neq \emptyset\} \\
&= \{f\,|\,\forall M \ \dim_K(M) \text{ finite}, \ \exists g \in U \ g|_M = f|_M\} \\
&= \{f\,|\,\forall M \ K \subseteq M \text{ Galois extension}, \ \dim_K(M) \text{ finite}, \\
&\qquad\qquad\qquad\qquad \exists g \in U \ g|_M = f|_M\}.
\end{aligned}
$$

The first equality follows from corollary 3.2.5, the second one from the definition of $V_M(f)$ and the third one, via proposition 3.1.4, from the fact that every finite dimensional subextension is contained in a finite dimensional Galois subextension. $\qquad\square$

3.3 Classical infinitary Galois theory

We shall now generalize theorem 1.4.5 to the case of an arbitrary Galois extension $K \subseteq L$. The Galois group $\mathsf{Gal}\,[L : K]$ is always considered as provided with its topology described in section 3.2.

Proposition 3.3.1 *Let $K \subseteq L$ be a Galois extension of fields. Let $K \subseteq M \subseteq L$ be a finite dimensional intermediate Galois extension. The canonical restriction morphism of definition 3.2.2*

$$p_M \colon \mathsf{Gal}\,[L : K] \longrightarrow \mathsf{Gal}\,[M : K]; f \mapsto f|_M$$

is a topological quotient for the equivalence relation determined by the subgroup $\mathsf{Gal}\,[L : M] \subseteq \mathsf{Gal}\,[L : K]$.

Proof   By proposition 3.1.2, the restriction map is surjective. One observes at once that

$$
\begin{aligned}
p_M(f) = p_M(g) \quad &\Leftrightarrow \quad f|_M = g|_M \\
&\Leftrightarrow \quad \forall m \in M \ \ f(m) = g(m) \\
&\Leftrightarrow \quad \forall m \in M \ \ m = (f^{-1} \circ g)(m) \\
&\Leftrightarrow \quad f^{-1} \circ g \in \mathsf{Gal}\,[L : M]
\end{aligned}
$$

which proves that the set theoretical quotient is induced by the subgroup $\mathsf{Gal}\,[L : K]$.

  The quotient topology on $\mathsf{Gal}\,[M : K]$ is the finest one making continuous the surjection p_M. Since the discrete topology makes p_M continuous, it is necessarily the finest one with that property.   □

Proposition 3.3.2 *Let $K \subseteq L$ be an arbitrary Galois extension of fields. For every finite dimensional intermediate extension $K \subseteq M \subseteq L$,*

$$\mathsf{Gal}\,[L : M] = \left\{ f \in \mathsf{Gal}\,[L : K] \,\middle|\, \forall m \in M \ \ f(m) = m \right\}$$

is an open and closed subgroup of $\mathsf{Gal}\,[L : K]$.

Proof We refer to proposition 3.1.4 and consider

$$K \subseteq M \subseteq N \subseteq L, \quad \dim{}_K N \text{ finite}, \quad K \subseteq N \text{ Galois extension}.$$

It follows at once that

$$\mathrm{id}_L \in \mathsf{Gal}\,[L : N] \subseteq \mathsf{Gal}\,[L : M]$$

and by lemma 3.2.3, we know that $\mathsf{Gal}\,[L:N]$ is an open and closed neighbourhood of id_L. This forces the conclusion, by elementary properties of topological groups, namely: every subgroup of a topological group containing an open subgroup is itself open, and every open subgroup is closed.

Indeed, in the special case we are handling here, for every $f \in \mathsf{Gal}\,[L:M]$, multiplication by f is a homeomorphism mapping id_L onto f. This homeomorphism also maps $\mathsf{Gal}\,[L:N]$ onto some open and closed subset $U_f \subseteq \mathsf{Gal}\,[L:M]$, because $\mathsf{Gal}\,[L:M]$ is a subgroup. Thus $\mathsf{Gal}\,[L:M]$ is a join of open subsets and therefore is open. Now if $g \notin \mathsf{Gal}\,[L:M]$, one must have $U_g \cap \mathsf{Gal}\,[L:M] = \emptyset$; otherwise, choosing h in this intersection, one would have $h = g \circ h'$ with $h, h' \in \mathsf{Gal}\,[L:M]$, thus $g \in \mathsf{Gal}\,[L:M]$. This proves that the set theoretical complement of $\mathsf{Gal}\,[L:M]$ is open.

| $\mathsf{Gal}\,[L:K]$ | $\mathsf{Gal}\,[L:M]$ | $\mathsf{Gal}\,[L:N]$ | $\mathrm{id}_L \bullet$ | $\bullet f$ | U_f | $\bullet g$ | U_g | $\square$ |

Corollary 3.3.3 *Let $K \subseteq L$ be an arbitrary Galois extension of fields. For every arbitrary intermediate extension $K \subseteq M \subseteq L$,*

$$\mathsf{Gal}\,[L:M] = \big\{ f \in \mathsf{Gal}\,[L:K] \,\big|\, \forall m \in M \quad f(m) = m \big\}$$

is a closed subgroup of $\mathsf{Gal}\,[L:K]$.

Proof One has

$$
\begin{aligned}
\mathsf{Gal}\,[L:M] &= \big\{ f \in \mathsf{Gal}\,[L:K] \,\big|\, \forall m \in M \quad f(m) = m \big\} \\
&= \big\{ f \in \mathsf{Gal}\,[L:K] \,\big|\, \forall m \in M \quad f \in \mathsf{Gal}\,[L:K(m)] \big\} \\
&= \bigcap_{m \in M} \mathsf{Gal}\,[L:K(m)].
\end{aligned}
$$

We recall that $K(m)$ is finite dimensional over K (see proposition 1.1.4). This forces the conclusion by proposition 3.3.2 and the fact that an intersection of closed subsets is closed. $\square$

Lemma 3.3.4 *Let $K \subseteq L$ be an arbitrary Galois extension of fields and $G \subseteq \mathsf{Gal}\,[L:K]$ a closed subgroup. Moreover, let us suppose that*

$$K = \mathsf{Fix}\,(G) = \big\{ l \in L \,\big|\, \forall g \in G \quad g(l) = l \big\}.$$

In these conditions, $G = \mathsf{Gal}\,[L:K]$.

Proof Let us first consider the subgroup

$$H_M = \{f|_M \,|\, f \in G\} \subseteq \mathsf{Gal}\,[M : K]$$

for every finite dimensional Galois extension $K \subseteq M \subseteq L$. By assumption,

$$
\begin{aligned}
\mathsf{Fix}\,(H_M) &= \{m \in M \,|\, \forall h \in H_M \ h(m) = m\} \\
&= \{m \in M \,|\, \forall f \in G \ f(m) = m\} \\
&= K.
\end{aligned}
$$

Applying the classical Galois theorem (see 1.4.5)

$$H_M = \mathsf{Gal}\,\big[M : \mathsf{Fix}\,(H_M)\big] = \mathsf{Gal}\,[M : K].$$

In other terms

$$\forall f \in \mathsf{Gal}\,[L : K] \ \ \forall M \ K \subseteq M \subseteq L \text{ Galois extension,}$$

$$\dim {}_K M \text{ finite,} \quad f|_M \in H_M$$

or also, with shortened notation,

$$\forall f \in \mathsf{Gal}\,[L : K] \ \ \forall M \ \exists g \in G \ g|_M = f|_M.$$

By corollary 3.2.6, this reduces to

$$\forall f \in \mathsf{Gal}\,[L : K] \ \ f \in \overline{G},$$

that is, finally, $\mathsf{Gal}\,[L : K] = \overline{G}$. Thus $G = \mathsf{Gal}\,[L : K]$ since G is closed. $\qquad\square$

Theorem 3.3.5 (Galois theorem) *Let $K \subseteq L$ be an arbitrary Galois extension of fields. The correspondences*

$$
\begin{aligned}
K \subseteq M \subseteq L &\ \mapsto\ \mathsf{Gal}\,[L : M], \\
G \subseteq \mathsf{Gal}\,[L : K] &\ \mapsto\ \mathsf{Fix}\,(G)
\end{aligned}
$$

induce a contravariant isomorphism between the lattice of arbitrary extensions $K \subseteq M \subseteq L$ and the lattice of closed subgroups $G \subseteq \mathsf{Gal}\,[L : K]$.

Proof By corollary 3.3.3, $\mathsf{Gal}\,[L : M]$ is a closed subgroup. On the other hand since the elements of G are field homomorphisms, it follows at once that $\mathsf{Fix}\,(G)$ is a field.

Let us consider a closed subgroup $G \subseteq$ Gal $[L : K]$. Since $K \subseteq L$ is a Galois extension, Fix $(G) \subseteq L$ is a Galois extension as well (see 1.4.2). On the other hand one has trivially

$$G \subseteq \text{Gal}\,[L : \text{Fix}\,(G)] \subseteq \text{Gal}\,[L : K]$$

with again G closed by assumption. Lemma 3.3.4 applies with Fix (G) instead of K, proving that $G = \text{Gal}\,[L : \text{Fix}\,(G)]$.

Conversely, consider an intermediate extension $K \subseteq M \subseteq L$. Trivially,

$$M \subseteq \text{Fix}\,\big(\text{Gal}\,[L : M]\big) \subseteq L$$

and $M \subseteq L$ is a Galois extension, since so is $K \subseteq L$ (see 1.4.2). Given $l \in \text{Fix}\,\big(\text{Gal}\,[L : M]\big)$, consider by proposition 3.1.4 a finite dimensional Galois extension $M \subseteq M'$, with

$$M \subseteq M' \subseteq L, \quad l \in M'.$$

By proposition 3.1.2, every M-automorphism of M' is the restriction of an M-automorphism of L. Since l is fixed by all elements of Gal $[L : M]$, it is thus also fixed by all elements of Gal $[M' : M]$. Again the classical Galois theorem (see 1.4.5) implies

$$l \in \text{Fix}\,\big(\text{Gal}\,[M' : M]\big) = M. \qquad \square$$

3.4 Profinite topological spaces

To treat Grothendieck infinitary Galois theory for fields, it is useful to study more extensively the profinite topological spaces, already mentioned after definition 3.2.2, in the special case of profinite groups.

Definition 3.4.1 A topological space is profinite when it is the projective limit, indexed by a cofiltered poset, of finite discrete topological spaces.

In fact, the requirement that the limit is indexed by a cofiltered poset can equivalently be omitted in definition 3.4.1, as shown by corollary 3.4.8.

Definition 3.4.2 A topological space is totally disconnected when two distinct points admit disjoint neighbourhoods which are both open and closed. Equivalently,

$$x \neq y \Rightarrow \exists U \subseteq X, \ U \text{ open and closed}, \ x \in U, \ y \notin U.$$

In particular, a totally disconnected space is a Hausdorff space. The terminology "totally disconnected" is justified by corollary 5.7.10.

Lemma 3.4.3 *In the category of topological spaces and continuous mappings, a projective limit of totally disconnected spaces is again totally disconnected.*

Proof We recall that given a diagram $\mathcal{D}$ of topological spaces, its projective limit L is the space

$$L = \left\{ (x_X)_{X \in \mathcal{D}} \in \prod_{X \in \mathcal{D}} X \,\middle|\, \forall f \in \mathcal{D}, \ f \colon X \longrightarrow Y, \ f(x_X) = x_Y \right\}$$

where the topology of the product induces that of L. In other words, the limit is given by the pullback below, where as above f is a morphism $f \colon X \longrightarrow Y$. This diagram presents the limit as an inverse image of a diagonal:

$$
\begin{array}{ccc}
L & \longrightarrow & \prod_{f \in \mathcal{D}} Y \\
\downarrow & & \downarrow \Delta \\
\prod_{X \in \mathcal{D}} X & \xrightarrow{\ \alpha\ } & \left(\prod_{f \in \mathcal{D}} Y \right) \times \left(\prod_{f \in \mathcal{D}} Y \right) \cong \prod_{f \in \mathcal{D}} (Y \times Y)
\end{array}
$$

where α is defined by

$$\alpha\big((x_X)_{X \in \mathcal{D}}\big) = \big(f(x_X), x_Y\big)_{f \in \mathcal{D}}.$$

First, a topological product $\prod_{i \in I} X_i$ of totally disconnected spaces is totally disconnected. Indeed, if $(x_i)_{i \in I} \neq (y_i)_{i \in I}$, there exists an index i_0 with $x_{i_0} \neq y_{i_0}$. In X_{i_0} one can find disjoint open and closed neighbourhoods $x_{i_0} \in U_{i_0}$, $y_{i_0} \in V_{i_0}$. Putting $U_i = X_i = V_i$ for $i \neq i_0$, we get two disjoint open and closed neighbourhoods

$$(x_i)_{i \in I} \in \prod_{i \in I} U_i, \quad (y_i)_{i \in I} \in \prod_{i \in I} V_i,$$

which proves the total disconnectedness of $\prod_{i \in I} X_i$. On the other hand it is immediate that a subspace of a totally disconnected space is itself totally disconnected. $\qquad \square$

The following corollary is interesting, even if not useful in this chapter.

Corollary 3.4.4 *The category of totally disconnected spaces is reflective both in the category of topological spaces and in the category of Hausdorff spaces.*

Proof The proof of lemma 3.4.3 applies to show that a projective limit of Hausdorff spaces is again Hausdorff: it suffices to work with arbitrary neighbourhoods instead of open and closed ones. Thus the category of totally disconnected spaces is closed under projective limits both in the category of topological and in that of Hausdorff spaces. In particular, monomorphisms of totally disconnected spaces are continuous injections and therefore the category is well-powered.

On the other hand the discrete two point space is trivially totally disconnected; it is in fact a cogenerator. Indeed given $x \neq y$ in X, we can choose an open and closed subset $U \subseteq X$, with $x \in U$ and $y \notin U$. We consider the equivalence relation $u \sim v$ when $u, v \in U$ or $u, v \notin U$. The topological quotient $q \colon X \longrightarrow\!\!\!\!\!\rightarrow X/\!\!\sim$ is homeomorphic to the discrete two point space and $q(x) \neq q(y)$. We conclude the proof by the special adjoint functor theorem (see [29]). $\qquad\square$

Lemma 3.4.5 *A projective limit of compact and totally disconnected spaces is again compact and totally disconnected.*

Proof By lemma 3.4.3, it suffices to prove that a projective limit of compact Hausdorff spaces is compact Hausdorff. By the Tychonoff theorem, a product of compact Hausdorff spaces is compact. Now in the diagram of lemma 3.4.3, the diagonal is closed because the spaces are Hausdorff, thus the limit L is closed in the product, which is compact Hausdorff; thus L is compact Hausdorff.

A more direct argument is the fact that the category of compact Hausdorff spaces is reflective in that of all topological spaces: the reflection is the Stone–Čech compactification. $\qquad\square$

Again the following corollary is interesting, even if not immediately useful for our purpose. Part of it will be studied more intensively in section 5.7 and in particular in proposition 5.7.12.

Corollary 3.4.6 *The category of compact and totally disconnected spaces is reflective in the categories of compact Hausdorff spaces, totally disconnected spaces, Hausdorff spaces and topological spaces.*

Proof The discrete two point space is compact and totally disconnected. So the proof of corollary 3.4.4 applies. $\qquad\square$

Theorem 3.4.7 *A topological space is profinite if and only if it is compact and totally disconnected.*

Proof Every finite discrete space is compact and totally disconnected, thus every projective limit of finite discrete spaces is still compact and totally disconnected by lemma 3.4.5.

Conversely, let X be compact and totally disconnected. We consider the poset $\mathcal{R}$ of all equivalence relations R on X such that the topological quotient X/R is discrete and finite; the ordering is inclusion. We write $[x]_R$ for the R-equivalence class of $x \in X$. We prove first the cofilteredness of $\mathcal{R}$. Let $R, S \in \mathcal{R}$ correspond respectively to the partitions $X = V_1 \cup \cdots \cup V_n$ and $X = W_1 \cup \cdots \cup W_m$. Let T be the equivalence relation corresponding to the partition

$$X = \bigcup_{1 \leq i \leq n, \ 1 \leq j \leq m} V_i \cap W_j.$$

Trivially $T \subseteq R$ and $T \subseteq S$ and the quotient X/T is finite. Since X/R and X/S are discrete, all V_i and W_j are both open and closed in X. Therefore each $V_i \cap W_j$ is open and closed in X and X/T is discrete as well. This proves the cofilteredness of $\mathcal{R}$.

When $R \subseteq S$ in $\mathcal{R}$, we get a factorization $X/R \longrightarrow X/S$ between the corresponding quotients and we shall prove that $X \cong \lim_{R \in \mathcal{R}} X/R$. To achieve this, we consider the canonical factorization

$$\lambda \colon X \longrightarrow \lim_{R \in \mathcal{R}} X/R, \quad x \mapsto \big([x]_R\big)_{R \in \mathcal{R}};$$

we must prove this is a homeomorphism. But X is compact Hausdorff and totally disconnected by assumption, and the same holds for $\lim_{R \in \mathcal{R}}$ by lemma 3.4.5. Since the spaces are compact Hausdorff, it suffices to prove that λ is bijective. But again by compactness and Hausdorffness, $\lambda(X)$ is compact, thus closed in $\lim_{R \in \mathcal{R}}$. It suffices therefore to prove that λ is injective and $\lambda(X)$ is dense.

To prove the injectivity, consider $x \neq y$ in X and $U \subseteq X$ open and closed, with $x \in U$ and $y \notin U$. The topological quotient X/U is the discrete two point space, thus the corresponding equivalence relation R is in $\mathcal{R}$ and, of course, $[x]_R \neq [y]_R$ since $x \in U$ and $y \notin U$. Therefore $\lambda(x) \neq \lambda(y)$.

To prove the density of $\lambda(X)$, let us fix an element $\big([x_R]_R\big)_{R \in \mathcal{R}}$ in $\lim_{R \in \mathcal{R}} X/R$. For every finite choice $R_1, \ldots, R_n$ in $\mathcal{R}$,

$$U_{R_1, \ldots, R_n} = \left\{ \big([y_R]_R\big)_{R \in \mathcal{R}} \middle| \forall i = 1, \ldots, n \ \ [x_{R_i}]_{R_i} = [y_{R_i}]_{R_i} \right\}$$

is a neighbourhood of $\left([x_R]_R\right)_{R \in \mathcal{R}}$ in $\lim_{R \in \mathcal{R}} X/R$ since it can be written

$$U_{R_1,\dots,R_n} = p_{R_1}^{-1}\left([x_{R_1}]_{R_1}\right) \cap \dots \cap p_{R_n}^{-1}\left([x_{R_n}]_{R_n}\right)$$

where the p_{R_i} are the canonical projections of the limit. The subsets $U_{R_1,\dots,R_n}$ even constitute a fundamental system of neighbourhoods of $\left([x_R]_R\right)_{R \in \mathcal{R}}$, since every elementary neighbourhood of this point has the form

$$V = p_{R_1}^{-1}(V_1) \cap \dots \cap p_{R_n}^{-1}(V_n)$$

for some finite family $R_1,\dots,R_n$ in $\mathcal{R}$ and subsets $V_i \subseteq X/R_i$ such that $[x_{R_i}]_{R_i} \in V_i$. Thus such a V indeed contains $U_{R_1,\dots,R_n}$. Now for each choice $R_1,\dots,R_n$ in $\mathcal{R}$, we can choose $R_0 \in \mathcal{R}$ with $R_0 \subseteq R_i$ for each $i = 1,\dots,n$, by cofilteredness of $\mathcal{R}$. For each index i, the relation $R_0 \subseteq R_i$ and the compatibility of the family $\left([x_R]_R\right)_{R \in \mathcal{R}}$ imply $[x_{R_i}]_{R_i} = [x_{R_0}]_{R_i}$. This proves that

$$\lambda(x_{R_0}) = \left([x_{R_0}]_R\right)_{R \in \mathcal{R}} \in U_{R_1,\dots,R_n}.$$

Every fundamental neighbourhood $U_{R_1,\dots,R_n}$ of $\left([x_R]_R\right)_{R \in \mathcal{R}}$ thus meets $\lambda(X)$, proving that $\lambda(X)$ is dense. $\qquad\square$

Corollary 3.4.8 *A topological space is profinite when it is homeomorphic to a projective limit of finite discrete spaces.*

Proof Finite discrete spaces are compact and totally disconnected; thus a projective limit of finite discrete spaces is compact and totally disconnected (see 3.4.5) and therefore profinite (see 3.4.7). $\qquad\square$

Corollary 3.4.9 *For a compact Hausdorff space X, the following conditions are equivalent:*

(i) *X is profinite;*
(ii) *the topology of X has a basis constituted of clopens (= simultaneously closed and open subsets);*
(iii) *X is totally disconnected.*

Proof (i) $\Leftrightarrow$ (iii) is part of theorem 3.4.7 and (ii) $\Rightarrow$ (iii) is obvious. Now assume (iii). The set of clopens in X is closed under finite intersections, and it suffices to prove that an open subset $U \subseteq X$ is a union of clopens. Choosing $x \in U$, we shall prove the existence of a clopen V such that $x \in V \subseteq U$, which will yield the result. For each $y \notin U$, choose a clopen

V_y such that $x \in V_y$ and $y \notin V_y$, thus $y \in \complement V_y$, where $\complement$ indicates the set-theoretical complement. The clopens $\complement V_y$ cover the compact subset $\complement U$, thus a finite number of them already covers $\complement U$. But

$$\complement U \subseteq \complement V_{y_1} \cup \cdots \cup \complement V_{y_n}$$

implies

$$x \in V_{y_1} \cap \cdots \cap V_{y_n} \subseteq U$$

with $V = V_{y_1} \cap \cdots \cap V_{y_n}$ a clopen. □

Lemma 3.4.10 *Let* $X = \lim_{i \in I} X_i$ *be a profinite space, presented as cofiltered limit of finite discrete spaces* X_i. *The following conditions are equivalent:*

(i) *X is not empty;*

(ii) *for each index $i \in I$, X_i is not empty.*

Proof The existence of the canonical projection $p_i \colon X \longrightarrow X_i$ of the limit forces (i) $\Rightarrow$ (ii).

Conversely, consider the product $\prod_{i \in I} X_i$ which is a non empty compact Hausdorff space by the Tychonoff theorem. For each fixed index $k \in I$,

$$C_k = \left\{ (x_i)_{i \in I} \in \prod_{i \in I} X_i \,\middle|\, \forall i \geq k \;\; x_i = f_{ki}(x_k) \right\}$$

where $f_{ki} \colon X_k \longrightarrow X_i$ is the morphism of the diagram corresponding to $k \leq i$. Each subset C_k is closed as intersection of the subsets

$$C_{k,j} = \{ (x_i)_{i \in I} \,|\, x_j = f_{kj}(x_k) \}, \quad j \geq k,$$

which are themselves closed, as inverse images of the diagonal by the continuous maps

$$X \xrightarrow{\left(\begin{array}{c} p_j \\ f_{kj} \circ p_k \end{array} \right)} X_j \times X_j, \quad x \mapsto (x_j, f_{kj}(x_k)).$$

Each C_k is non empty; indeed choosing an arbitrary $x \in X_k$, it suffices to put $x_i = f_{ki}(x)$ for $i \geq k$ and $x_i \in X_i$ arbitrary for the other indices; it follows at once that this family $(x_i)_{i \in I}$ is in C_k. If $k \leq l$, one has $C_k \subseteq C_l$ and therefore the family $(C_k)_{k \in I}$ is cofiltered, because so is I.

Since in the compact Hausdorff space $\prod_{i \in I} X_i$, a cofiltered intersection of non empty closed subsets is again non empty, we deduce that

$$X = \lim_{i \in I} X_i = \bigcap_{k \in I} C_k$$

is non empty. $\square$

Lemma 3.4.11 *Let $X = \lim_{i \in I} X_i$ be a profinite space, presented as cofiltered limit of finite discrete spaces. Let us write $p_i \colon X \longrightarrow X_i$ for the projections of this limit and $f_{ji} \colon X_j \longrightarrow X_i$ for the morphism of the diagram corresponding to $j \leq i$. For every index $i \in I$, there exists an index $j \leq i$ such that $\operatorname{Im} f_{ji} = \operatorname{Im} p_i$, where Im stands for the image.*

Proof We fix the index $i \in I$. Since the limit is cofiltered, it is equivalent to compute it on the indices $j \leq i$. In other terms, we can assume that i is the terminal object of the poset I.

Given $x \in X_i$ and an index $j \leq i$, let us put $Y_j(x) = f_{ji}^{-1}(\{x\}) \subseteq X_j$. Applying lemma 3.4.10 and the obvious equality $p_i^{-1}(x) = \lim_{j \in I} Y_j(x)$, we get

$$
\begin{aligned}
x \in \bigcap_{j \in I} \operatorname{Im} f_{ji} \quad &\Leftrightarrow \quad \forall j \leq i \; Y_j(x) \neq \emptyset \\
&\Leftrightarrow \quad \lim_{j \in I} Y_j(x) \neq \emptyset \\
&\Leftrightarrow \quad p_i^{-1}(x) \neq \emptyset \\
&\Leftrightarrow \quad x \in \operatorname{Im} p_i.
\end{aligned}
$$

This proves that

$$\operatorname{Im} p_i = \bigcap_{j \in I} \operatorname{Im} f_{ji}.$$

The right hand side of this equality is a cofiltered intersection of finite subsets of X_i, thus by finiteness, is one of these subsets. In other words

$$\exists j \in I \; \operatorname{Im} p_i = \operatorname{Im} f_{ji}. \qquad \square$$

Lemma 3.4.12 *Let $X = \lim_{i \in I} X_i$ be a profinite space, presented as a cofiltered limit of finite discrete spaces X_i. For every finite discrete space Y, the canonical comparison morphism*

$$\alpha \colon \operatorname{colim} \operatorname{Cont}(X_i, Y) \overset{\cong}{\longrightarrow} \operatorname{Cont}\left(\lim_{i \in I} X_i, Y\right)$$

is a bijection, with Cont denoting the set of continuous maps.

Proof   Observe first that since X_i is discrete, $\mathsf{Cont}(X_i, Y)$ is the set of all maps from X_i to Y. To fix notation, we recall that α is the unique factorization making the following diagrams commutative:

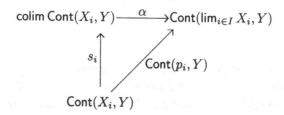

with s_i the canonical morphisms of the colimit and p_i the canonical morphisms of the limit. We must prove that α is bijective. Since the limit is cofiltered, the colimit is filtered. We recall that this filtered colimit is the quotient of the disjoint union $\coprod_{i \in I} \mathsf{Cont}(X_i, Y)$ by the equivalence relation $\approx$ given by

$$\big(u \in \mathsf{Cont}(X_i, Y)\big) \approx \big(v \in \mathsf{Cont}(X_j, Y)\big)$$
$$\Leftrightarrow \exists k \in I,\ k \leq i,\ k \leq j,\ \mathsf{Cont}(f_{ki}, Y)(u) = \mathsf{Cont}(f_{kj}, Y)(v).$$

In other words

$$\big(u \in \mathsf{Cont}(X_i, Y)\big) \approx \big(v \in \mathsf{Cont}(X_j, Y)\big)$$
$$\Leftrightarrow \exists k \in I,\ k \leq i,\ k \leq j,\ u \circ f_{ki} = v \circ f_{kj}.$$

We first prove the injectivity of α. Two elements of the colimit are thus equivalences classes $[u]$, $[v]$ of elements which, by filteredness, we can choose in the same factor; let us say, $u, v \in \mathsf{Cont}(X_i, Y)$. We assume that $\alpha\big([u]\big) = \alpha\big([v]\big)$. This implies at once

$$u \circ p_i = \mathsf{Cont}(p_i, Y)(u) = (\alpha \circ s_i)(u) = \alpha\big([u]\big)$$
$$= \alpha\big([v]\big) = (\alpha \circ s_i)(v) = \mathsf{Cont}(p_i, Y)(v) = v \circ p_i.$$

But by lemma 3.4.11, we know the existence of an index $j \leq i$ with corresponding morphism $f_{ji} \colon X_j \longrightarrow X_i$ in the diagram and such that $\mathsf{Im}\, f_{ji} = \mathsf{Im}\, p_i$. The previous equalities show that u and v coincide on $\mathsf{Im}\, p_i$, thus on $\mathsf{Im}\, f_{ji}$. But then $u \circ f_{ji} = v \circ f_{ji}$, that is both elements u, v are identified in the factor $\mathsf{Cont}(X_j, Y)$ of the colimit. Thus u, v are identified in the colimit and $[u] = [v]$.

The surjectivity of α is harder to prove. Let us thus consider a continuous map $f \colon \lim_{i \in I} X_i \longrightarrow Y$. We consider first its kernel pair in the category of topological spaces

$$R \underset{r_2}{\overset{r_1}{\rightrightarrows}} X = \lim_{i \in I} X_i \overset{f}{\longrightarrow} Y, \quad R = \{(x, x') \mid f(x) = f(x')\}$$

and the kernel pair of each projection

$$R_i \underset{r_2^i}{\overset{r_1^i}{\rightrightarrows}} X = \lim_{i \in I} X_i \overset{p_i}{\longrightarrow} Y, \quad R_i = \{(x, x') \mid p_i(x) = p_i(x')\}.$$

These spaces R and R_i are thus obtained as inverse images of the diagonals of Y or X_i.

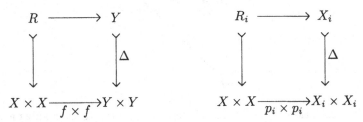

Since Y and X_i are discrete, their diagonal is both open and closed and therefore R and R_i are open and closed subspaces of $X \times X$.

Moreover if $j \leq i$, the commutativity of the triangle

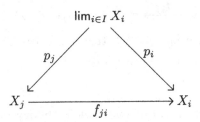

implies $R_j \subseteq R_i$. Since I is a cofiltered poset, the poset $(R_i)_{i \in I}$ of equivalence relations, ordered by inclusion, is cofiltered as well.

We observe also that the diagonal $\Delta_X \subseteq X \times X$ of X is closed, because X is a Hausdorff space.

Since R is an equivalence relation, $\Delta_X \subseteq R$. On the other hand, by definition of R_i, one has $\Delta_X = \bigcap_{i \in I} R_i$. Therefore

$$\emptyset = \Delta_X \cap \mathsf{C}\Delta_X = \left(\bigcap_{i \in I} R_i\right) \cap \mathsf{C}\Delta_X \supseteq \left(\bigcap_{i \in I} R_i\right) \cap \mathsf{C}R = \bigcap_{i \in I} (R_i \cap \mathsf{C}R),$$

which proves that $\bigcap_{i \in I} (R_i \cap \mathsf{C}R) = \emptyset$. Since the $(R_i)_{i \in I}$ constitute a cofiltered poset, so do the $R_i \cap \mathsf{C}R$. We have thus a cofiltered family of closed subsets of the compact Hausdorff space $X \times X$, whose intersection is empty; by compactness, one of these subsets is already empty. Thus

$$\exists j \in I \ \ R_j \cap \complement R = \emptyset, \quad \text{that is,} \quad \exists j \in I \ \ R_j \subseteq R.$$

In other words

$$\exists j \in I \ \ \forall x, x' \in X \ \ p_j(x) = p_j(x') \Rightarrow f(x) = f(x').$$

This shows precisely the existence of a factorization g in the following diagram:

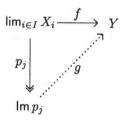

This factorization is continuous since $\operatorname{Im} p_j$ is discrete, as a subspace of X_j.

Applying once more lemma 3.4.10, let us choose $k \leq j$ such that $\operatorname{Im} p_j = \operatorname{Im} f_{kj}$. This yields the composite

$$X_k \xrightarrow{\ f_{kj}\ } \operatorname{Im} f_{kj} = \operatorname{Im} p_j \xrightarrow{\ g\ } Y$$

which is such that $g \circ f_{kj} \circ p_k = f$, that is $\operatorname{Cont}(p_k, Y)(g \circ f_{kj}) = f$. By definition of α, this implies $\alpha([g \circ f_{kj}]) = f$. $\qquad\square$

3.5 Infinitary extension of the Galois theory of Grothendieck

Definition 3.5.1 Let G be a topological group. A topological G-space is a topological space provided with a continuous action of G; a morphism of topological G-spaces is a continuous morphism of G-sets. A topological G-space is profinite when it is a projective limit, indexed by a cofiltered poset, of finite discrete topological G-spaces.

It is immediate to observe that projective limits of topological G-spaces are computed as in the category of topological spaces, with the corresponding componentwise action of G. Given a topological group G, we shall write G-**Prof** for the category of profinite G-spaces and their continuous homomorphisms.

Notice that given a topological group, a discrete G-set X, even finite, has no reason to be a topological G-space, since the action $G \times X \longrightarrow X$ is no longer defined on a discrete space. In fact a G-set X equipped with the discrete topology is a topological G-space if and only if the stabilizer $G_x = \{g \in G \mid gx = x\}$ is open for every $x \in X$.

Lemma 3.5.2 *Let K be a field. Every algebraic K-algebra A is the set-theoretical filtered union of its finite dimensional subalgebras.*

Proof Every element $a \in A$ is algebraic, thus the subalgebra $K(a) \subseteq A$ is finite dimensional (see proposition 2.1.10). This implies immediately that $A = \bigcup_B B$, for $B \subseteq A$ running through its finite dimensional subalgebras.

This union is filtered. Indeed given two such B and B', they have the form $B = K(a_1, \dots, a_n)$, $B' = K(c_1, \dots, c_m)$, for finite sets of generators. It suffices to observe that $K(a_1, \dots, a_n, c_1, \dots, c_m)$ is still finite dimensional. Trivially, $K(a_1, \dots, a_n, c_1, \dots, c_m)$ can be described as

$$\{p(a_1, \dots, a_n, c_1, \dots, c_m) \mid$$
$$p(X_1, \dots, X_n, Y_1, \dots, Y_m) \in K[X_1, \dots, X_n, Y_1, \dots, Y_m]\}.$$

But if $a_i \in A$ has a minimal polynomial $p(X) \in K[X]$ of degree n_i,

$$p(X) = X^{n_i} + k_{n_i - 1} X^{n_i - 1} + \cdots + k_0,$$

every occurrence of $a_i^{n_i}$ can be replaced by

$$a_i^{n_i} = -k_{n_i - 1} a_i^{n_i - 1} - \cdots - k_0.$$

An analogous argument holds for all a_i and c_j. Thus in the description of $K(a_1, \dots, a_n, c_1, \dots, c_m)$, there is no restriction in bounding the degree of the polynomial p in each variable, specifically, forcing this degree to be strictly less than the degree of the corresponding minimal polynomial. But these polynomials with bounded degrees have only a bounded number of coefficients, thus the space is finite dimensional. $\quad\square$

Lemma 3.5.3 *Let $K \subseteq L$ be an arbitrary Galois extension of fields. For every K-algebra A which is split by L, there is a bijection*

$$\mathsf{Hom}_K(A, L) \cong \lim_B \mathsf{Hom}_K(B, L)$$

where the limit is cofiltered and indexed by the finite dimensional subalgebras $B \subseteq A$. Moreover, each $\mathsf{Hom}_K(B, L)$ is finite; in particular *the above limit formula provides* $\mathsf{Hom}_K(A, L)$ *with the structure of a profinite space.*

Proof By lemma 3.5.2, A is the filtered union, or equivalently, the filtered colimit, of its finite dimensional subalgebras. This yields

$$\mathsf{Hom}_K(A, L) \cong \mathsf{Hom}_K \left(\operatorname*{colim}_B B, L \right) \cong \lim_B \mathsf{Hom}_K(B, L).$$

It remains to prove that $\mathsf{Hom}_K(B, L)$ is finite, for each finite dimensional $B \subseteq A$. By proposition 3.1.5, there exists a finite dimensional Galois extension $K \subseteq M \subseteq L$ which splits B. Every K-homomorphism $f \colon B \longrightarrow L$ is such that, for every element $b \in B$ with minimal polynomial $p(X) \in K[X]$,

$$p\big(f(b)\big) = f\big(p(b)\big) = f(0) = 0,$$

thus $f(b)$ is a root of $p(X)$ in L and therefore $f(b) \in M$. This proves the isomorphism $\mathsf{Hom}_K(B, L) \cong \mathsf{Hom}_K(B, M)$, and this last set is finite by theorem 2.3.3. $\square$

Lemma 3.5.4 *Let $K \subseteq L$ be an arbitrary Galois extension of fields. For every K-algebra A which is split by L, the map*

$$\mu \colon \mathsf{Gal}\,[L : K] \times \mathsf{Hom}_K(A, L) \longrightarrow \mathsf{Hom}_K(A, L), \quad (g, f) \mapsto g \circ f$$

is a continuous action of the topological group $\mathsf{Gal}\,[L : K]$ on the topological space $\mathsf{Hom}_K(A, L)$, when these are provided with the profinite topologies inherited from definition 3.2.2 and lemma 3.5.3.

Proof By associativity of the composition, μ is a group action. Proving its continuity reduces to proving that for every finite dimensional subalgebra $B \subseteq A$, the composite $p_B \circ \mu$ is continuous, where p_B is the canonical projection of the limit:

$$p_B \colon \mathsf{Hom}_K(A, L) \cong \lim_B \mathsf{Hom}_K(B, L) \longrightarrow \mathsf{Hom}_K(B, L).$$

By proposition 3.1.5, we choose a finite dimensional Galois extension $K \subseteq M \subseteq L$ which splits B. As observed in the proof of lemma 3.5.3, $\mathsf{Hom}_K(B, L) \cong \mathsf{Hom}_K(B, M)$ and is a finite discrete space. This allows consideration of diagram 3.1. Since p_M and p_B are continuous by definition, the left vertical composite is continuous. The bottom arrow is the composition, that is maps (f, g) onto $f \circ g$; it is of course continuous, since it is defined between finite discrete spaces. $\square$

$$\text{Gal}\,[L:K] \times \text{Hom}_K(A,L) \xrightarrow{\quad \mu \quad} \text{Hom}_K(A,L)$$

$$p_M \times \text{id} \downarrow \qquad\qquad\qquad\qquad \downarrow p_B$$

$$\text{Gal}\,[M:K] \times \text{Hom}_K(A,L)$$

$$\text{id} \times p_B \downarrow$$

$$\text{Gal}\,[M:K] \times \text{Hom}_K(B,M) \longrightarrow \text{Hom}_K(B,M)$$

Diagram 3.1

Lemma 3.5.5 *Let $K \subseteq L$ be an arbitrary Galois extension of fields. Consider a homomorphism $f\colon A \longrightarrow B$ of K-algebras, where A and B are split by L. The map*

$$\Gamma(f)\colon \text{Hom}_K(B,L) \longrightarrow \text{Hom}_K(A,L), \quad h \mapsto h \circ f$$

is a continuous homomorphism of $\text{Gal}\,[L:K]$-sets, when $\text{Hom}_K(A,L)$ and $\text{Hom}_K(B,L)$ are provided with the profinite topology of lemma 3.5.3.

Proof For every finite dimensional subalgebra $C \subseteq A$, we must prove that the composite

$$\text{Hom}_K(B,L) \xrightarrow{\ \Gamma(f)\ } \text{Hom}_K(A,L) = \varprojlim_C \text{Hom}_K(C,L) \xrightarrow{\ p_C\ } \text{Hom}_K(C,L)$$

is continuous. Since C is finite dimensional, $f(C) \subseteq B$ is a finite dimensional K-algebra. The following diagram is commutative:

$$\text{Hom}_K(B,L) \xrightarrow{\quad \Gamma(f) \quad} \text{Hom}_K(A,L)$$

$$p_C \downarrow \qquad\qquad\qquad\qquad \downarrow p_C$$

$$\text{Hom}_K\big(f(C),L\big) \xrightarrow[\quad \gamma(f) \quad]{} \text{Hom}_K(C,L)$$

where $\big(\gamma(f)(h)\big)(c) = h\big(f(c)\big)$ for all $c \in C$. The morphisms p_C are continuous by definition and $\gamma(f)$ is continuous since it is defined between finite discrete spaces. $\qquad\square$

Lemma 3.5.6 *Let K be a field and A an algebraic K-algebra. As in lemma 3.5.2, let us write $A = \operatorname{colim} B$ where B runs through the finite dimensional subalgebras of A. For every finite dimensional K-algebra C, the canonical morphism*

$$\rho\colon \operatorname*{colim}_{B} \operatorname{Hom}_K(C, B) \xrightarrow{\;\cong\;} \operatorname{Hom}_K(C, A)$$

is bijective.

Proof The canonical morphism ρ of the statement is that induced by the canonical inclusions $\operatorname{Hom}_K(C, B) \subseteq \operatorname{Hom}_K(C, A)$.

Let $c_1, \ldots, c_n$ be a basis of C as K-vector space. A morphism of K-algebras $f\colon C \longrightarrow A \cong \operatorname{colim}_B B$ is such that

$$\forall i = 1, \ldots, n \quad f(c_i) = b_i \quad \text{with} \quad b_i \in B_i.$$

By filteredness, there is no restriction in choosing all elements b_i in the same factor B_0 of the colimit. This implies the existence of a K-linear factorization

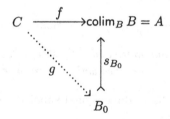

where

$$g\left(\sum_{i=1,\ldots,n} k_i c_i \right) = \sum_{i=1,\ldots,n} k_i b_i.$$

Since s_{B_0} is injective and f is an algebra homomorphism, it follows at once that g is an algebra homomorphism. The above diagram shows that $f = \rho([g])$, proving the surjectivity of ρ.

To prove the injectivity of ρ, consider a homomorphism $g'\colon C \longrightarrow B_1$ such that $f = \rho([g'])$. By filteredness, we can view g and g' as having values in the same B_2, from which $g = g'$ as maps with values in B_2, since s_{B_2} is injective. This means $[g] = [g']$ in the colimit. $\square$

Lemma 3.5.7 *Let $G = \lim_{i \in I} G_i$ be a profinite group, expressed as a cofiltered projective limit of finite discrete groups. Let us assume that the projections $p_i\colon G \longrightarrow G_i$ are surjective. We write G_i-Set$_f$ for the* *category of finite G_i-sets and G-Top$_f$ for the category of discrete finite topological G-spaces. For every index $i \in I$, there is a functor*

$$\gamma_i \colon G_i\text{-Set}_f \longrightarrow G\text{-Top}_f, \quad X \mapsto X$$

where the G-set action on X is given by $g \cdot x = p_i(g) \cdot x$. This functor γ_i identifies G_i-Set$_f$ with a full subcategory of G-Top$_f$. Moreover the category G-Top$_f$ is the set theoretical filtered union of the full subcategories G_i-Set$_f$.

Proof For $X \in G_i$-Set$_f$, the action of $\gamma_i(X)$ is continuous since it is the composite

$$G \times X \xrightarrow{\; p_i \times \text{id} \;} G_i \times X \xrightarrow{\;\mu_i\;} X$$

where G_i and X are discrete, μ_i is the group action and p_i is continuous by definition.

To prove the full- and faithfulness, consider $f \colon X \longrightarrow Y$, an arbitrary map between finite G_i-sets. Observe that in the diagram

$$
\begin{array}{ccccc}
G \times X & \xrightarrow{\; p_i \times \text{id} \;} & G_i \times X & \xrightarrow{\;\mu_i\;} & X \\
{\scriptstyle \text{id} \times f}\big\downarrow & & {\scriptstyle \text{id} \times f}\big\downarrow & & \big\downarrow{\scriptstyle f} \\
G \times Y & \xrightarrow[\; p_i \times \text{id} \;]{} & G_i \times Y & \xrightarrow[\;\mu_i\;]{} & Y
\end{array}
$$

the left hand square is commutative. Since p_i is surjective, the commutativity of the outer diagram is equivalent to that of the right hand square. In other words, f is a morphism of G-sets if and only if it is a morphism of G_i-sets.

To prove the last assertion, consider a finite G-space X. The canonical comparison morphism

$$\left(\lim_{i \in I} G_i \right) \times X \xrightarrow{\;\cong\;} \lim_{i \in I} (G_i \times X)$$

is a homeomorphism. Indeed, products commute with limits in every category; moreover $X = \lim_{i \in I} X$ because I is connected. Therefore

$$\lim_{i \in I} (G_i \times X) \cong \left(\lim_{i \in I} G_i \right) \times \left(\lim_{i \in I} X \right) \cong \left(\lim_{i \in I} G_i \right) \times X.$$

By lemma 3.4.12, the composite

$$\lim_i (G_i \times X) \xrightarrow{\;\cong\;} \left(\lim_i G_i \right) \times X \xrightarrow{\;\cong\;} G \times X \xrightarrow{\;\mu\;} X$$

factors as

$$\lim_{i \in I}(G_i \times X) \xrightarrow{\;p_i \times \mathrm{id}\;} G_{i_0} \times X \xrightarrow{\;\mu_{i_0}\;} X.$$

By surjectivity of $p_i \times \mathrm{id}$, this presents X as a G_{i_0}-set.   □

Theorem 3.5.8 (Galois theorem) *Let $K \subseteq L$ be an arbitrary Galois extension of fields. We write $\mathsf{Split}_K(L)$ for the category of K-algebras split by L and $\mathsf{Gal}\,[L : K]$-Prof for the category of profinite $\mathsf{Gal}\,[L : K]$-spaces. The functor*

$$\Gamma \colon \mathsf{Split}_K(L) \longrightarrow \mathsf{Gal}\,[L : K]\text{-}\mathsf{Prof}, \quad A \mapsto \mathsf{Hom}_K(A, L)$$

described in lemmas 3.5.3, 3.5.4, 3.5.5 is a contravariant equivalence of categories.

Proof   We prove first that Γ is full and faithful. For this consider $A, B \in \mathsf{Split}_K(L)$. By definition 2.3.1 and lemma 3.5.2, let us write

$$A = \mathrm{colim}\, C, \quad C \subseteq A; \quad B = \mathrm{colim}\, D, \quad D \subseteq B$$

where the colimits are filtered and C, D run respectively through the finite dimensional subalgebras of A and B. For each pair C, D, we can by proposition 3.1.5 choose finite dimensional Galois extensions M_C, M_D which split respectively C and D. By proposition 3.1.4, we can even choose a finite dimensional Galois extension M_{CD} which splits both C and D and yields

$$K \subseteq M_C \subseteq M_{CD} \subseteq L, \quad K \subseteq M_D \subseteq M_{CD} \subseteq L.$$

As observed in the proof of lemma 3.5.3

$$\mathsf{Hom}_K(C, L) \cong \mathsf{Hom}_K(C, M_{CD}), \quad \mathsf{Hom}_K(D, L) \cong \mathsf{Hom}_K(D, M_{CD}).$$

We have then, using 3.4.12,

$$
\begin{aligned}
\mathsf{Hom}\big(\Gamma(A), \Gamma(B)\big) \\
&\cong \mathsf{Hom}\big(\mathsf{Hom}_K(A, L), \mathsf{Hom}_K(B, L)\big) \\
&\cong \mathsf{Hom}\big(\mathsf{Hom}_K(\mathrm{colim}_C\, C, L), \mathsf{Hom}_K(\mathrm{colim}_D\, D, L)\big) \\
&\cong \mathsf{Hom}\big(\lim_C \mathsf{Hom}_K(C, L), \lim_D \mathsf{Hom}_K(D, L)\big) \\
&\cong \lim_D \mathsf{Hom}\big(\lim_C \mathsf{Hom}_K(C, L), \mathsf{Hom}_K(D, L)\big) \\
&\cong \lim_D \mathrm{colim}_C \mathsf{Hom}\big(\mathsf{Hom}_K(C, L), \mathsf{Hom}_K(D, L)\big) \\
&\cong \lim_D \mathrm{colim}_C \mathsf{Hom}\big(\mathsf{Hom}_K(C, M_{CD}), \mathsf{Hom}_K(D, M_{CD})\big) \\
&\cong \lim_D \mathrm{colim}_C \mathsf{Hom}(D, C) \quad \text{by 2.4.3}
\end{aligned}
$$

$$\cong \lim_D \mathsf{Hom}(D, \mathsf{colim}\, C) \quad \text{by 3.5.6}$$

$$\cong \mathsf{Hom}(\mathsf{colim}\, D, \mathsf{colim}\, C)$$

$$\cong \mathsf{Hom}(B, A).$$

This proves the full- and faithfulness of Γ.

Observe next that for every finite dimensional Galois extension $K \subseteq M \subseteq L$, the projection, acting by restriction,

$$\mathsf{Gal}\,[L : K] = \lim_M \mathsf{Gal}\,[M : K] \longrightarrow \mathsf{Gal}\,[M : K]$$

is surjective, by definition 3.2.2 and proposition 3.1.2. Thus lemma 3.5.7 applies.

It remains to prove that Γ is essentially surjective on the objects. A profinite $\mathsf{Gal}\,[L : K]$-space is a cofiltered projective limit $X \cong \lim_{i \in I} X_i$ of finite discrete topological $\mathsf{Gal}\,[L : K]$-spaces. By lemma 3.5.7, each X_i is a finite $\mathsf{Gal}\,[M_i : K]$-set for some finite dimensional Galois extension $K \subseteq M_i \subseteq L$. By theorem 2.4.3, $X_i = \mathsf{Hom}_K(C_i, M_i)$ for some finite dimensional K-algebra C_i which is split by M_i. As already observed in lemma 3.5.3

$$X_i = \mathsf{Hom}_K(C_i, M_i) \cong \mathsf{Hom}_K(C_i, L).$$

We observe moreover that given $f_{ij} \colon X_i \longrightarrow X_j$ in the diagram defining X, the space X_i is a finite discrete $\mathsf{Gal}\,[M_i : K]$-space and the space X_j is a finite discrete $\mathsf{Gal}\,[M_j : K]$-space. By proposition 3.1.4, we choose a finite dimensional Galois extension $K \subseteq M \subseteq L$ such that $M_i \subseteq M$ and $M_j \subseteq M$. As above, this yields $X_i = \mathsf{Hom}_K(C_i, M)$ and $X_j = \mathsf{Hom}_K(C_j, M)$, where C_i and C_j are finite dimensional K-algebras split by M. Again by theorem 2.4.3

$$f_{ij} \colon \mathsf{Hom}_K(C_i, M) = X_i \longrightarrow X_j = \mathsf{Hom}_K(C_j, M)$$

is induced by a morphism $h_{ij} \colon C_j \longrightarrow C_i$ of K-algebras. In conclusion, from the cofiltered diagram constituted by the X_i, we have constructed a corresponding filtered diagram constituted by the C_i. We put $A = \mathsf{colim}_{i \in I} C_i$; since this colimit of algebras is filtered, it is computed as in the category of sets.

Each element $a \in A$ has the form $[a_i]$ for some index $i \in I$ and some element $a_i \in C_i$. The minimal polynomial $p(X)$ of a_i factors in $L[X]$ into factors of degree 1 (see definition 2.3.1). But from $p(a_i) = 0$ in C_i we deduce $p(a) = 0$ in A, thus a admits a minimal polynomial $q(X)$ which is a factor of $p(X)$ (see proposition 2.1.7). Therefore $q(X)$ factors in $L[X]$ into factors of degree 1. This proves that L splits A.

Finally one has

$$\operatorname{Hom}_K(A, L) \cong \operatorname{Hom}_K \left(\operatorname*{colim}_{i \in I} C_i, L \right) \cong \varprojlim_{i \in I} \operatorname{Hom}_K(C_i, L) \cong \varprojlim_{i \in I} X_i \cong X.$$

$\square$

4

Categorical Galois theory
of commutative rings

Convention *In this chapter, all rings and algebras are commutative and have a unit.*

In this chapter, we propose an approach to Galois theory of rings originally inspired by the work of A.R. Magid [67], but put in a categorical presentation which will make transparent the generalization of chapter 5. In particular we make an extensive use of the Pierce spectrum functor and its adjoint, which relate the category of commutative rings with that of profinite topological spaces, and then naturally bring in profinite topological groupoids.

4.1 Stone duality

We shall now prove that the category of profinite topological spaces is dual to the category of boolean algebras. We first fix the terminology and recall some elementary facts.

Definition 4.1.1 Let B be a boolean algebra.

(i) A filter in B is a subset $F \subseteq B$ such that

(F1) $1 \in F$,

(F2) $x \in F$ and $y \in F \Rightarrow x \wedge y \in F$,

(F3) $x \in F$ and $x \leq y \Rightarrow y \in F$.

(ii) The filter F is proper when, moreover,

(F4) $0 \notin F$.

(iii) An ultrafilter is a maximal element of the poset of proper filters, ordered by inclusion.

The notion of *ideal* is dual to that of filter. More precisely,

Definition 4.1.2 Let B be a boolean algebra.

(i) An ideal in B is a subset $I \subseteq B$ such that

    (I1) $0 \in I$,

    (I2) $x \in I$ and $y \in I \Rightarrow x \vee y \in I$,

    (I3) $x \in I$ and $x \geq y \Rightarrow y \in I$.

(ii) The ideal I is proper when, moreover,

    (I4) $1 \notin I$.

(iii) An ideal is maximal when it is a maximal element in the poset of proper ideals, ordered by inclusion.

In a boolean algebra B, we shall write Cx for the complement of an element $x \in B$.

Proposition 4.1.3 *Let $F \subseteq B$ be a proper filter of a boolean algebra B. The following conditions are equivalent.*

(i) *F is an ultrafilter;*

(ii) *$\forall x \in B \quad x \in F$ or $Cx \in F$;*

(iii) *$\forall x, y \in B \quad x \vee y \in F \Rightarrow x \in F$ or $y \in F$;*

(iv) *there exists a homomorphism of boolean algebras $f \colon B \longrightarrow \{0,1\}$ such that $F = f^{-1}(1)$.*

Proof (i) $\Rightarrow$ (ii) If $x \notin F$, then $G = \{z \mid x \vee z \in F\}$ is a proper filter which contains F, thus $G = F$ and $Cx \in G = F$.

(ii) $\Rightarrow$ (iii) If $x \vee y \in F$ and $x \notin F$, one has $Cx \in F$ and thus $y \wedge Cx = (x \vee y) \wedge Cx \in F$, from which $y \in F$.

(iii) $\Rightarrow$ (iv) Putting $f(x) = 1$ iff $x \in F$, f preserves $\wedge$, $\leq$, 1 because F is a filter and 0 because F is proper. Condition (iii) expresses the preservation of $\vee$. The preservation of C follows at once from that of 0, 1, $\wedge$ and $\vee$.

(iv) $\Rightarrow$ (i) Since f preserves C, condition (ii) is satisfied. But then if G is a filter containing F, given $x \in G \setminus F$ one has $Cx \in F$ and thus $0 = x \wedge Cx \in G$, which proves $G = B$. $\qquad\square$

Proposition 4.1.4 *In a boolean algebra B,*

(i) *every proper filter is contained in an ultrafilter,*

(ii) *every non-zero element belongs to an ultrafilter,*

(iii) *$x \not\leq y \Rightarrow \exists F$ ultrafilter, with $x \in F$, $y \notin F$,*

(iv) *every filter is the intersection of the ultrafilters which contain it.*

Proof Proper filters containing a given filter constitute an inductive poset, from which condition (i) holds by Zorn's lemma. Applying condition (i) to

$$\uparrow x = \{y \mid y \geq x\}$$

one gets condition (ii). Applying condition (ii) to $x \wedge \complement y$, one gets condition (iii); this implies condition (iv) for proper filters. Condition (iv) for the trivial filter B is obvious, since the intersection becomes that of the empty family and thus is equal to the whole space. □

Proposition 4.1.5 *Given a boolean algebra B, let us write $\mathsf{Spec}(B)$ for the set of its ultrafilters. For every filter H on B, consider*

$$\mathcal{O}_H = \{F \in \mathsf{Spec}(B) \mid H \nsubseteq F\}.$$

The subsets $\mathcal{O}_H \subseteq \mathsf{Spec}(B)$ constitute a topology $\mathcal{T}$ on $\mathsf{Spec}(B)$. The map

$$\mathcal{O} \colon \mathsf{Filters}(B) \longrightarrow \mathcal{T}, \quad H \mapsto \mathcal{O}_H$$

is an isomorphism of posets.

Proof The map $\mathcal{O}$ is surjective by definition and injective by condition (iv) of proposition 4.1.4. It is a homomorphism of posets, since given filters $H \subseteq H'$, one has at once $\mathcal{O}_H \subseteq \mathcal{O}_{H'}$. Conversely $\mathcal{O}_H \subseteq \mathcal{O}_{H'}$, for arbitrary filters H and H', implies $H \subseteq H'$, again by condition (iv) of proposition 4.1.4. □

Corollary 4.1.6 *Given a boolean algebra B, the following conditions hold:*

(i) $\mathcal{O}_{\{1\}} = \emptyset$ *and* $\mathcal{O}_B = \mathsf{Spec}(B)$,

(ii) $\mathcal{O}_{H \cap H'} = \mathcal{O}_H \cap \mathcal{O}_{H'}$,

(iii) $\mathcal{O}_{\langle \bigcup_{i \in I} H_i \rangle} = \bigcup_{i \in I} \mathcal{O}_{H_i}$,

where H, H' and the H_i are filters in B, and $\langle \bigcup_{i \in I} H_i \rangle$ denotes the filter generated by the set theoretical union of the filters H_i.

Proof Any isomorphism of posets preserves the top and bottom elements, arbitrary infima and suprema. In $\mathcal{T}$, finite infima and arbitrary suprema are computed set theoretically. □

Definition 4.1.7 Given a boolean algebra B, the space $(\mathsf{Spec}(B), \mathcal{T})$ of proposition 4.1.5 is called the spectrum of B.

Lemma 4.1.8 *Let B be a boolean algebra. The subsets*

$$\forall b \in B \quad \mathcal{O}_b = \mathcal{O}_{\uparrow b} = \{F \in \mathsf{Spec}(B) | b \notin F\}$$

are both open and closed and constitute a base of open subsets for the topology of $\mathsf{Spec}(B)$. We write $U_b = \mathcal{O}_{\complement b}$ for the complement of $\mathcal{O}_b$.

Proof For every filter $H \subseteq B$, $H = \bigcup_{b \in H} \uparrow b$, from which $\mathcal{O}_H = \bigcup_{b \in H} \mathcal{O}_b$ by condition (iii) of corollary 4.1.6. Moreover condition (iii) of proposition 4.1.3 can be rephrased as $\mathcal{O}_b \cap \mathcal{O}_{b'} = \mathcal{O}_{b \vee b'}$, proving that the $\mathcal{O}_b$ are stable under finite intersections. Finally condition (ii) of proposition 4.1.3 indicates that $U_b = \mathcal{O}_{\complement b} = \complement \mathcal{O}_b$, proving that each open subset $\mathcal{O}_b$ is also closed. $\qquad\square$

Lemma 4.1.9 *Let B be a boolean algebra. One has*

(i) $b \leq b' \Rightarrow \mathcal{O}_b \supseteq \mathcal{O}_{b'}$,
(ii) $b \neq b' \Rightarrow \mathcal{O}_b \neq \mathcal{O}_{b'}$,
(iii) $\mathcal{O}_{b \wedge b'} = \mathcal{O}_b \cup \mathcal{O}_{b'}$,
(iv) $\mathcal{O}_{b \vee b'} = \mathcal{O}_b \cap \mathcal{O}_{b'}$,
(v) $\mathcal{O}_1 = \emptyset$, $\mathcal{O}_0 = B$,
(vi) $\mathcal{O}_{\complement b} = \complement \mathcal{O}_b$,

for all elements $b, b' \in B$.

Proof With the notation of 4.1.6 and by proposition 4.1.3, observe that

$$\langle \uparrow b \cup \uparrow b' \rangle = \uparrow (b \wedge b'), \quad \uparrow b \cap \uparrow b' = \uparrow (b \vee b')$$

and apply proposition 4.1.5. $\qquad\square$

Corollary 4.1.10 *Let B be a boolean algebra. Using the alternative notation $U_b = \mathcal{O}_{\complement b}$, one gets the corresponding covariant formulæ between basic open-closed subsets:*

(i) $b \leq b' \Rightarrow U_b \subseteq U_{b'}$;
(ii) $b \neq b' \Rightarrow U_b \neq U_{b'}$;
(iii) $U_{b \wedge b'} = U_b \cap U_{b'}$;
(iv) $U_{b \vee b'} = U_b \cup U_{b'}$;
(v) $U_1 = B$, $U_0 = \emptyset$;
(vi) $U_{\complement b} = \complement U_b$;

for all elements $b, b' \in B$. $\qquad\square$

Proposition 4.1.11 *The spectrum of a boolean algebra is a profinite space.*

Proof If $F \neq F'$ are distinct ultrafilters, the maximality of F implies at once $F \not\subseteq F'$. Choosing $b \in F \setminus F'$, we get $F' \in \mathcal{O}_b$ and $F \notin \mathcal{O}_b$. The open subset $\mathcal{O}_b$ is also closed by lemma 4.1.8. This proves that $\mathsf{Spec}(B)$ is totally disconnected (see definition 3.4.2).

By theorem 3.4.7, it remains to prove that $\mathsf{Spec}(B)$ is compact. For this apply lemma 4.1.8 and consider $\mathsf{Spec}(B) = \bigcup_{i \in I} \mathcal{O}_{b_i}$, for elements $b_i \in B$. We have thus

$$\forall F \in \mathsf{Spec}(B) \quad \exists i \in I \quad F \in \mathcal{O}_{b_i}$$

or, equivalently,

$$\forall F \in \mathsf{Spec}(B) \quad \exists i \in I \quad b_i \notin F.$$

Consider now

$$G = \{y \in B \mid \exists i_1, \dots, i_n \in I \quad b_{i_1} \wedge \cdots \wedge b_{i_n} \leq y\}.$$

We argue by reduction *ad absurdum*. If we cannot extract a finite covering from the original one, for every finite sequence $b_{i_1}, \dots, b_{i_n}$, there exists a corresponding ultrafilter $F_{i_1, \dots, i_n}$ such that

$$F_{i_1, \dots, i_n} \notin \mathcal{O}_{b_{i_1}} \cup \cdots \cup \mathcal{O}_{b_{i_n}} = \mathcal{O}_{b_{i_1} \wedge \cdots \wedge b_{i_n}},$$

that is, $b_{i_1} \wedge \cdots \wedge b_{i_n} \in F_{i_1, \dots, i_n}$. In particular $b_{i_1} \wedge \cdots \wedge b_{i_n} \neq 0$, which proves that G, which is obviously a filter, is in fact a proper filter. By proposition 4.1.4, this proper filter G is contained in an ultrafilter F. But G, and thus F, contains all the elements b_i, $i \in I$, which is a contradiction. $\qquad\square$

Corollary 4.1.12 *A subset $U \subseteq \mathsf{Spec}(B)$ of the spectrum of a boolean algebra is a clopen (= closed and open) iff it has the form U_b, for some element $b \in B$.*

Proof By lemma 4.1.8, all U_b are clopen. Conversely if U is a clopen, U is compact in $\mathsf{Spec}(B)$ by proposition 4.1.11. Moreover, by lemma 4.1.8, $U = \bigcup_{i \in I} U_{b_i}$, for elements $b_i \in B$. By compactness,

$$U = U_{b_1} \cup \cdots \cup U_{b_n} = U_{b_1 \vee \cdots \vee b_n}. \qquad\square$$

Corollary 4.1.13 *Every boolean algebra B is isomorphic to the boolean algebra of clopens in $\mathsf{Spec}(B)$.* $\qquad\square$

Corollary 4.1.14 *Every finite boolean algebra is isomorphic to the boolean algebra of subsets of a finite set.*

Proof The spectrum of a finite boolean algebra is compact Hausdorff and finite, thus discrete. Its clopens are all its subsets.   □

Proposition 4.1.15 *Every profinite space is homeomorphic to the spectrum of the boolean algebra of its clopens.*

Proof Let X be a profinite space. Given $x \in X$, we put

$$F_x = \{U \subseteq X \mid U \text{ clopen}, \, x \in U\}.$$

Each F_x is trivially a proper filter in the boolean algebra $\mathsf{Clopen}(X)$ of clopens of X. Since moreover

$$\forall U \in \mathsf{Clopen}(X) \quad x \in U \text{ or } x \in \complement U,$$

the filter F_x is an ultrafilter (see proposition 4.1.3). We consider now the map

$$\varphi \colon X \longrightarrow \mathsf{Spec}\big(\mathsf{Clopen}(X)\big), \quad x \mapsto F_x$$

and we shall prove it is a homeomorphism.

If $x \neq y$, by theorem 3.4.7 there exists a clopen U with $x \in U$ and $y \in \complement U$. Thus $U \in F_x$ and $U \notin F_y$, from which $F_x \neq F_y$ and φ is injective.

If $\mathcal{F} \subseteq \mathsf{Clopen}(X)$ is an ultrafilter, consider

$$V = \bigcap \{U \subseteq X \mid U \in \mathcal{F}\}.$$

Since $\mathcal{F}$ is a proper filter, a finite intersection of elements in $\mathcal{F}$ is never empty. By compactness of X, the subset V is thus non empty and we fix $x \in V$. The proper filter F_x contains $\mathcal{F}$, thus $\mathcal{F} = F_x = \varphi(x)$ by maximality of $\mathcal{F}$. This proves the surjectivity of φ.

Since φ is a bijection between compact Hausdorff spaces, it remains to prove its continuity. Applying lemma 4.1.8, we consider a basic open subset of $\mathsf{Spec}\big(\mathsf{Clopen}(X)\big)$, which has the form U_W for some $W \in \mathsf{Clopen}(X)$. It follows at once that

$$\varphi^{-1}(U_W) = \{x \in X \mid W \in F_x\} = \{x \in X \mid x \in W\} = W.   □$$

Theorem 4.1.16 (Stone duality) *The maps*

$$B \mapsto \mathsf{Spec}(B), \quad X \mapsto \mathsf{Clopen}(X)$$

extend to a contravariant equivalence between the categories of boolean algebras and profinite spaces.

Proof If $f: B \longrightarrow B'$ is a homomorphism of boolean algebras, composition with f

$$\mathsf{Hom}(B', \{0,1\}) \longrightarrow \mathsf{Hom}(B, \{0,1\})$$

yields, by proposition 4.1.3, a map

$$\mathsf{Spec}(f): \mathsf{Spec}(B') \longrightarrow \mathsf{Spec}(B).$$

This map is continous since given $b \in B$

$$
\begin{aligned}
\mathsf{Spec}(f)^{-1}(\mathcal{O}_b) &= \{F \in \mathsf{Spec}(B') \,|\, b \notin \mathsf{Spec}(f)(F)\} \\
&= \{F \in \mathsf{Spec}(B') \,|\, b \notin f^{-1}(F)\} \\
&= \{F \in \mathsf{Spec}(B') \,|\, f(b) \notin F\} \\
&= \mathcal{O}_{f(b)}.
\end{aligned}
$$

This yields at once a functor from the category **Bool** of boolean algebras to the category **Prof** of profinite spaces –

$$\mathsf{Spec}: \mathbf{Bool} \longrightarrow \mathbf{Prof}.$$

On the other hand every continuous map $g: X \longrightarrow X'$ between (profinite) topological spaces induces a homomorphism of boolean algebras

$$g^{-1}: \mathsf{Clopen}(X') \longrightarrow \mathsf{Clopen}(X)$$

which yields at once a functor

$$\mathsf{Clopen}: \mathbf{Prof} \longrightarrow \mathbf{Bool}.$$

Corollary 4.1.13 and proposition 4.1.15 imply already that **Spec** and **Clopen** are mutually inverse (up to isomorphisms) at the level of objects. Moreover given a homomorphism $f: B \longrightarrow B'$ of boolean algebras,

$$(\mathsf{Clopen} \circ \mathsf{Spec})(f)(\mathcal{O}_b) = \mathsf{Spec}(f)^{-1}(\mathcal{O}_b) = \mathcal{O}_{f(b)}$$

from which $\mathsf{Clopen} \circ \mathsf{Spec} \cong \mathrm{id}$ since $U_b = \mathcal{O}_{\mathfrak{C}b}$. On the other hand, for a continuous map $g: X \longrightarrow X'$ between profinite spaces,

$$
\begin{aligned}
(\mathsf{Spec} \circ \mathsf{Clopen})(g)(F_x) &= \{W \in \mathsf{Clopen}(X') \,|\, g^{-1}(W) \in F_x\} \\
&= \{W \in \mathsf{Clopen}(X') \,|\, x \in g^{-1}(W)\} \\
&= \{W \in \mathsf{Clopen}(X') \,|\, g(x) \in W\} \\
&= F_{g(x)},
\end{aligned}
$$

which proves $\mathsf{Spec} \circ \mathsf{Clopen} \cong \mathrm{id}$.   □

4.2 Pierce representation of a commutative ring

Let us recall that an element e of a ring R is idempotent when $e^2 = e$. We also recall that all our rings are commutative and have a unit.

Proposition 4.2.1 *The idempotents of a ring R, with the operations*

$$e \wedge e' = ee', \quad e \vee e' = e + e' - ee',$$

constitute a boolean algebra.

Proof If e, e' are idempotents,

$$
\begin{aligned}
(ee')^2 &= e^2 e'^2 = ee', \\
(e + e' - ee')^2 &= e^2 + e'^2 + e^2 e'^2 + 2ee' - 2e^2 e' - 2ee'^2 \\
&= e + e' + ee' + 2ee' - 2ee' - 2ee' \\
&= e + e' - ee'
\end{aligned}
$$

from which ee' and $e + e' - ee'$ are again idempotents.

Observe next that

$$e \wedge e' = e \Leftrightarrow ee' = e \Leftrightarrow e' = e + e' - ee' \Leftrightarrow e' = e \vee e'.$$

This shows the existence of a unique relation $\leq$ defined by

$$e \leq e' \text{ iff } e \wedge e' = e \text{ iff } e \vee e' = e'.$$

This yields a poset structure on the set of idempotents:

- $e \leq e$ because $ee = e$;
- $e \leq e' \leq e''$ yields $ee' = e$ and $e'e'' = e'$, from which $e = ee' = ee'e'' = ee''$ and thus $e \leq e''$;
- $e \leq e'$ and $e' \leq e$ yield $e = ee'$ and $e' = e'e$, from which $e = e'$.

It follows at once, by definition of the poset structure, that $\wedge$ and $\vee$ are the infimum and the supremum in the poset of idempotents. Clearly 0 is the bottom element and 1 is the top element. We prove next that every idempotent has a complement.

If e is idempotent,

$$(1 - e)^2 = 1 - 2e + e^2 = 1 - 2e + e = 1 - e$$

proving that $1 - e$ is idempotent as well. Moreover

$$e(1 - e) = e - e^2 = e - e = 0$$

proving $e \wedge (1 - e) = 0$, while

$$e + (1 - e) - e(1 - e) = e + 1 - e - e + e^2 = e + 1 - e - e + e = 1$$

proving that $e \vee (1 - e) = 1$. Thus $1 - e$ is the complement of e.

The distributivity law

$$e \wedge (e' \vee e'') = (e \wedge e') \vee (e \wedge e'')$$

reduces to

$$e(e' + e'' - e'e'') = ee' + ee'' - e^2 e' e''$$

which holds since $e = e^2$. Thus the idempotents of R constitute a boolean algebra. □

Corollary 4.2.2 *If $I \triangleleft R$ is an ideal of the ring R, the idempotents of I constitute an ideal in the boolean algebra of idempotents of R.*

Proof Of course $0 \in I$. Write e, e' for idempotents in R. If $e, e' \in I$, then $e \vee e' = e + e' - ee' \in I$. If $e' \in I$ and $e \leq e'$, then $e = e \wedge e' = ee' \in I$. □

Proposition 4.2.3 *The construction of proposition 4.2.1 extends to a functor defined on the category* Ring *of rings –*

$$\mathsf{Idemp} \colon \mathsf{Ring} \longrightarrow \mathsf{Bool}.$$

Proof A ring homomomorphism $f \colon R \longrightarrow R'$ maps an idempotent of R onto an idempotent of R' and preserves operations like $e \wedge e' = ee'$, $e \vee e' = e + e' - ee'$, $\complement e = 1 - e$. □

Definition 4.2.4 The Pierce spectrum (or boolean spectrum) of a ring R is the spectrum of the boolean algebra of its idempotents.

The composite functor

$$\mathsf{Ring} \xrightarrow{\ \mathsf{Idemp}\ } \mathsf{Bool} \xrightarrow{\ \mathsf{Spec}\ } \mathsf{Prof}$$

will be written $\mathsf{Sp} \colon \mathsf{Ring} \longrightarrow \mathsf{Prof}$; it is a contravariant functor.

Lemma 4.2.5 *In the Pierce spectrum of a ring R, a partition of a clopen U_e into non-empty clopens has the form*

$$U_e = U_{e_1} \cup \cdots \cup U_{e_n}$$

where

(i) *each $e_i \in R$ is a non-zero idempotent,*
(ii) $e_1 + \cdots + e_n = e$,
(iii) $i \neq j \Rightarrow e_i e_j = 0$

and $U_{e'} = \{F \in \mathsf{Sp}(R) | e' \in F\}$ is defined as in corollary 4.1.10. In particular, putting $e = 1$, this applies to $\mathsf{Sp}(R) = U_1$.

Proof Since U_e is compact, every partition into non-empty clopens is necessarily finite. And each clopen has the form $U_{e'}$ for some idempotent $e' \in R$, by corollary 4.1.12. By corollary 4.1.10, a partition of U_e into clopens thus has the form

$$U_e = U_{e_1} \cup \cdots \cup U_{e_n}$$

with

$$
\begin{aligned}
e_1 \vee \cdots \vee e_n &= e, \\
e_i e_j &= 0 \quad \text{for } i \neq j.
\end{aligned}
$$

Notice next that $e_i \wedge e_j = e_i e_j = 0$ implies at once

$$e_i \vee e_j = e_i + e_j - e_i e_j = e_i + e_j.$$

It is straightforward to extend this formula by induction to an n term partition; this yields the required result. $\qquad\square$

Definition 4.2.6 In a ring R, an ideal $I \lhd R$ is regular when, as an ideal, it is generated by its idempotent elements.

An ideal $I \lhd R$ is thus regular when every element $i \in I$ can be written

$$i = r_1 e_1 + \cdots + r_n e_n$$

with each $r_i \in R$, $e_i \in I$ and $e_i^2 = e_i$.

Lemma 4.2.7 *The following conditions are equivalent, for an ideal $I \lhd R$:*

(i) *I is regular;*
(ii) *$\forall i \in I \; \exists e \in I \;\; e = e^2, \;\; i = ie$;*
(iii) *$\forall i_1, \ldots, i_n \in I \;\; \exists e \in I \;\; e = e^2, \;\; \forall k = 1, \ldots, n \; i_k = i_k e$.*

Proof Trivially (iii) $\Rightarrow$ (ii) $\Rightarrow$ (i). Let us prove (i) $\Rightarrow$ (iii). Each i_k can be written

$$i_k = r_1^{(k)} e_1^{(k)} + \cdots + r_{m_k}^{(k)} e_{m_k}^{(k)}$$

with each $r_i^{(j)} \in R$, $e_i^{(j)} \in I$ and $\left(e_i^{(j)}\right)^2 = e_i^{(j)}$. It suffices clearly to find $e \in I$ such that $e^2 = e$ and $e_i^{(j)}e = e_i^{(j)}$ for all possible indices i, j.

This reduces the problem to the following statement. Given idempotents $e_1, \ldots, e_m$ in I, there exists an idempotent $e \in I$ such that $e_ie = e_i$, that is $e_i \le e$, for each index i. By corollary 4.2.2, it suffices to take $e = e_1 \vee \cdots \vee e_n$. $\qquad\square$

Lemma 4.2.8 *In every ring,*

(i) *the ring itself is a regular ideal,*

(ii) *a finite non-empty intersection of regular ideals is again regular and coincides with the product of the ideals,*

(iii) *an arbitrary sum of regular ideals is again a regular ideal.*

Proof The ring itself is regular, since $r = r1$ for each $r \in R$ and $1 \in R$ is idempotent. This is in fact the case of an empty intersection of ideals, which is the ring itself.

Let us treat the case of a binary intersection, which will imply the case of a finite non-empty intersection. If I, J are regular ideals, one has at once $IJ \subseteq I \cap J$. Conversely if $r \in I \cap J$, by lemma 4.2.7 write $r = re = (re)e$ with $e = e^2 \in J$; this proves at once $r \in IJ$ and therefore, $I \cap J = IJ$. It remains to prove that IJ is regular. Indeed an element of IJ has the form

$$r = i_1 j_1 + \cdots + i_n j_n, \quad i_k \in I, \ j_k \in J.$$

By lemma 4.2.7, choose $e = e^2 \in I$ such that $i_k e = i_k$ for all indices k, and $e' = e'^2 \in J$ such that $j_k e' = j_k$ for all indices k. It follows immediately that $e \wedge e' = ee' \in IJ$ is an idempotent such that $ree' = r$.

Next choose a family $(I_k)_{k \in K}$ of regular ideals. An element $r \in I = \sum_{k \in K} I_k$ has the form $r = r_1 + \cdots + r_n$ with each $r_i \in I_{k_i}$. By lemma 4.2.7, for each index i choose $e_i = e_i^2 \in I_{k_i} \subseteq I$ such that $r_i e_i = r_i$. It remains to find $e = e^2 \in I$ such that $e_i e = e_i$, that is $e_i \le e$, for each index i. By corollary 4.2.2, it suffices to choose $e = e_1 \vee \cdots \vee e_n$. $\qquad\square$

Let us recall a classical definition.

Definition 4.2.9 A locale is a complete lattice in which the infinite distributivity law

$$a \wedge \left(\bigvee_{i \in I} b_i \right) = \bigvee_{i \in I} (a \wedge b_i)$$

holds for all elements a, b_i and every set I of indices.

A typical example of a locale is the lattice of open subsets of a topological space: the distributivity law holds since finite meets and arbitrary joins of open subsets are just set theoretical intersections and unions.

Proposition 4.2.10 *The regular ideals of a ring R constitute a locale isomorphic to the locale of open subsets of the Pierce spectrum of R.*

Proof   By proposition 4.1.5, the locale of open subsets of $\mathsf{Sp}(R)$ is isomorphic to the poset of filters in the boolean algebra of idempotents of R. But in every boolean algebra B the poset of filters is isomorphic to the poset of ideals: the correspondence between an ideal I and a filter F is simply given by

$$I = \{b \in B \mid \complement b \in F\}, \quad F = \{b \in B \mid \complement b \in I\}.$$

It remains to prove that the lattice of regular ideals of R is isomorphic to the lattice of ideals in the boolean algebra $\mathsf{Idemp}(R)$ of idempotents of R.

By corollary 4.2.2 we know already that the idempotents of a regular ideal I constitute an ideal $\mathsf{Idemp}(I)$ in the boolean algebra $\mathsf{Idemp}(R)$. By definition of a regular ideal, distinct regular ideals yield distinct ideals in $\mathsf{Idemp}(R)$, proving the injectivity of the correspondence. Moreover it is obvious that this correspondence preserves and respects the ordering. It remains to prove that every ideal J of $\mathsf{Idemp}(R)$ is the ideal of idempotents of a regular ideal $I \lhd R$. Of course, we define I as the ideal of R generated by all the elements of J, which implies at once that I is regular and $J \subseteq \mathsf{Idemp}(I)$. Conversely, if $e \in \mathsf{Idemp}(I)$, then e can be written $e = r_1 e_1 + \cdots + r_n e_n$ with each $r_i \in R$ and $e_i \in J$. Since J is an ideal in $\mathsf{Idemp}(R)$, $e' = e_1 \vee \ldots \vee e_n \in J$ and $e_i = e_i \wedge e' = e_i e'$ for each index i. In particular $e = ee'$, that is $e \leq e'$ with $e' \in J$. Since J is an ideal, $e \in J$.   □

Corollary 4.2.11 *Let R be a ring. The Pierce spectrum of R is homeomorphic to the set of maximal regular ideals of R (that is, maximal elements in the poset of proper regular ideals), provided with the topology constituted of the subsets*

$$\mathcal{O}_I = \{M \mid I \not\subseteq M\}$$

for every regular ideal $I \lhd R$. We keep writing $\mathsf{Sp}(R)$ for the Pierce spectrum described in this way.

Proof By definition 4.1.7 and proposition 4.2.10. □

Given an idempotent $e \in R$, we shall generally write

$$\mathcal{O}_e = \mathcal{O}_{Re} = \{M \mid Re \not\subseteq M\} = \{M \mid e \notin M\}.$$

Definition 4.2.12 The Pierce structural space of a ring R is the disjoint union of all the quotients R/M, where M runs through all the maximal regular ideals of R; this set is provided with the final topology for all the maps

$$s_r^I \colon \mathcal{O}_I \longrightarrow \coprod_M R/M, \quad N \mapsto [r] \in R/N$$

where I runs through the regular ideals of R and r through the elements of R.

We recall another classical concept.

Definition 4.2.13 A map $f \colon X \longrightarrow Y$ between topological spaces is étale when, for every point $x \in X$, there exist open neighbourhoods $x \in U \subseteq X$ and $y \in V \subseteq Y$ such that f restricts as a homeomorphism $f \colon U \longrightarrow V$.

Of course, an étale map is both continuous and open.

Theorem 4.2.14 *Given a ring R, the projection*

$$p \colon \coprod_M R/M \longrightarrow \mathsf{Sp}(R), \quad [r] \in R/N \mapsto N$$

is étale, when the structural space $\coprod_M R/M$ is provided with its topology from definition 4.2.12.

Proof We coinsider first an element $r \in R$ and prove that

$$U_r = \{M \in \mathsf{Sp}(R) \mid r \in M\}$$

is open in $\mathsf{Sp}(R)$. For this consider

$$
\begin{aligned}
J &= \{e \in R \mid e = e^2, \ \forall N \in \mathsf{Sp}(R) \ r \notin N \Rightarrow e \in N\} \\
&= \bigcap \{\mathsf{Idemp}(N) \mid N \in \mathsf{Sp}(R), \ r \notin N\}.
\end{aligned}
$$

J is trivially an ideal in the boolean algebra $\mathsf{Idemp}(R)$. The proof of proposition 4.2.10 shows that $J = \mathsf{Idemp}(I)$ for a unique regular ideal $I \lhd R$. Observe moreover that given $N \in \mathsf{Sp}(R)$, one has immediately

$$N \in \mathcal{O}_I \iff I \not\subseteq N$$

$$\Leftrightarrow \quad \exists e \ \ e = e^2, \ \ e \in I, \ \ e \notin N$$

$$\Leftrightarrow \quad \exists e \ \ e \in J, \ \ e \notin N$$

$$\Rightarrow \quad r \in N$$

$$\Leftrightarrow \quad N \in U_r.$$

In fact, the implication in the previous formula is itself an equivalence. Indeed, if $r \in N$, then $r = re'$ with $e' = e'^2 \in N$ by lemma 4.2.7. If $M \in \mathsf{Sp}(R)$ and $r \notin M$, then $e' \notin M$ since $r = re'$. Thus $1 - e' \in M$ by maximality of M (propositions 4.1.3 and 4.2.10). This proves that $e = 1 - e' \in J$ with $e = 1 - e' \notin N$, since $e' \in N$. This concludes the proof that $U_r = \mathcal{O}_I$, thus U_r is open.

Now if s_r^I and $s_t^{I'}$ are two sections as in definition 4.2.12,

$$
\begin{aligned}
(s_t^{I'})^{-1}(s_r^I(\mathcal{O}_I)) &= \{M \in \mathcal{O}_{I'} | s_t^{I'}(M) \in s_r^I(\mathcal{O}_I)\} \\
&= \{M \in \mathcal{O}_I \cap \mathcal{O}_{I'} | [t] = [r] \in R/M\} \\
&= \{M \in \mathcal{O}_{I \cap I'} | [t - r] = 0 \in R/M\} \\
&= \{M \in \mathcal{O}_{I \cap I'} | t - r \in M\} \\
&= \mathcal{O}_{I \cap I'} \cap U_{t-r}
\end{aligned}
$$

and this is an open subset. By definition of a final topology, $s_r^I(\mathcal{O}_I)$ is thus an open subset of $\coprod_M R/M$.

Again by definition of a final topology, the continuity of p reduces to that of all composites $p \circ s_r^I$, which are the canonical inclusions of the open subsets $\mathcal{O}_I$ in $\mathsf{Sp}(R)$. As a consequence, for every element $[r] \in R/M$, we obtain reciprocal homeomorphisms

$$\mathsf{Sp}(R) \underset{s_r^R}{\overset{p}{\rightleftarrows}} s_r^R(\mathsf{Sp}(R))$$

with moreover $s_r^R(\mathsf{Sp}(R))$ a neighbourhood of $[r]$. $\qquad\square$

Theorem 4.2.15 *Every ring R is isomorphic to the ring of continuous sections of the projection p,*

$$p \colon \coprod_M R/M \longrightarrow \mathsf{Sp}(R), \quad [r] \in R/N \mapsto N,$$

defined on the Pierce structural space of R. More generally, the ring of continous sections of p defined on the open subset $\mathcal{O}_e$, for an idempotent element $e \in R$, is isomorphic to the principal ideal Re.

Proof Let us fix an idempotent element $e \in R$; the first part of the statement is the special case $e = 1$. For every element $r \in R$ we have a continuous section

$$s_r^e = s_r^{Re} : \mathcal{O}_e \longrightarrow \coprod_M R/M, \quad N \mapsto [r] \in R/N.$$

Writing $\mathsf{Sec}_e(p)$ for the set of continuous sections of p on $\mathcal{O}_e$, we get a map

$$\varphi_e : R \longrightarrow \mathsf{Sec}_e(p), \quad r \mapsto s_r^e$$

It is obvious that the image of φ_e is a ring for the fibrewise operations and that φ_e, corestricted to its image $\varphi_e(R)$, is a ring homomorphism.

Now observe that $s_{1-e}^e = 0$ since

$$M \in \mathcal{O}_e \Leftrightarrow e \notin M \Leftrightarrow 1 - e \in M$$

by propositions 4.1.3 and 4.2.10. This implies the existence of a ring homomorphism factorization of φ_e as

$$R \longrightarrow\!\!\!\!\rightarrow R/R(1-e) \xrightarrow{\ \psi_e\ } \varphi_e(R), \quad \psi_e\big([r]\big) = \varphi_e(r).$$

This factorization ψ_e is injective because when $s_r^e = 0$, by propositions 4.1.4 and 4.2.10,

$$r \in \bigcap_{M \in \mathcal{O}_e} M = \bigcap_{e \notin M} M = \bigcap_{1 - e \in M} M = \bigcap_{R(1-e) \subseteq M} M = R(1-e).$$

The factorization ψ_e is also surjective, by definition of its codomain.

Next we observe that $R/R(1-e) \cong Re$. Indeed we have a direct sum $R = Re \oplus R(1-e)$ since this sum contains $e + (1-e) = 1$ and $Re \cap R(1-e) = Re \cdot R(1-e) = 0$ (see lemma 4.2.8). Thus ψ_e yields an isomorphism between Re and $\varphi_e(R)$.

It remains to prove that $\varphi_e(R) = \mathsf{Sec}_e(p)$. For this choose a continuous section σ of p on $\mathcal{O}_e$. Given $M \in \mathcal{O}_e$, let us write $\sigma(M) = [r_M] \in R/M$. The proof of theorem 4.2.14 shows that $s_{r_M}^e(\mathcal{O}_e)$ is open, from which $\sigma^{-1}\big(s_{r_M}^e(\mathcal{O}_e)\big)$ is open. In other words

$$W_M = \big\{ N \in \mathcal{O}_e \,\big|\, \sigma(N) = [r_M] \in R/N \big\} \subseteq \mathcal{O}_e$$

is open and contains M. This open subset W_M is a union of clopens of the form $\mathcal{O}_{e_i}$, with each e_i idempotent. In particular there exists an idempotent e_M such that $M \in \mathcal{O}_{e_M} \subseteq W_M \subseteq \mathcal{O}_e$. These various $\mathcal{O}_{e_M}$ constitute a covering of $\mathcal{O}_e$, since each $M \in \mathcal{O}_e$ belongs to $\mathcal{O}_{e_M}$. But $\mathcal{O}_e$ is closed and thus compact, from which we get finitely many $M_i \in \mathcal{O}_e$ such that the corresponding clopens $\mathcal{O}_{e_{M_i}}$ cover $\mathcal{O}_e$. A finite intersection of clopens of the form $\mathcal{O}_{e_{M_i}}$ is a clopen of the form $\mathcal{O}_{e'}$ for some idempotent e' (see lemma 4.1.9). Thus from the finite covering by clopens we get, by considering the various intersections, a finite partition $\mathcal{O}_e = \mathcal{O}_{e_1} \cup \cdots \cup \mathcal{O}_{e_n}$ with each e_i an idempotent. Now each $\mathcal{O}_{e_i}$ is by construction contained in some W_{M_i} and therefore $\sigma(N) = [r_{M_i}]$ for each $N \in \mathcal{O}_{e_i}$. For simplicity, write $r_i = r_{M_i}$, yielding $\sigma(N) = [r_i] \in R/N$ for each $N \in \mathcal{O}_{e_i}$.

We now put $r = r_1 e_1 + \cdots + r_n e_n$ and we shall prove that $\sigma = s_r^e$. If $M \in \mathcal{O}_{e_i}$, then $e_i \notin M$ and thus $1 - e_i \in M$. The fact of working with a partition also yields $M \notin \mathcal{O}_{e_j}$ for $j \neq i$, thus $e_j \in M$. Therefore, in R/M,

$$
\begin{aligned}
s_r^e(M) &= [r_1 e_1 + \cdots + r_n e_n] \\
&= [r_1][e_1] + \cdots + [r_n][e_n] \\
&= [r_i][e_i] \quad \text{since } e_j \in M \text{ for } j \neq i \\
&= [r_i] \quad \text{since } 1 - e_i \in M, \text{ thus } [e_i] = [1] \text{ in } R/M \\
&= \sigma(M).
\end{aligned}
$$

This concludes the proof. $\qquad\square$

Proposition 4.2.16 *Given a maximal regular ideal M of a ring R, the quotient ring R/M admits only 0 and 1 as idempotents.*

Proof The ring R/M is the filtered colimit of the rings $R/\langle e_1, \ldots, e_n \rangle$, where $\langle e_1, \ldots, e_n \rangle$ is the ideal generated by the idempotents $e_1, \ldots, e_n$ of M. If $[r] \in R/M$ is idempotent, then r and r^2 are identified in R/M, thus are already identified in a quotient $R/\langle e_1, \ldots, e_n \rangle = R/\langle e_1 \vee \cdots \vee e_n \rangle$. But $R/\langle e_1 \vee \cdots \vee, e_n \rangle = R\big(1 - (e_1 \vee \cdots \vee e_n)\big)$ (see proof of theorem 4.2.15), thus $[r] = [e] \in R/M$, where $e = e^2 \in R\big(1 - (e_1 \vee \cdots \vee e_n)\big)$. If $e \in M$, then $[e] = 0$; otherwise $1 - e \in M$ and $[e] = 1$. $\qquad\square$

4.3 The adjoint of the 'spectrum' functor

We recall once more that our rings and algebras are commutative and have a unit.

Lemma 4.3.1 *The category R-Alg of R-algebras on a ring R is isomorphic to the category R/Ring of rings under R. In other words, the dual of the category of R-algebras is the slice category $(\mathsf{Ring})^{\mathrm{op}}/R$. More generally, if S is an R-algebra, the dual of the category of S-algebras is the slice category $(R\text{-}\mathsf{Alg})^{\mathrm{op}}/S$.*

Proof An R-algebra A is a ring A provided with a ring homomorphism

$$\rho_A \colon R \longrightarrow A, \quad r \mapsto r1.$$

Conversely a ring A provided with such a ring homomorphism ρ_A is an R-algebra, with scalar multiplication defined by $ra = \rho_A(r)a$. The rest is obvious.

More generally, if S is an R-algebra and A is an S-algebra, the morphism

$$\rho_A \colon S \longrightarrow A, \quad s \mapsto s1$$

is now a homomorphism of R-algebras and, conversely, the existence of such a homomorphism ρ_A on an R-algebra A induces the structure of an S-algebra on A via the formula $sa = \rho_A(s)a$. □

Theorem 4.3.2 *Let R be a ring. The Pierce spectrum functor*

$$\mathsf{Sp} \colon (R\text{-}\mathsf{Alg})^{\mathrm{op}} \longrightarrow \mathsf{Prof}, \quad A \mapsto \mathsf{Sp}(A)$$

admits as right adjoint the functor

$$\mathcal{C}(-, R) \colon \mathsf{Prof} \longrightarrow (R\text{-}\mathsf{Alg})^{\mathrm{op}}, \quad X \mapsto \mathcal{C}(X, R)$$

where R is provided with the discrete topology and $\mathcal{C}(X, R)$ denotes the ring of continuous functions.

Proof The functor $\mathcal{C}(-, R)$ acts by composition on the arrows of Prof. To prove the adjunction, we shall construct the corresponding unit and counit and prove the required triangular identities. For the sake of clarity, we work in $R\text{-}\mathsf{Alg}$ instead of its dual.

Given a profinite space X, let us construct

$$\alpha_X \colon \mathsf{Sp}\big(\mathcal{C}(X, R)\big) \longrightarrow X.$$

By the Stone duality theorem (see 4.1.16), this reduces to constructing a homomorphism of boolean algebras

$$\widetilde{\alpha}_X \colon \mathsf{Clopen}(X) \longrightarrow \mathsf{Idemp}\big(\mathcal{C}(X, R)\big), \quad U \mapsto \widetilde{\alpha}_X(U) = f_U,$$

where we define

$$f_U \colon X \longrightarrow R, \quad f_U(x) = \begin{cases} 1 & \text{if} \quad x \in U, \\ 0 & \text{if} \quad x \notin U. \end{cases}$$

One has trivially

$$f_{U \cap V} = f_U \cdot f_V, \quad f_{U \cup V} = f_U + f_V - f_{U \cap V} = f_U + f_V - f_U \cdot f_V$$

from which it follows at once that $\widetilde{\alpha}_X$ is a homomorphism of boolean algebras.

The naturality of α is easy. If $h \colon X \longrightarrow Y$ is a morphism in Prof, for every $V \in \mathsf{Clopen}(Y)$ and $x \in X$,

$$f_{h^{-1}(U)}(x) = 1 \Leftrightarrow x \in h^{-1}(U) \Leftrightarrow h(x) \in U \Leftrightarrow f_U\big(h(x)\big) = 1$$

from which $\widetilde{\alpha}_X\big(h^{-1}(U)\big) = \mathcal{C}(h, -)\big(\widetilde{\alpha}_Y(U)\big)$. This expresses the naturality of $\widetilde{\alpha}$, thus of α.

Next for an R-algebra A, we construct

$$\beta_A \colon \mathcal{C}\big(\mathsf{Sp}(A), R\big) \longrightarrow A.$$

For every continuous map $f \colon \mathsf{Sp}(A) \longrightarrow R$, we shall prove the existence of a finite partition $\mathsf{Sp}(A) = U_{e_1} \cup \cdots \cup U_{e_n}$ of $\mathsf{Sp}(A)$ into clopens U_{e_i}, with $e_i \in A$ idempotent, and such that f is constant on each U_{e_i}, let us say, with value $r_i \in R$. We shall define $\beta_A(f) = \sum_{i=1}^{n} r_i e_i$ and we shall prove that this definition is independent of the choice of the partition.

To prove the existence of such a partition, observe we have a covering $\mathsf{Sp}(A) = \bigcup_{r \in R} f^{-1}(r)$. Clearly $r \neq r'$ implies that $f^{-1}(r)$ and $f^{-1}(r')$ are disjoint, thus the covering is a partition. Moreover since R is discrete, each $f^{-1}(r)$ is open and closed. And finally since $\mathsf{Sp}(A)$ is compact, we can extract a finite subpartition of $\mathsf{Sp}(A)$, that is, only finitely many $f^{-1}(r_i)$ are non empty. Let us write

$$\mathsf{Sp}(A) = f^{-1}(r_1) \cup \cdots \cup f^{-1}(r_n).$$

Since each $f^{-1}(r_i)$ is a clopen in the profinite space $\mathsf{Sp}(A)$, it has the form U_{e_i}, for an idempotent $e_i \in A$ (see corollary 4.1.12). And of course for each $a \in U_{e_i}$, we have $f(a) = r_i$.

Let us prove now that the definition $\beta_A(f) = r_1 e_1 + \cdots + r_n e_n$ depends only on f, not on the choice of the partition. The fact that f is constant on each piece of the partition, with the r_i distinct, implies that every possible partition is necessarily a refinement of the partition we have just exhibited. But, with the previous notation, if $U_{e_i} = U_{e_1^i} \cup \cdots \cup U_{e_{k_i}^i}$, by lemma 4.2.5 we get $e_i = e_1^i + \cdots + e_{k_i}^i$ and obviously

$$r_i e_i = r_i(e_1^i + \cdots + e_{k_i}^i) = r_i e_1^i + \cdots + r_i e_{k_i}^i$$

which shows that the definition of $\beta_A(f)$ does not depend on the choice of the partition.

It is now easy to prove that β_A is a homomorphism of R-algebras. For example, given $f, g \in \mathcal{C}\big(\mathsf{Sp}(A), R\big)$, let us prove that $\beta_A(f+g) = \beta_A(f) + \beta_A(g)$. Using the formula $U_e \cap U_{e'} = U_{e \wedge e'} = U_{ee'}$ of corollary 4.1.10, one can choose a partition into clopens U_{e_i} such that both $f(a) = r_i$ and $g(a) = s_i$ hold for all elements $a \in U_{e_i}$. Then $(f + g)(a) = r_i + s_i$ for all $a \in U_{e_i}$. Therefore

$$\beta_A(f + g) = \sum_{i=1}^{n}(r_i + s_i)e_i = \sum_{i=1}^{n} r_i e_i + \sum_{i=1}^{n} s_i e_i = \beta_A(f) + \beta_A(g).$$

To prove the naturality of β, choose a morphism $k \colon A \longrightarrow B$ of R-algebras and $f \in C(\mathsf{Sp}(A), R)$. With the above notation, the conditions

$$e_1 + \cdots + e_n = 1, \quad i \neq j \Rightarrow e_i e_j = 0$$

imply at once

$$k(e_1) + \cdots + k(e_n) = 1, \quad i \neq j \Rightarrow k(e_i)k(e_j) = 0.$$

But

$$
\begin{aligned}
U_{k(e)} &= \{F \in \mathsf{Sp}(B) \,|\, k(e) \in F\} \\
&= \{F \in \mathsf{Sp}(B) \,|\, e \in k^{-1}(F)\} \\
&= \mathsf{Sp}(k)^{-1}(U_e),
\end{aligned}
$$

proving that $C(\mathsf{Sp}(k), \mathsf{id})(f) = f \circ \mathsf{Sp}(k)$ maps $U_{k(e)}$ onto $f(e)$. Therefore

$$(k \circ \beta_A)(f) = k\left(\sum_{i=1}^{n} r_i e_i\right) = \sum_{i=1}^{n} r_i k(e_i) = \beta_B\Big(C(\mathsf{Sp}(k), \mathsf{id})\Big)(f).$$

We must now prove the triangular identities of the adjunction. The first identity is the commutativity of the left hand triangle below,

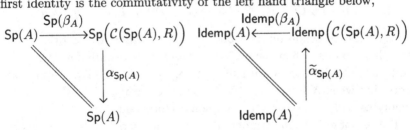

which is equivalent to the commutativity of the right hand triangle, by identifying the clopens of $\mathsf{Sp}(A)$ with the idempotents of A. An idempotent $e \in A$ is mapped by $\tilde{\alpha}_{\mathsf{Sp}(A)}$ onto

$$f_e \colon \mathsf{Sp}(A) \longrightarrow R, \quad \begin{cases} f_e(e') = 1 & \text{if} \quad e' \in U_e, \\ f_e(e') = 0 & \text{if} \quad e' \notin U_e, \text{ i.e. } e' \in U_{1-e}, \end{cases}$$

from which $\beta_A(f_e) = 1e + 0(1 - e) = e$.

The second triangular identity is the commutativity of

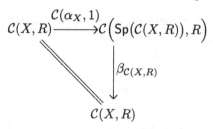

Choose $g \in \mathcal{C}(X, R)$ and consider the composite

$$\mathsf{Sp}\big(\mathcal{C}(X, R)\big) \xrightarrow{\ \alpha_X\ } X \xrightarrow{\ g\ } R.$$

Write $X = \bigcup_{r \in R} g^{-1}(r)$, which is a partition into clopens, since R is discrete. By compactness, we extract a finite partition $X = g^{-1}(r_1) \cup \cdots \cup g^{-1}(r_n)$ and, obviously, $g = \sum_{i=1}^{n} r_i f_{g^{-1}(r_i)}$. Now the subsets $\alpha_X^{-1}\big(g^{-1}(r_i)\big)$ constitute a partition of $\mathsf{Sp}\big(\mathcal{C}(X, R)\big)$ into clopens and by lemma 4.2.5, $\alpha_X^{-1}\big(g^{-1}(r_i)\big) = U_{g_i}$ for some idempotent $g_i \in \mathcal{C}(X, R)$. By definition of α_X, we have $g_i = f_{g^{-1}(r_i)}$, from which $g = \sum_{i=1}^{n} r_i g_i$ and therefore

$$\beta_{\mathcal{C}(X,R)}(g \circ \alpha_X) = \sum_{i=1}^{n} r_i g_i = g. \qquad \square$$

Proposition 4.3.3 *In the conditions and with the notation of theorem 4.3.2, the following conditions are equivalent:*

(i) *the ring R has exactly two distinct idempotents, 0 and 1;*

(ii) *the functor $\mathcal{C}(-, R) \colon \mathsf{Prof} \longrightarrow (R\text{-}\mathsf{Alg})^{\mathrm{op}}$ is full and faithful.*

Proof Assume condition (i). Condition (ii) reduces to the fact that each morphism α_X is a homeomorphism. Since α_X is continuous between compact Hausdorff spaces, it suffices to prove its bijectivity. Observe at once that when X is empty, $\mathcal{C}(X, R)$ is the zero ring and $\mathsf{Sp}\big(\mathcal{C}(X, R)\big)$ is empty as well, from which α_X is a homeomorphism.

If X is not empty, an idempotent continuous map $f \colon X \longrightarrow R$, under assumption (i), takes only the values 0 and 1. Since R is discrete, f takes value 1 on a clopen of X and 0 on its complement. Therefore $\mathsf{Idemp}\big(\mathcal{C}(X, R)\big) \cong \mathsf{Clopen}(X)$ and $\tilde{\alpha}_X$ is an isomorphism.

Conversely if α_X is an isomorphism for each X, choosing for X the singleton yields

$$\mathsf{Sp}(R) \cong \mathsf{Sp}\big(\mathcal{C}(\{*\}, R)\big) \cong \{*\},$$

which proves that $\mathsf{Idemp}(R) = \{0, 1\}$, with $0 \neq 1$. $\qquad \square$

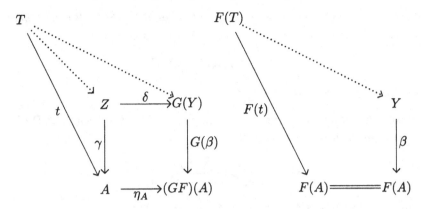

Diagram 4.1

The condition "R admits 0 and 1 as its only idempotents" is very strong, and will now be avoided via a localization process over $\mathsf{Sp}(R)$.

Lemma 4.3.4 *Consider an adjunction* $A \underset{F}{\overset{G}{\rightleftarrows}} B$, $F \dashv G$. *Assume that* A *has pullbacks. In those conditions, for every object* $A \in A$, *the functor*

$$F_A\colon A/A \longrightarrow B/F(A), \quad (X,\alpha) \mapsto \big(F(X), F(\alpha)\big)$$

has a right adjoint functor

$$G_A\colon B/F(A) \longrightarrow A/A, \quad (Y,\beta) \mapsto (Z,\gamma)$$

where (Z,γ) *is defined by the pullback*

$$
\begin{array}{ccc}
Z & \xrightarrow{\ \delta\ } & G(Y) \\
{\scriptstyle\gamma}\downarrow & & \downarrow{\scriptstyle G(\beta)} \\
A & \xrightarrow[\eta_A]{} & (GF)(A)
\end{array}
$$

in which η *is the unit of the adjunction* $G \dashv F$.

Proof Considering diagram 4.1, we observe that

$$A/A\big((T,t), G_A(Y,\beta)\big) \cong A/GF(A)\big((T, \eta_A \circ t), (G(Y), G(\beta))\big)$$

$$\cong B/F(A)\big(F_A(T,t), (Y,\beta)\big).$$

This exhibits the required adjunction.   □

Corollary 4.3.5 *Let R be a ring and S an R-algebra. When S has exactly two distinct idempotents 0 and 1, the isomorphism*

$$S \otimes_R C(X, R) \cong C(X, S)$$

holds for every profinite space X.

Proof By assumption, $\mathsf{Sp}(S)$ is a singleton, thus $\mathsf{Prof}/\mathsf{Sp}(S) \cong \mathsf{Prof}$. The adjunction of theorem 4.3.2

$$(R\text{-}\mathsf{Alg})^{\mathrm{op}} \underset{\mathsf{Sp}}{\overset{C(-, R)}{\rightleftarrows}} \mathsf{Prof}$$

induces, via lemmas 4.3.1 and 4.3.4, an adjunction

$$(S\text{-}\mathsf{Alg})^{\mathrm{op}} \cong (R\text{-}\mathsf{Alg})^{\mathrm{op}}/S \underset{\mathsf{Sp}_S}{\overset{C(-, R)_S}{\rightleftarrows}} \mathsf{Prof}/\mathsf{Sp}(S) \cong \mathsf{Prof}$$

where $C(-, R)_S$ is defined via the following pushout diagram in R-Alg:

$$
\begin{array}{ccc}
C(X, R)_S \cong S \otimes_R C(X, R) & \longleftarrow & C(X, R) \\
\uparrow & & \uparrow \\
\\
S & \underset{\beta_S}{\longleftarrow} & C(\mathsf{Sp}(S), R) \cong C(\{*\}, S) \cong R
\end{array}
$$

Applying theorem 4.3.2 to the ring S, we conclude that

$$S \otimes_R C(X, R) \cong C(X, S). \qquad \qquad \square$$

Theorem 4.3.6 *For every ring R, the right adjoint of the functor*

$$\mathsf{Sp}_R \colon (R\text{-}\mathsf{Alg})^{\mathrm{op}} \longrightarrow \mathsf{Prof}/\mathsf{Sp}(R), \quad A \mapsto (\mathsf{Sp}(A) \longrightarrow \mathsf{Sp}(R))$$

is the functor

$$C_R \colon \mathsf{Prof}/\mathsf{Sp}(R) \longrightarrow (R\text{-}\mathsf{Alg})^{\mathrm{op}}, \quad (X, f) \mapsto \mathsf{Hom}\Big((X, f), \big(\textstyle\coprod_M R/M, p\big)\Big)$$

where $p \colon \coprod_M R/M \longrightarrow \mathsf{Sp}(R)$ is the projection of the Pierce structural space of the ring R (see theorem 4.2.14). This right adjoint is full and faithful.

Proof The case $R = \{0\}$ is trivial since $(R\text{-}\mathsf{Alg})^{\mathrm{op}} \cong \{\{0\}\}$ and $\mathsf{Prof}/\mathsf{Sp}(R) \cong \{\emptyset = \!\!=\!\!= \emptyset\}$. So we shall assume $0 \neq 1$ in R.

The set $\mathsf{Hom}\big((X,f),(\coprod_M R/M,p)\big)$ is provided pointwise with the structure of R-algebra, via the structure of R-algebra on each fibre R/M. Indeed, checking that

$$g, g' \text{ continuous} \Rightarrow g + g', \ gg' \text{ continuous}$$

is routine. An idempotent in this ring is a function $g\colon X \longrightarrow \coprod_M R/M$ such that $p \circ g = f$ and each $g(x) \in R/f(x)$ is idempotent, that is, equal to 0 or 1 by proposition 4.2.16. The subsets

$$s_0^{\mathsf{Sp}(R)}\big(\mathsf{Sp}(R)\big) = \{[0] \in R/M | M \in \mathsf{Sp}(R)\},$$
$$s_1^{\mathsf{Sp}(R)}\big(\mathsf{Sp}(R)\big) = \{[1] \in R/M | M \in \mathsf{Sp}(R)\}$$

are open (proof of 4.2.14) and disjoint (because $1 \notin M$).Therefore the idempotents of $\mathsf{Hom}\big((X,f),(\coprod_M R/M)\big)$ are given by

$$\mathsf{Hom}\Big((X,f),(\{0,1\} \times \mathsf{Sp}(R),p)\Big)$$

where $\{0,1\}$ is provided with the discrete topology and p is the projection onto the factor $\mathsf{Sp}(R)$. This is the same thing as $\mathcal{C}(X,\{0,1\})$ or, equivalently, the boolean algebra of clopens in X. The Stone duality theorem (see 4.1.16) then yields

$$\mathsf{Sp}\Big(\mathsf{Hom}\big((X,f),(\coprod_M R/M,p)\big)\Big) \cong X.$$

When the adjunction property is established, this will prove that the counit of the adjunction is an isomorphism, thus $\mathcal{C}_R$ is full and faithful.

To prove the adjointness property, we fix an R-algebra A; we must prove the existence of natural bijections

$$\mathsf{Hom}\Big(\mathsf{Hom}\big((X,f),(\coprod_M R/M,p)\big),A\Big) \cong \mathsf{Hom}\Big((\mathsf{Sp}(A),\alpha),(X,f)\Big)$$

where $\alpha\colon \mathsf{Sp}(A) \longrightarrow \mathsf{Sp}(R)$ is the canonical morphism corresponding, by the Stone duality, to the morphism

$$\mathsf{Idemp}(R) \longrightarrow \mathsf{Idemp}(A), \quad e \mapsto e \cdot 1$$

of boolean algebras.

Given $\varphi\colon \mathsf{Hom}\big((X,f),(\coprod_M R/M,p)\big) \longrightarrow A$ and a clopen $U \subseteq X$, we consider the continuous function

$$\chi_U \colon X \longrightarrow \coprod_M R/M, \quad \begin{cases} x \mapsto [1] \in R/f(x) & \text{if} \quad x \in U, \\ x \mapsto [0] \in R/f(x) & \text{if} \quad x \notin U. \end{cases}$$

The map χ_U is idempotent and $\varphi(\chi_U)$ is thus an idempotent of A. This yields a homomorphism

$$\varphi' \colon \mathsf{Clopen}(X) \longrightarrow \mathsf{Idemp}(A), \quad U \mapsto \varphi'(U) = \varphi(\chi_U);$$

this homomorphism corresponds by the Stone duality to a continuous map $\varphi'' \colon \mathsf{Sp}(A) \longrightarrow X$. To have a morphism $\varphi'' \colon (\mathsf{Sp}(A), \alpha) \longrightarrow (X, f)$, we must still prove the equality $f \circ \varphi'' = \alpha$. This reduces, for every idempotent $e \in R$, to the equality $\varphi'(f^{-1}(\mathcal{O}_e)) = e \cdot 1$. But

$$\chi_{f^{-1}(\mathcal{O}_e)}(x) = \begin{cases} [1] & \text{if } x \in f^{-1}(\mathcal{O}_e), \quad \text{i.e.} \quad e \notin f(x), \\ [0] & \text{if } x \notin f^{-1}(\mathcal{O}_e), \quad \text{i.e.} \quad e \in f(x). \end{cases}$$

But $e \notin f(x)$ implies $1-e \in f(x)$, thus $[1] = [e] \in R/f(x)$, while $e \in f(x)$ implies $[0] = [e] \in R/f(x)$. This proves that $\chi_{f^{-1}(\mathcal{O}_e)}(x) = [e] = e \cdot [1]$ for all x. Since φ is a homomorphism of R-algebras,

$$\varphi'(f^{-1}(\mathcal{O}_e)) = \varphi(\chi_{f^{-1}(\mathcal{O}_e)}) = e \cdot 1 = e.$$

It is obvious to observe that the correspondence $\varphi \mapsto \varphi''$ is natural.

To prove the injectivity of this correspondence, observe first that given $h \colon (X, f) \longrightarrow (\coprod_M R/M, p)$, the subsets $h^{-1}(s_r^{\mathsf{Sp}(R)}(\mathsf{Sp}(R)))$, for $r \in R$, constitute an open covering of X (see proof of 4.2.14). Since X has a base of clopens (see 3.4.9), there is a refinement of this covering constituted of clopens and, by compactness, we extract from it a finite covering. This thus yields clopens $U_1, \ldots, U_n \subseteq X$ and elements $r_1, \ldots, r_n \in R$ such that for each index i and each element $x \in U_i$, $h(x) = [r_i] \in R/f(x)$. This proves that $h = \sum_{i=1}^n r_i \cdot \chi_{U_i}$. Now choose $\varphi, \psi \colon \mathsf{Hom}((X, f), (\coprod_M R/M, p)) \rightrightarrows A$ such that $\varphi'' = \psi''$; proving $\varphi(\chi_U) = \psi(\chi_U)$ for each clopen $U \subseteq X$ will imply $\varphi(h) = \psi(h)$, by R-linearity of φ and ψ. But $\varphi(\chi_U) = \varphi'(U) = \psi'(U) = \psi(\chi_U)$, since $\varphi'' = \psi''$ implies $\varphi' = \psi'$ by the Stone duality.

To prove the surjectivity of the correspondence $\varphi \mapsto \varphi''$, let us consider a morphism $g \colon (\mathsf{Sp}(A), \alpha) \dashrightarrow (X, f)$, which yields by the Stone duality a morphism $\bar{g} \colon \mathsf{Clopen}(X) \longrightarrow \mathsf{Idemp}(A)$ of boolean algebras, such that $\bar{g}(f^{-1}(\mathcal{O}_e)) = e \cdot 1$ for each idempotent $e \in R$. We have already observed that $h \colon (X, f) \longrightarrow (\coprod_M R/M)$ can be written $h = \sum_{i=1}^n r_i \chi_{U_i}$ with $r_i \in R$ and $X = U_1 \cup \cdots \cup U_n$ a partition into clopens such that $h(x) = [r_i] \in R/f(x)$ for $x \in U_i$. We define

$$\varphi \colon \mathsf{Hom}\big((X, f), (\coprod_M R/M, p)\big) \longrightarrow A, \quad h \mapsto \varphi(h) = \sum_{i=1}^n r_i \cdot \bar{g}(U_i).$$

If we can prove that this definition is independent of the decomposition $h = \sum_{i=1}^{n} r_i \chi_{U_i}$, the relation $\varphi(\chi_U) = \overline{g}(U)$ will yield $\varphi'' = g$ and prove the surjectivity.

So let us consider another decomposition $h = \sum_{j=1}^{m} s_j \chi_{V_j}$ as above. Putting $W_{ij} = U_i \cap V_j$, we obtain a new partition of X into clopens with the property

$$x \in W_{ij} \Rightarrow [r_i] = h(x) = [s_j] \in R/f(x).$$

On the other hand fixing the index i, the clopens W_{ij} for $j = 1, \dots, m$ constitute a finite partition of U_i into clopens, thus the elements $\overline{g}(W_{ij})$ constitute a partition of the idempotent $\overline{g}(U_i)$; by lemma 4.2.5, this implies $\overline{g}(U_i) = \sum_{j=1}^{m} \overline{g}(W_{ij})$. Consequently

$$r_i \overline{g}(U_i) = r_i \sum_{j=1}^{m} \overline{g}(W_{ij})$$

and analogously

$$s_j \overline{g}(V_j) = s_j \sum_{i=1}^{n} \overline{g}(W_{ij}).$$

It suffices now to prove that $r_i \overline{g}(W_{ij}) = s_j \overline{g}(W_{ij})$ for all indices i, j, to imply the independence with respect to the decomposition of h. But for $x \in W_{ij}$, the relation $[r_i] = [s_j] \in R/f(x)$ implies $r_i - s_j \in f(x)$. Thus it suffices to prove that given a clopen $U \subseteq X$ and an element $r \in R$

$$\big(\forall x \in U \quad r \in f(x) \big) \Rightarrow \big(r\overline{g}(U) = 0 \big).$$

Let us prove this last fact. With the notation of theorem 4.2.14 and referring freely to its proof,

$$f(U) \subseteq \{ M \in \mathsf{Sp}(R) | r \in M \} = U_r$$

with U_r open. By 3.4.9, U_r is a union of clopens, from which by finite intersections we deduce a partition $U_r = \bigcup_{i \in I} \mathcal{O}_{e_i}$ into clopens $\mathcal{O}_{e_i}$, with each $e_i \in R$ idempotent. Therefore the clopens $f^{-1}(\mathcal{O}_{e_i})$ cover the clopen $U \subset X$; but U is closed, thus compact, and therefore we can extract a finite covering by clopens

$$U \subseteq f^{-1}(\mathcal{O}_{e_{i_1}}) \cup \cdots \cup f^{-1}(\mathcal{O}_{e_{i_k}}).$$

This implies

$$r \cdot \overline{g}(U) \;=\; r \cdot \overline{g} \left(U \cap \bigcup_{j=1}^{k} f^{-1}(\mathcal{O}_{e_{i_j}}) \right)$$

$$= r \cdot \bar{g}\left(\bigcup_{j=1}^{k} U \cap f^{-1}(\mathcal{O}_{e_{i_j}})\right)$$

$$= r \cdot \sum_{j=1}^{k} \left(\bar{g}(U) \cdot \bar{g}(f^{-1}(\mathcal{O}_{e_{i_j}}))\right)$$

$$= \sum_{j=1}^{k} r\bar{g}(U) \cdot e_i$$

and it remains to prove that each $re_i = 0$ for each index i. But since $\mathcal{O}_{e_i} \subseteq U_r$,

$$r \in \bigcap \{M \in \mathsf{Sp}(R) | M \in \mathcal{O}_{e_i}\} = \bigcap \{M \in \mathsf{Sp}(R) | e_i \notin M\}$$
$$= \bigcap \{M \in \mathsf{Sp}(R) | 1 - e_i \in M\}.$$

Applying proposition 4.2.10 and condition (iv) in 4.1.4, this proves $r \in R(1 - e_i)$. But then $r = r(1 - e_i)$, that is $re_i = 0$. $\qquad\square$

Let us conclude this section with a last observation on the functor Sp.

Proposition 4.3.7 *For every ring R, the functor*

$$\mathsf{Sp} \colon (R\text{-}\mathsf{Alg})^{\mathrm{op}} \longrightarrow \mathsf{Prof}$$

preserves cofiltered limits.

Proof Applying the Stone duality, we must verify that the functor

$$\mathsf{Idemp} \colon R\text{-}\mathsf{Alg} \longrightarrow \mathsf{Bool}$$

preserves filtered colimits. But both categories are algebraic, thus their filtered colimits are computed as in the category of sets. Now the functor Idemp is the factorization, through the category of boolean algebras, of the representable functor

$$\mathsf{Hom}\left(R[X]/\langle X^2 \rangle, - \right) \colon R\text{-}\mathsf{Alg} \longrightarrow \mathsf{Set}$$

with values in the category Set of sets. This representable functor preserves filtered colimits, because $R[X]/\langle X^2 \rangle$ is finitely presentable as an R-algebra (see [1]). $\qquad\square$

$$\begin{array}{ccc}
X \times_Y A & \xrightarrow{p_A} & A \\
{\scriptstyle \mathsf{id}_X \times a}\downarrow & & \downarrow{\scriptstyle a} \\
X \times_Y X & \xrightarrow{p_2} & X \\
{\scriptstyle p_1}\downarrow & & \downarrow{\scriptstyle f} \\
X & \xrightarrow[f]{} & Y
\end{array}$$

Diagram 4.2

4.4 Descent morphisms

This section presents a rough introduction to descent theory along pullbacks in a given category.

Definition 4.4.1 Let $\mathcal{C}$ be a category with pullbacks. A morphism $f\colon X \longrightarrow Y$ is an effective descent morphism when the functor "pullback along f"

$$f^*\colon \mathcal{C}/Y \longrightarrow \mathcal{C}/X$$

is monadic.

Let us recall the precise form of the monad. The functor f^* admits the left adjoint functor

$$\Sigma_f\colon \mathcal{C}/X \longrightarrow \mathcal{C}/Y, \quad (A, a) \mapsto (A, f \circ a)$$

of composition with f. This yields a composite functor

$$T = f^* \circ \Sigma_f\colon \mathcal{C}/X \longrightarrow \mathcal{C}/X, \quad (A, a) \mapsto \big(X \times_Y A, p_1 \circ (\mathsf{id}_X \times a)\big)$$

described by diagram 4.2, where both squares are pullbacks. We get at once natural transformations

$$\eta\colon \mathsf{id} \Longrightarrow T, \quad \mu\colon T \circ T \Longrightarrow T$$

defined by

$$\eta_{(A,a)} = \begin{pmatrix} a \\ \mathsf{id}_A \end{pmatrix} : A \longrightarrow X \times_Y A$$

and

$$\mu_{(A,a)} = p_1 \times \mathrm{id}_A : X \times_Y X \times_Y A \longrightarrow X \times_Y A,$$

and these are the unit and the multiplication of the monad $\mathbb{T} = (T, \varepsilon, \mu)$.

By the Beck criterion (see [66]), the monadicity of the functor f^* reduces to the following properties:

(i) the functor f^* reflects isomorphisms;

(ii) the functor f^* creates the coequalizers of those pairs (u, v) such that $(f^*(u), f^*(v))$ has a split coequalizer.

For the sake of completeness, we also recall the precise meaning of this condition (ii). A split coequalizer of $(f^*(u), f^*(v))$ consists in three arrows q, r, s,

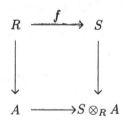

such that

$$q \circ f^*(u) = q \circ f^*(v), \quad q \circ r = \mathrm{id}_Q, \quad f^*(u) \circ s = \mathrm{id}_{f^*(V)}, \quad f^*(v) \circ s = r \circ q;$$

this implies immediately that q is the coequalizer of $(f^*(u), f^*(v))$ and that this coequalizer is preserved by every functor defined on $\mathcal{C}/X$. Condition (ii) requires that for every such pair (u, v), the coequalizer of (u, v) exists in $\mathcal{C}/Y$ and is preserved by f^*.

Our interest will be in descent morphisms in the dual of the category of rings (commutative, with unit). We must therefore study the co-monadicity of the "pushout" functor in the category of rings. Let us recall that in the category of rings, the pushout is given by the tensor product

$$
\begin{array}{ccc}
R & \xrightarrow{\ f\ } & S \\
\downarrow & & \downarrow \\
A & \longrightarrow & S \otimes_R A
\end{array}
$$

Let us also recall a classical notion in commutative algebra.

Definition 4.4.2 A module M over a ring R is flat when the functor

$$M \otimes_R - : \mathsf{Mod}_R \longrightarrow \mathsf{Mod}_R$$

preserves exact sequences. The module M is faithfully flat when it preserves and reflects exact sequences.

Proposition 4.4.3 *Let $f : R \longrightarrow S$ be a morphism of rings. When S is faithfully flat as an R-module, f is an effective descent morphism in the dual of the category of rings.*

Proof A morphism $g : A \longrightarrow B$ of R-modules is an isomorphism if and only if the sequence

$$0 \longrightarrow A \xrightarrow{g} B \longrightarrow 0$$

is exact. Since $S \otimes_R - : \mathsf{Mod}_R \longrightarrow \mathsf{Mod}_R$ reflects exact sequences, it thus reflects isomorphisms. Since the forgetful functors $R\text{-}\mathsf{Alg} \longrightarrow \mathsf{Mod}_R$ and $S\text{-}\mathsf{Alg} \longrightarrow \mathsf{Mod}_S$ have "all properties" we need (preserve, reflect, create exact sequences), the functor $S \otimes_R - : R\text{-}\mathsf{Alg} \longrightarrow S\text{-}\mathsf{Alg}$, which is the restriction of the previous functor, reflects isomorphisms as well, since these are bijective homomorphisms.

Since $R\text{-}\mathsf{Alg}$ and $S\text{-}\mathsf{Alg}$ have equalizers computed as in Mod_R and Mod_S, the second condition of the Beck criterion is satisfied because $S \otimes_R - : \mathsf{Mod}_R \longrightarrow \mathsf{Mod}_R$ preserves equalizers, by flatness of S. $\quad\square$

Proposition 4.4.4 *Let $f : R \longrightarrow S$ be a morphism of rings admitting an R-linear retraction $g : S \longrightarrow R$; that is, $g \circ f = \mathrm{id}_R$. Then f is an effective descent morphism in the dual of the category of rings.*

Proof If $u, v : A \rightrightarrows B$ are such that $S \otimes u = S \otimes v$, the consideration of the diagram

$$
\begin{array}{ccc}
R \otimes_R A \cong A & \begin{array}{c} \underrightarrow{\ \ u\ \ } \\[-4pt] \underrightarrow{\ \ v\ \ } \end{array} & B \cong R \otimes_R B \\[6pt]
{\scriptstyle f \otimes A} \Big\downarrow\Big\uparrow {\scriptstyle g \otimes A} & & {\scriptstyle f \otimes B} \Big\downarrow\Big\uparrow {\scriptstyle g \otimes B} \\[6pt]
S \otimes_R A & \begin{array}{c} \underrightarrow{\ \ S \otimes u\ \ } \\[-4pt] \underrightarrow{\ \ S \otimes v\ \ } \end{array} & S \otimes_R B
\end{array}
$$

yields

$$(f \otimes B) \circ u = (S \otimes u) \circ (f \otimes A) = (S \otimes v) \circ (f \otimes A) = (f \otimes B) \circ v,$$

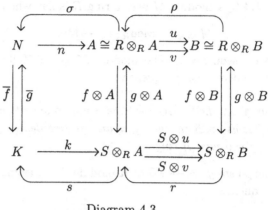

Diagram 4.3

from which $u = v$ since $f \otimes B$ admits the retraction $g \otimes B$ and therefore is injective. Thus the functor $S \otimes_R - : \mathsf{Mod}_R \longrightarrow \mathsf{Mod}_R$ is faithful, and therefore reflects monomorphisms and epimorphisms. It thus reflects isomorphisms, since in Mod_R being an isomorphism reduces to being both a monomorphism and an epimorphism. And finally the functor $S \otimes_R - : R\text{-}\mathsf{Alg} \longrightarrow S\text{-}\mathsf{Alg}$ reflects isomorphisms, since in all those categories, isomorphisms are just bijective homomorphisms.

Let us check now the condition on split equalizers. We thus consider two morphisms $u, v : A \rightrightarrows B$ of R-algebras, such that the pair $(S \otimes u, S \otimes v)$ has a split equalizer in $S\text{-}\mathsf{Alg}$. This yields a part of diagram 4.3 where

$$(S \otimes u) \circ k = (S \otimes v) \circ k, \quad s \circ k = \mathsf{id}, \quad r \circ (S \otimes u) = \mathsf{id}, \quad r \circ (S \otimes v) = k \circ s.$$

We put $n = \mathsf{Ker}\,(u, v)$ in $R\text{-}\mathsf{Alg}$ or Mod_R: this is the same map. We must prove that $(S \otimes_R -)$ preserves this equalizer. Considering both equalizers

$$n = \mathsf{Ker}\,(u, v), \quad k = \mathsf{Ker}\,(S \otimes u, S \otimes v)$$

we conclude that there exist morphisms $\overline{f}, \overline{g}$ inducing corresponding commutativities in the diagram in Mod_R. We put then

$$\sigma = \overline{g} \circ s \circ (f \otimes A), \quad \rho = (g \otimes A) \circ r \circ (f \otimes B).$$

It remains to observe that ρ and σ present n as the split equalizer of the pair (u, v), from which this coequalizer will be preserved by every functor and in particular by the functor $S \otimes_R -$. And indeed (omitting for short the composition symbols)

$$\sigma n = \overline{g}s(f \otimes A)n = \overline{g}sk\overline{f} = \overline{g}\overline{f} = \mathsf{id},$$
$$\rho u = (g \otimes A)r(f \otimes B)u = (g \otimes A)r(S \otimes u)(f \otimes A)$$
$$= (g \otimes A)(f \otimes A) = \mathsf{id},$$
$$\rho v = (g \otimes A)r(f \otimes B)v = (g \otimes A)r(S \otimes v)(f \otimes A)$$
$$= (g \otimes A)ks(f \otimes A) = n\overline{g}s(f \otimes A) = n\sigma.$$

This concludes the proof.   □

Corollary 4.4.5 *Every morphism of fields* $f\colon K \longrightarrow L$ *is an effective descent morphism in the dual of the category of rings.*

Proof   The field L is a K-vector space and the K-linear map f is injective, as a field homomorphism. Thus K is a sub-K-vector-space of L and has therefore a complementary sub-K-vector-space V. So $L \cong K \oplus V$ and the projection of the direct sum onto K is the expected retraction.

An alternative proof follows easily from proposition 4.4.3.   □

The last result of this section exhibits an interesting class of effective descent morphisms; this is a special case of a more general result valid for exact categories (see [50]). This result will be useful in section 5.2 to apply Galois theory to the study of central extensions of groups.

Lemma 4.4.6 *In a category which is monadic over the category of sets, every regular epimorphism is an effective descent morphism.*

Proof   Let $\mathbb{T}$ be a monad over **Set** and consider the corresponding category $\mathbf{Set}^{\mathbb{T}}$ of $\mathbb{T}$-algebras. In this category, a regular epimorphism $\sigma\colon (S,\xi) \longrightarrow (R,\zeta)$ is exactly a surjective morphism (see [8], volume 2). Via the axiom of choice, we fix a section $\tau\colon R \longrightarrow S$ of σ in the category of sets; thus $\sigma \circ \tau = \mathsf{id}_R$.

We shall use the Beck criterion to prove the statement. In $\mathbf{Set}^{\mathbb{T}}$, the functor σ^* admits as left adjoint the functor Σ_σ of composition with σ, which is one of the conditions of this criterion.

To prove that σ^* reflects isomorphisms, consider diagram 4.4, where the quadrilaterals containing σ are pullbacks. Since σ is surjective, so are σ_A and σ_B. If $\sigma^*(\gamma)$ is an isomorphism, then $\sigma_B \circ \sigma^*(\gamma) = \gamma \circ \sigma_A$ is surjective, from which γ is surjective. To prove that γ is also injective, consider $x, y \in A$ such that $\gamma(x) = \gamma(y)$. Since σ is surjective, we choose

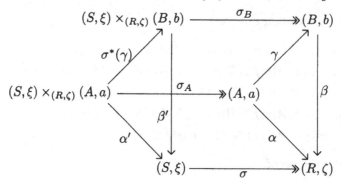

Diagram 4.4

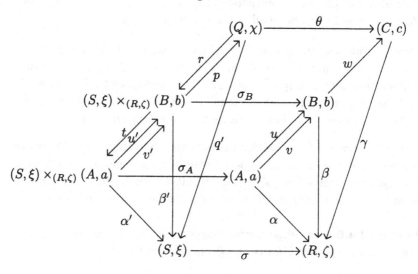

Diagram 4.5

$s \in S$ such that $\sigma(s) = \beta\gamma(x) = \beta\gamma(y)$. We have then

$$\sigma(s) = \beta\gamma(x) = \alpha(x), \quad \sigma(s) = \beta\gamma(y) = \alpha(y);$$

thus $(s, x) \in S \times_R A$ and $(s, y) \in S \times_R A$. But

$$\sigma^*(\gamma)(s, x) = \big(s, \gamma(x)\big) = \big(s, \gamma(y)\big) = \sigma^*(\gamma)(s, y)$$

and thus $(s, x) = (s, y)$ since $\sigma^*(\gamma)$ is an isomorphism. Finally $x = y$, which proves the injectivity of γ. Thus γ is an isomorphism.

To check the last condition in the Beck criterion, we refer to diagram 4.5. Consider two morphisms $u, v \colon \big((A, a), \alpha\big) \rightrightarrows \big((B, b), \beta\big)$ in

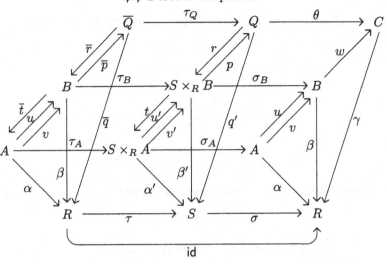

Diagram 4.6

$\mathsf{Set}^{\mathbf{T}}/(R,\zeta)$ such that the pair $\big(\sigma^*(u),\sigma^*(v)\big)$ admits in $\mathsf{Set}^{\mathbf{T}}/(S,\xi)$ the coequalizer p which is split by morphisms r and t. To shorten notation, we write $u' = \sigma^*(u)$ and $v' = \sigma^*(v)$. Putting $w = \mathsf{Coker}\,(u,v)$ in $\mathsf{Set}^{\mathbf{T}}$, from $\beta \circ u = \alpha = \beta \circ v$ we get γ such that $\gamma \circ w = \beta$, yielding $w = \mathsf{Coker}\,(u,v)$ in $\mathsf{Set}^{\mathbf{T}}/(R,\zeta)$. We must prove that $p = \sigma^*(w)$. First, there is a factorization θ between the two coequalizer diagrams. An easy diagram chasing shows $\gamma \circ \theta \circ p = \sigma \circ q' \circ p$, from which $\gamma \circ \theta = \sigma \circ q'$, because p is a surjection. In the category of sets, let us further compute the pullbacks along the map $\tau\colon R \longrightarrow S$, as in diagram 4.6. The split coequalizer p is preserved by every functor, thus

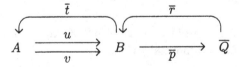

is a split coequalizer in Set. By the Beck criterion applied to the morphisms $u,v\colon (A,a) \rightrightarrows (B,b)$ and the forgetful functor $\mathsf{Set}^{\mathbf{T}} \longrightarrow \mathsf{Set}$, the coequalizer $w = \mathsf{Coker}\,(u,v)$ in $\mathsf{Set}^{\mathbf{T}}$ is also the coequalizer in Set, that is, $\theta \circ \tau_Q$ is a bijection. But since $\overline{p} \cong w$ is a split coequalizer of (u,v) in Set, the diagram

$$S \times_R A \underset{v'}{\overset{u'}{\rightrightarrows}} S \times_R B \overset{\sigma^*(w)}{\longrightarrow} S \times_R C$$

is a coequalizer in Set. This proves that in Set, the factorization ψ in the diagram

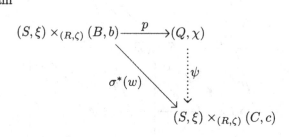

$$(S,\xi) \times_{(R,\zeta)} (B,b) \xrightarrow{\ p\ } (Q,\chi)$$

$$\sigma^*(w) \qquad \psi$$

$$(S,\xi) \times_{(R,\zeta)} (C,c)$$

is bijective. Therefore ψ is an isomorphism in $\mathsf{Set}^{\mathbb{T}}$.   □

4.5 Morphisms of Galois descent

This section generalizes, to the case of rings, the notion of "Galois extension of fields".

Definition 4.5.1 Let $\sigma: R \longrightarrow S$ be a morphism of rings. Write β for the unit of the adjunction

$$(S\text{-Alg})^{\mathrm{op}} \overset{\mathcal{C}_S}{\underset{\mathsf{Sp}_S}{\rightleftarrows}} \mathsf{Prof}/\mathsf{Sp}(S)$$

described in theorem 4.3.6. For the sake of clarity, we express each β_A in the category S-Alg, not in its dual. An R-algebra A is split by σ when the morphism

$$\beta_{S\otimes_R A}: \mathcal{C}_S\mathsf{Sp}_S(S \otimes_R A) \longrightarrow S \otimes_R A$$

is an isomorphism.

Definition 4.5.2 A morphism $\sigma: R \longrightarrow S$ of rings is of (effective) Galois descent (also called 'normal') when

  (i) σ is an effective descent morphism in the dual of the category of rings,
  (ii) for every object $(X,\varphi) \in \mathsf{Prof}/\mathsf{Sp}(S)$, the R-algebra $\mathcal{C}_S(X,\varphi)$ is split by σ.

Let us mention that the second condition could be required only for $X = \{*\}$. Let us also mention that with every ring R can be associated a "separable closure" $\overline{R}$ of R, with the property that the inclusion $R \hookrightarrow \overline{R}$ is a Galois descent morphism. We shall not need this result in this book.

We prove now that the previous notions extend the situation we have studied in the special case of fields (see also 4.7.16).

Lemma 4.5.3

(i) *Every finite dimensional Galois extension of fields is a Galois descent morphism.*

(ii) *In these conditions, every finite dimensional K-algebra which is split by L in the sense of definition 2.3.1 is also split by σ.*

Proof We recall that a field admits only 0 and 1 as idempotents, thus its Pierce spectrum is a singleton. By corollary 4.4.5, σ is an effective descent morphism, and corollary 4.3.5 holds in the present case. We shall prove that for every finite dimensional K-algebra A which is split by L in the sense of 2.3.1 and every profinite space X, the canonical morphism

$$\mathcal{C}\Big(\mathsf{Sp}\big(A \otimes_K \mathcal{C}(X, L)\big), L\Big) \longrightarrow A \otimes_K \mathcal{C}(X, L)$$

is an isomorphism. Putting $A = L$ (see proposition 2.3.2) will yield that $\mathcal{C}(X, L)$ is split by σ. Putting $X = \{*\}$ will prove that A is split by σ.

By corollary 4.3.5, $\mathcal{C}(X, L) \cong L \otimes_K \mathcal{C}(X, K)$. Using proposition 2.3.2, we get

$$
\begin{aligned}
A \otimes_K \mathcal{C}(X, L) &\cong A \otimes_K L \otimes_K \mathcal{C}(X, K) \\
&\cong L^n \otimes_K \mathcal{C}(X, K) \\
&\cong \big(L \otimes_K \mathcal{C}(X, K)\big)^n \\
&\cong \mathcal{C}(X, L)^n.
\end{aligned}
$$

Let us recall that $\mathsf{Sp} \colon K\text{-}\mathsf{Alg} \longrightarrow \mathsf{Prof}$ transforms limits into colimits, because it has an adjoint functor. Therefore

$$
\begin{aligned}
\mathcal{C}\Big(\mathsf{Sp}\big(A \otimes_K \mathcal{C}(X, L)\big), L\Big) &\cong \mathcal{C}\Big(\mathsf{Sp}\big(\mathcal{C}(X, L)^n\big), L\Big) \\
&\cong \mathcal{C}\left(\coprod_{i=1}^{n} \mathsf{Sp}\,\mathcal{C}(X, L), L\right) \\
&\cong \mathcal{C}\left(\coprod_{i=1}^{n} X, L\right) \\
&\cong \prod_{i=1}^{n} \mathcal{C}(X, L) \\
&\cong A \otimes_K \mathcal{C}(X, L)
\end{aligned}
$$

where we have used proposition 4.3.3 for the isomorphism $\mathsf{Sp}\,\mathcal{C}(X,L) \cong X$. □

Proposition 4.5.4

(i) *Every Galois extension of fields* $\sigma \colon K \longrightarrow L$ *is a Galois descent morphism.*

(ii) *In these conditions, every K-algebra which is split by L in the sense of definition 2.3.1 is also split by σ.*

Proof As in the proof of 4.5.3, we recall that a field admits only 0 and 1 as idempotents, thus its Pierce spectrum is a singleton. By corollary 4.4.5, σ is an effective descent morphism; corollary 4.3.5 and proposition 4.3.3 hold in the present case. We shall prove that for every K-algebra A which is split by L in the sense of 2.3.1 and every profinite space X, the canonical morphism

$$\mathcal{C}\Big(\mathsf{Sp}\big(A \otimes_K \mathcal{C}(X,L)\big), L\Big) \longrightarrow A \otimes_K \mathcal{C}(X,L)$$

is an isomorphism. Putting $A = L$ (see proposition 2.3.2) will yield that $\mathcal{C}(X,L)$ is split by σ. Putting $X = \{*\}$ will prove that A is split by σ.

By proposition 3.1.5, every algebra A which is split by L in the sense of 2.3.1 is a filtered colimit of its finite dimensional subalgebras $B \subseteq A$, each of these being itself split, in the sense of 2.3.1, by a finite dimensional Galois extension $K \subseteq M_B \subseteq L$. By proposition 3.1.4, the extension L is itself the filtered colimit of its finite dimensional Galois subextensions $K \subseteq M \subseteq L$. We thus have

$$A \otimes_K L \cong \Big(\operatorname*{colim}_{B} B\Big) \otimes_K \Big(\operatorname*{colim}_{M} M\Big) \cong \operatorname*{colim}_{(B,M)} B \otimes_K M$$

where B and M are as above. We can equivalently compute the last filtered colimit on a cofinal subset of indices, for example, via proposition 3.1.5, by restricting our attention to those pairs (B,M) such that B is split by M in the sense of 2.3.1. Having made this choice, we get the following isomorphisms, using the same arguments as in the proof of lemma 4.5.3 and the fact that the functor $\mathsf{Sp} \colon K\text{-}\mathsf{Alg} \longrightarrow \mathsf{Prof}$ transforms filtered colimits into cofiltered limits (see proposition 4.3.7).

$$\mathcal{C}\Big(\mathsf{Sp}\big(A \otimes_K \mathcal{C}(X,L)\big), L\Big)$$
$$\cong L \otimes_K \mathcal{C}\Big(\mathsf{Sp}\big(A \otimes_K L \otimes_K \mathcal{C}(X,K)\big), K\Big)$$
$$\cong L \otimes_K \mathcal{C}\Big(\mathsf{Sp}\big((\operatorname*{colim}_{(B,M)} B \otimes_K M) \otimes_K \mathcal{C}(X,K)\big), K\Big)$$

$$\cong L \otimes_K \mathcal{C}\Big(\mathsf{Sp}(\mathrm{colim}_{(B,M)} B \otimes_K M \otimes_K \mathcal{C}(X,K)), K\Big)$$

$$\cong L \otimes_K \mathcal{C}\Big(\mathrm{lim}_{(B,M)} \mathsf{Sp}(B \otimes_K M \otimes_K \mathcal{C}(X,K)), K\Big)$$

$$\cong L \otimes_K \mathrm{colim}_{(B,M)} \mathcal{C}\Big(\mathsf{Sp}(B \otimes_K M \otimes_K \mathcal{C}(X,K)), K\Big) \quad (*)$$

$$\cong \Big(\mathrm{colim}_{M'} M'\Big) \otimes_K \mathrm{colim}_{(B,M)} \mathcal{C}\Big(\mathsf{Sp}(B \otimes_K M \otimes_K \mathcal{C}(X,K)), K\Big)$$

$$\cong \mathrm{colim}_{(B,M,M')} M' \otimes_K \mathcal{C}\Big(\mathsf{Sp}(B \otimes_K M \otimes_K \mathcal{C}(X,K)), K\Big)$$

$$\cong \mathrm{colim}_{(B,M)} M \otimes_K \mathcal{C}\Big(\mathsf{Sp}(B \otimes_K M \otimes_K \mathcal{C}(X,K)), K\Big) \quad (**)$$

$$\cong \mathrm{colim}_{(B,M)} \mathcal{C}\Big(\mathsf{Sp}(B \otimes_K \mathcal{C}(X,M)), M\Big)$$

$$\cong \mathrm{colim}_{(B,M)} B \otimes_K \mathcal{C}(X,M) \quad \text{see } 4.5.3$$

$$\cong \mathrm{colim}_{(B,M)} B \otimes_K M \otimes_K \mathcal{C}(X,K)$$

$$\cong \big(\mathrm{colim}_B B\big) \otimes_K \big(\mathrm{colim}_M M\big) \otimes_K \mathcal{C}(X,K)$$

$$\cong L \otimes_K A \otimes_K \mathcal{C}(X,K)$$

$$\cong A \otimes_K \mathcal{C}(X,L).$$

The argument $(*)$ holds since the functor $\mathcal{C}(-,K)\colon \mathsf{Prof} \longrightarrow (K\text{-}\mathsf{Alg})^{\mathrm{op}}$ has the left adjoint Sp, thus preserves limits. The argument $(**)$ reduces to the computation of the colimit on a cofinal subset of indices. $\qquad\square$

We conclude this section with a useful technical result, showing how the spectrum construction is naturally present in the theory developed in the previous chapters.

Proposition 4.5.5 *Let* $\sigma\colon K \longrightarrow L$ *be a Galois extension of fields. The functors, defined on the category of K-algebras split by L in the sense of definition 2.3.1,*

$$\mathsf{Split}_K(L) \longrightarrow \mathsf{Prof}, \qquad A \mapsto \mathrm{Hom}_K(A, L),$$

$$\mathsf{Split}_K(L) \longrightarrow \mathsf{Prof}, \qquad A \mapsto \mathsf{Sp}(A \otimes_K L)$$

(see lemma 3.5.3) are isomorphic.

Proof When the K-algebra A is finite dimensional, we know by proposition 3.1.5 that $\mathrm{Hom}_K(A, L) \cong \mathrm{Hom}_K(A, M)$ for some finite dimensional Galois extension $K \subseteq M \subseteq L$, and this isomorphism is clearly natural in A. Moreover, by theorem 2.3.3, $\#\mathrm{Hom}_K(A, M) = n = \dim A$ while $A \otimes_K L \cong L^n$, again with naturality properties, hidden in particular in the cardinality argument. A finite profinite space is discrete, thus

$\mathsf{Hom}_K(A, L)$ is the n point discrete space. On the other hand $\mathsf{Sp}(L^n)$ is the discrete n point space too, since L has only 0 and 1 as idempotents.

Now when A is infinite dimensional, by lemma 3.5.3

$$\mathsf{Hom}_K(A, L) \cong \lim_B \mathsf{Hom}_K(B, L)$$

where B runs through the finite dimensional subalgebras of A. On the other hand, applying proposition 4.3.7, we obtain

$$\mathsf{Sp}(A \otimes_K L) \cong \mathsf{Sp}((\mathrm{colim}_B B) \otimes_K L) \cong \mathsf{Sp}(\mathrm{colim}_B (B \otimes_K L))$$
$$\cong \lim_B \mathsf{Sp}(B \otimes_K L) \cong \lim_B \mathsf{Hom}_K(B, L)$$
$$\cong \mathsf{Hom}_K(\mathrm{colim}_B B, L) \cong \mathsf{Hom}_K(A, L).$$

This concludes the proof.   □

4.6 Internal presheaves

In this section, "presheaf" will always mean "covariant presheaf". A covariant presheaf on a small category $\mathcal{C}$ is thus, classically, a functor $P \colon \mathcal{C} \longrightarrow \mathsf{Set}$. And a small category has a *set* of objects and a *set* of morphisms. It is well known that this situation can be generalized by replacing all *sets* by objects of an arbitrary category $\mathcal{X}$ with pullbacks. To avoid losing ourselves through heavy technical considerations, we recall only the spirit of the definitions and refer to section 7.1 or [8], volume 1, for the details.

An internal category in $\mathcal{X}$ consists in giving the following situation:

$$C_2 \xrightarrow{\ m\ } C_1 \underset{\substack{\longrightarrow \\ d_1 \\ \longrightarrow}}{\overset{\substack{d_0 \\ \longrightarrow \\ n}}{\longleftarrow}} C_0$$

where C_2 is defined by the following pullback:

$$
\begin{array}{ccc}
C_2 & \xrightarrow{\ p_2\ } & C_1 \\
{\scriptstyle p_1}\downarrow & & \downarrow{\scriptstyle d_0} \\
C_1 & \xrightarrow[\ d_1\ ]{} & C_0
\end{array}
$$

One should think of

- C_0 as the "object of objects" of $\mathcal{C}$,

- C_1 as the "object of arrows" of $\mathcal{C}$,
- d_0 as the "domain morphism",
- d_1 as the "codomain morphism",
- n as the "identity morphism",
- C_2 as the "object of composable pairs",
- m as the "composition morphism".

It remains to impose on these data the diagrammatical transcription of the axioms of units and associativity.

The internal category is an internal groupoid when there exists an additional morphism $s: C_1 \longrightarrow C_1$ with axioms indicating that s formally inverts the arrows of $\mathcal{C}$.

We leave to the reader the definitions of an internal functor between two internal categories in $\mathcal{X}$ and of an internal natural transformation between internal functors: we shall not need these notions.

An internal covariant presheaf on the internal category $\mathcal{C}$ is a triple (P, p, π) where

(i) $p: P \longrightarrow C_0$ is a morphism of $\mathcal{X}$,
(ii) $\pi: P_1 \longrightarrow P$ is a morphism of $\mathcal{X}$, where P_1 is defined by the first square below, which is a pullback,

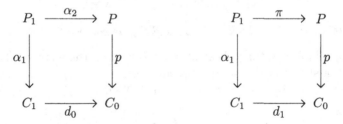

while the second square above is commutative.

To get an intuition of this notion, consider the case $\mathcal{X} = \mathsf{Set}$. Giving $p: P \longrightarrow C_0$ is equivalent to giving the family $\left(p^{-1}(C)\right)_{C \in C_0}$; for each $C \in C_0$, we put $P(C) = p^{-1}(C)$. This defines the functor P on the objects. Next, we have by definition

$$P_1 = \{(f, a) \mid f \text{ arrow of } \mathcal{C}, \ a \in P(\text{domain of } f)\}.$$

Giving π as indicated in condition (ii) consists, for each pair (f, a) as above, in giving an element of $P(\text{codomain of } f)$, which we choose as $P(f)(a)$. This defines the functor P on the morphisms. It remains to impose axioms expressing respectively the compatibility of P with identities and composition.

Finally given two internal covariant presheaves (P, p, π), (P', p', π') on an internal category $\mathcal{C}$, an internal natural transformation

$$\gamma\colon (P, p, \pi) \Longrightarrow (P', p', \pi')$$

is a morphism $\alpha\colon P \longrightarrow P'$ such that $p' \circ \alpha = p$. In the case $\mathcal{X} = \mathsf{Set}$, the condition $p' \circ \alpha = p$ means precisely that γ splits as a family of maps

$$\gamma_C\colon P(C) = p^{-1}(C) \longrightarrow (p')^{-1}(C) = P'(C).$$

It remains to impose diagrammatically the naturality axiom.

It is straightforward to observe that internal covariant presheaves and internal natural transformations on an internal category $\mathcal{C}$ constitute a category $\mathcal{X}^{\mathcal{C}}$.

Proposition 4.6.1 *Let $\mathcal{X}$ be a category with pullbacks and $\mathcal{X}^{\mathcal{C}}$ the category of internal covariant presheaves on an internal category $\mathcal{C}$. The functor*

$$\mathcal{X}^{\mathcal{C}} \longrightarrow \mathcal{X}/C_0, \quad (P, p, \pi) \mapsto (P, p)$$

is monadic and the corresponding monad admits as functorial part the composite

$$\mathcal{X}/C_0 \xrightarrow{d_0^*} \mathcal{X}/C_1 \xrightarrow{\Sigma_{d_1}} \mathcal{X}/C_0$$

where d_0^* is pulling back along d_0 and Σ_{d_1} is composition with d_1.

Proof This is again a classical result (see [55]); therefore we shall only recall a sketch of the proof. One first writes $T = \Sigma_{d_1} \circ d_0^*$. In the case $\mathcal{X} = \mathsf{Set}$, an object of $\mathcal{X}/C_0$ is a family $(X_C)_{C \in C_0}$ of sets, and

$$T\Big((X_C)_{C \in C_0}\Big) = \left(\coprod_{d_1(f) = C} X_{d_0(f)}\right)_{C \in C_0}.$$

With this special case in mind, one observes that composition and identities in $\mathcal{C}$ induce at once the structure of a monad on T. An algebra for this monad is then a pair $((P, p), \pi)$ where $(P, p) \in \mathcal{X}/C_0$ and $\pi\colon T(P, p) \longrightarrow (P, p)$. Thus p and π make commutative the diagram:

$$
\begin{array}{ccccc}
P & \xleftarrow{\ \pi\ } & P_1 & \xrightarrow{\ \alpha_1\ } & P \\
{\scriptstyle p}\big\downarrow & & {\scriptstyle \alpha_1}\big\downarrow & {\scriptstyle \text{p.b.}} & \big\downarrow{\scriptstyle p} \\
C_0 & \xleftarrow{\ d_1\ } & C_1 & \xrightarrow{\ d_0\ } & C_0
\end{array}
$$

This yields the same data as in the description of an internal covariant presheaf. It remains to observe that the axioms for being a T-algebra are equivalent to those for being an internal covariant presheaf, and analogously for the morphisms.   □

It is probably more interesting to really grasp the intuitive meaning of this result in the case of $\mathcal{X} = \mathsf{Set}$. In this case, writing $\mathcal{C}_0$ for the discrete category on the set C_0, it follows at once that the category Set/C_0 is isomorphic to the category of covariant presheaves $\mathsf{Set}^{\mathcal{C}_0}$. Indeed, in both cases, we recapture the category of C_0-indexed families of sets and maps. The functor

$$\mathsf{Set}^{\mathcal{C}} \longrightarrow \mathsf{Set}^{\mathcal{C}_0}$$

we are interested in is composition with the inclusion $i \colon \mathcal{C}_0 \longrightarrow \mathcal{C}$ and we are interested in its monadicity. This functor has a left adjoint, namely the Kan extension along i. This same functor also reflects isomorphisms (that is, natural transformations all of whose components are bijective) simply because i is bijective on the objects (a natural transformation α is an isomorphism when each α_C is an isomorphisms). Finally both categories have coequalizers computed pointwise, thus preserved by composition with i. By the Beck criterion (see [66]), we get the expected monadicity.

Example 4.6.2 Let $\sigma \colon S \longrightarrow R$ be a morphism in a category $\mathcal{X}$ with pullbacks. The diagram

$$(S \times_R S) \times_S (S \times_R S) \xrightarrow{\ (p_1, p_4)\ } S \times_R S \overset{\underset{p_2}{\longrightarrow}}{\underset{\tau \uparrow}{\overset{\Delta}{\underset{\longleftarrow}{\xrightarrow{p_1}}}}} S$$

is an internal groupoid, with (p_0, p_1) the kernel pair of σ, Δ the diagonal of the pullback and τ the twisting isomorphism which interchanges factors.

Proof The proof is obvious.   □

In the previous example, the object $S \times_R S$ of morphisms is in fact a subobject of $S \times S$, where S is the object of arrows. Thus the internal category is in fact an internal preorder, thus a reflective and transitive relation. Being a groupoid expresses the symmetry of the relation. Thus the previous example is a reformulation of the well-known fact that the kernel pair of σ is an equivalence relation on S.

4.7 The Galois theorem for rings

Convention *Throughout this section,* $\sigma: R \longrightarrow S$ *indicates a Galois descent morphism of rings, in the sense of definition 4.5.2. This fact will generally not be recalled in the various statements of the present section.*

We shall now develop, in the special case of rings, the Galois theory of Janelidze. This is a special case of the more general theory of the same author, developed in section 5.1.

Lemma 4.7.1 *The following conditions are equivalent for an S-algebra A:*

   (i) *A is split by* $\mathrm{id}_S: S =\!\!=\!\!= S$*;*
   (ii) *the canonical morphism* $\mathcal{C}_S(\mathsf{Sp}_S(A)) \longrightarrow A$ *is an isomorphism.*

Proof One has $S \otimes_S A \cong A$ and thus $\mathcal{C}_S(\mathsf{Sp}_S(S \otimes_S A)) \cong \mathcal{C}_S(\mathsf{Sp}_S(A))$, from which the result follows by 4.5.1. □

Corollary 4.7.2 *The following conditions are equivalent for an R-algebra A:*

   (i) *A is split by* $\sigma: R \longrightarrow S$*;*
   (ii) *the S-algebra* $S \otimes_R A$ *is split by* $\mathrm{id}_S: S =\!\!=\!\!= S$*.* □

Lemma 4.7.3 *The following conditions are equivalent for an S-algebra A:*

   (i) *A is split by* $\mathrm{id}_S: S =\!\!=\!\!= S$*;*
   (ii) $A \cong \mathcal{C}_S(X, \varphi)$ *for some* $(X, \varphi) \in \mathsf{Prof}/\mathsf{Sp}(S)$*.*

Proof Putting $(X, \varphi) = \mathsf{Sp}_S(A)$, lemma 4.7.1 yields (i) ⇒ (ii). Conversely, by lemma 4.7.1 again and theorem 4.3.6

$$A \cong \mathcal{C}_S(X, \varphi) \cong \mathcal{C}_S(\mathsf{Sp}_S \mathcal{C}_S(X, \varphi)) \cong \mathcal{C}_S(\mathsf{Sp}_S(A)).$$ □

Corollary 4.7.4 *The following conditions are equivalent for an R-algebra A:*

   (i) *A is split by* $\sigma: R \longrightarrow S$*;*
   (ii) $S \otimes_R A \cong \mathcal{C}(X, \varphi)$ *for some* $(X, \varphi) \in \mathsf{Prof}/\mathsf{Sp}(S)$*.* □

Lemma 4.7.5 *Let us write* $\mathsf{Split}_S(S)$ *for the category of S-algebras split by* $\mathrm{id}_S\colon S \Longrightarrow S$. *The functors*

$$\left(\mathsf{Split}_S(S)\right)^{\mathrm{op}} \underset{\mathsf{Sp}_S}{\overset{\mathcal{C}_S}{\longleftarrow\!\!\!-\!\!\!\longrightarrow}} \mathsf{Prof}/\mathsf{Sp}(S)$$

constitute an equivalence of categories.

Proof By lemma 4.7.3, $\mathcal{C}_S$ indeed takes values in $\mathsf{Split}_S(S)$. One has $\mathsf{Sp}_S \circ \mathcal{C}_S \cong \mathrm{id}$ by theorem 4.3.6 and $\mathcal{C}_S \circ \mathsf{Sp}_S \cong \mathrm{id}$ by lemma 4.7.1. $\quad\square$

Lemma 4.7.6 *If A is an S-algebra split by* $\mathrm{id}_S\colon S \Longrightarrow S$, *then $S \otimes_R A$ is another S-algebra split by* $\mathrm{id}_S\colon S \Longrightarrow S$.

Proof By lemma 4.7.3, $A \cong \mathcal{C}_S(X, \varphi)$ for some $(X, \varphi) \in \mathsf{Prof}/\mathsf{Sp}(S)$. Since σ is of Galois descent, $\mathcal{C}_S(X, \varphi)$ is an R-algebra split by σ, thus, by corollary 4.7.2, $S \otimes_R \mathcal{C}_S(X, \varphi)$ is an S-algebra split by id_S. $\quad\square$

Lemma 4.7.7 *The ring S is an S-algebra which is split by* $\mathrm{id}_S\colon S \Longrightarrow S$.

Proof The algebra S is the initial object of the category S-Alg of S-algebras and $\mathsf{Sp}_S(S) = \left(\mathsf{Sp}(S) \Longrightarrow \mathsf{Sp}(S)\right)$ is the final object of the category $\mathsf{Prof}/\mathsf{Sp}(S)$. The functor $\mathcal{C}_S\colon \mathsf{Prof}/\mathsf{Sp}(S) \longrightarrow (S\text{-}\mathsf{Alg})^{\mathrm{op}}$ preserves the terminal object, since it has a left adjoint Sp_S (see 4.5.1), thus $\mathcal{C}_S\left(\mathsf{Sp}_S(S)\right) \cong S$. One concludes the proof by lemma 4.7.1. $\quad\square$

Corollary 4.7.8 *The ring R is an R-algebra split by* $\sigma\colon R \longrightarrow S$. $\quad\square$

Lemma 4.7.9 *For every integer $n \in \mathbb{N}$, consider $\otimes_{i=1}^{n} S = S \otimes_R \cdots \otimes_R S$ provided with the structure of S-algebra induced by the multiplication on the first factor. These S-algebras $\otimes_{i=1}^{n} S$ are split by* $\mathrm{id}_S\colon S \Longrightarrow S$.

Proof By lemma 4.7.7 and an iterated application of lemma 4.7.6. $\quad\square$

Let us generalize further some of the previous arguments. If A and B are S-algebras, $A \otimes_R B$ can be provided with two canonical structures of S-algebra:

$$s\left(\sum_i a_i \otimes b_i\right) = \sum_i (sa_i) \otimes b_i,$$

$$s\left(\sum_i a_i \otimes b_i\right) = \sum_i a_i \otimes (sb_i).$$

Let us write $A \otimes_R^1 B$ and $A \otimes_R^2 B$ to distinguish those two structures of S-algebra. One has obviously $A \otimes_R^1 B \cong B \otimes_R^2 A$. Observe that $\otimes^1$ was already used in the proof of lemmas 4.7.1 and 4.7.6.

Lemma 4.7.10 *If A and B are S-algebras split by id_S, then so are $A \otimes_R^1 B$ and $A \otimes_R^2 B$.*

Proof It suffices to develop the proof for $A \otimes_R^1 B$. Obviously $A \otimes_R^1 B \cong A \otimes_S (S \otimes_R^1 B)$, with $S \otimes_R^1 B$ split by id_S, by lemma 4.7.6. So putting $C = S \otimes_R^1 B$, it suffices to prove that

$$(A \text{ and } C \text{ split by } \mathrm{id}_S) \Rightarrow (A \otimes_S C \text{ split by } \mathrm{id}_S).$$

This implication follows from lemma 4.7.3 and the fact that the functor

$$C_S \colon \mathsf{Prof}/\mathsf{Sp}(S) \dashrightarrow (S\text{-}\mathsf{Alg})^{\mathrm{op}}$$

preserves products, since it is a right adjoint. □

Lemma 4.7.11 *Let us write $\mathsf{Split}_R(\sigma)$ for the category of R-algebras split by σ. The functor*

$$S \otimes_R - \colon \mathsf{Split}_R(\sigma) \longrightarrow \mathsf{Split}_S(S)$$

is comonadic.

Proof By corollary 4.7.2, this functor is correctly defined. Observe that given $A \in \mathsf{Split}_S(S)$, lemma 4.7.10 implies $S \otimes_R A \in \mathsf{Split}_S(S)$ and thus $A \in \mathsf{Split}_R(\sigma)$ by corollary 4.7.2. Therefore the classical adjunction

$$R\text{-}\mathsf{Alg} \underset{S \otimes_R -}{\overset{U}{\rightleftarrows}} S\text{-}\mathsf{Alg}, \quad U(A) = A$$

restricts, via corollary 4.7.2, to an adjunction

$$\mathsf{Split}_R(\sigma) \underset{S \otimes_R -}{\overset{U}{\rightleftarrows}} \mathsf{Split}_S(S), \quad S \otimes_R - \dashv U.$$

This is one of the conditions of the Beck criterion.

Since σ is an effective descent morphism,

$$S \otimes_R - \colon R\text{-}\mathsf{Alg} \longrightarrow S\text{-}\mathsf{Alg}$$

is comonadic (see 4.4.1 and 4.5.2), thus reflects isomorphisms. Therefore its restriction to split algebras reflects isomorphisms as well.

Finally consider two morphisms $u, v \colon A \rightrightarrows B$ in $\mathsf{Split}_R(\sigma)$ whose images under $(S \otimes_R -)$ admit a split equalizer

$$K \xrightarrow[k]{\quad r \quad} S \otimes_R A \underset{S \otimes v}{\overset{S \otimes u \quad t}{\rightrightarrows}} S \otimes_R B$$

in $\mathsf{Split}_S(S)$. This split equalizer is preserved by every functor, thus in particular by the inclusion in S-Alg. The Beck criterion in the case of the functor

$$S \otimes_R - : R\text{-}\mathsf{Alg} \longrightarrow S\text{-}\mathsf{Alg}$$

implies thus the existence of an equalizer

$$N \xrightarrow{\ n\ } A \underset{v}{\overset{u}{\rightrightarrows}} B$$

in R-Alg which is preserved by $S \otimes_R -$, thus mapped onto k. In particular, $S \otimes_R N \cong K \in \mathsf{Split}_S(S)$, which implies by corollary 4.7.2 that $N \in \mathsf{Split}_R(\sigma)$. □

Corollary 4.7.12 *The functor*

$$\left(\mathsf{Split}_R(\sigma)\right)^{\mathrm{op}} \longrightarrow \mathsf{Prof}/\mathsf{Sp}(S), \quad A \mapsto \mathsf{Sp}_S(S \otimes_R A)$$

is monadic.

Proof This functor is the composite of functors in lemmas 4.7.11 and 4.7.5. □

For clarity, let us present on a single diagram the various functors we have already studied.

$$\mathsf{Split}_R(\sigma) \underset{S \otimes_R -}{\overset{U}{\rightleftarrows}} \mathsf{Split}_S(S) \underset{\mathsf{Sp}_S}{\overset{\mathcal{C}_S}{\rightleftarrows}} \mathsf{Prof}/\mathsf{Sp}(S)$$

$$R\text{-}\mathsf{Alg} \underset{S \otimes_R -}{\overset{U}{\rightleftarrows}} S\text{-}\mathsf{Alg} \underset{\mathsf{Sp}_S}{\overset{\mathcal{C}_S}{\rightleftarrows}} \mathsf{Prof}/\mathsf{Sp}(S)$$

The next lemma is crucial, since it allows defining the Galois groupoid of the morphism $\sigma: R \longrightarrow S$.

Lemma 4.7.13 *Consider the cokernel pair of $\sigma: R \longrightarrow S$ in the category of R-algebras, viewed as a groupoid in the dual category (see example 4.6.2). The functor*

$$\mathsf{Sp}: (R\text{-}\mathsf{Alg})^{\mathrm{op}} \longrightarrow \mathsf{Prof}$$

transforms this groupoid into another groupoid Gal $[\sigma]$ *in the category of profinite spaces.*

Proof   The cogroupoid in the category of R-algebras (that is, the groupoid in the opposite category) is thus (see section 4.4)

$$(S \otimes_R S) \otimes_S (S \otimes_R S) \xleftarrow{\;s_1 \otimes s_2\;} S \otimes_R S \; \substack{\xleftarrow{\;\;s_1\;\;} \\ \xrightarrow{\;\;\mu\;\;} \\ \tau \bigcup \xleftarrow{\;\;s_2\;\;}} \; S$$

where

$$s_1(a) = a \otimes 1, \quad s_2(a) = 1 \otimes a, \quad \mu(a \otimes b) = ab, \quad \tau(a \otimes b) = b \otimes a.$$

Notice that S is canonically an S-algebra. The objects $S \otimes_R S$ and $(S \otimes_R S) \otimes_S (S \otimes_R S)$ can be provided with two distinct structures of S-algebra (see lemma 4.7.10). With the notation of 4.7.10, let us observe that the following morphisms are morphisms of S-algebras:

$$s_1 \colon S \longrightarrow S \otimes_R^1 S, \quad s_2 \colon S \longrightarrow S \otimes_R^2 S.$$

When we refer to $(S \otimes_R S) \otimes_S (S \otimes_R S)$ as an S-algebra, it will always be as the following pushout in S-Alg, which is by the way also a pushout in R-Alg:

$$\begin{array}{ccc}
S & \xrightarrow{\;\;s_1\;\;} & S \otimes_R^1 S \\
{\scriptstyle s_2}\big\downarrow & & \big\downarrow \\
S \otimes_R^2 S & \longrightarrow & (S \otimes_R^2 S) \otimes_S (S \otimes_R^1 S)
\end{array}$$

By lemma 4.7.9, the S-algebras S, $S \otimes_R^1 S$ and $S \otimes_R^2 S$ are split by $\mathrm{id}_S \colon S =\!=\!= S$. On the other hand we have the classical isomorphism

$$(S \otimes_R^2 S) \otimes_S (S \otimes_R^1 S) \cong S \otimes_R S \otimes_R S$$

where the S-module structure on $S \otimes_R S \otimes_R S$ is now given by

$$s \left(\sum_i u_i \otimes v_i \otimes w_i \right) = \sum_i u_i \otimes (sv_i) \otimes w_i.$$

Again by lemma 4.7.9, $(S \otimes_R^2 S) \otimes_S (S \otimes_R^1 S)$ is thus also an S-algebra split by id_S.

This proves in particular that the pushout defining the S-algebra structure of $(S \otimes_R^2 S) \otimes_S (S \otimes_R^1 S)$ lies entirely in the category $\mathsf{Split}_S(S)$, thus is a pushout in this category. Applying lemma 4.7.5, we get therefore a pullback in $\mathsf{Prof}/\mathsf{Sp}(S)$

$$
\begin{array}{ccc}
\mathsf{Sp}_S\big((S \otimes_R^2 S) \otimes_S (S \otimes_R^1 S)\big) & \longrightarrow & \mathsf{Sp}_S(S \otimes_R^2 S) \\
\downarrow & & \downarrow {\scriptstyle \mathsf{Sp}_S(s_2)} \\
\mathsf{Sp}_S(S \otimes_R^1 S) & \xrightarrow[\mathsf{Sp}_S(s_1)]{} & \mathsf{Sp}_S(S)
\end{array}
$$

and therefrom the following pullback in Prof:

$$
\begin{array}{ccc}
\mathsf{Sp}\big((S \otimes_R S) \otimes_S (S \otimes_R S)\big) & \longrightarrow & \mathsf{Sp}(S \otimes_R S) \\
\downarrow & & \downarrow {\scriptstyle \mathsf{Sp}(s_2)} \\
\mathsf{Sp}(S \otimes_R S) & \xrightarrow[\mathsf{Sp}(s_1)]{} & \mathsf{Sp}(S)
\end{array}
$$

This shows that applying the functor Sp to the cokernel pair of σ, viewed as a groupoid in $(R\text{-}\mathsf{Alg})^{\mathrm{op}}$, yields the following situation in Prof:

$$
\mathsf{Sp}(S \otimes_R S) \times_{\mathsf{Sp}(S)} \mathsf{Sp}(S \otimes_R S) \xrightarrow[\mathsf{Sp}(\tau)\,\cup]{\mathsf{Sp}(s_1 \otimes s_2)} \mathsf{Sp}(S \otimes_R S) \underset{\xrightarrow{\mathsf{Sp}(\mu)}}{\overset{\xrightarrow{\mathsf{Sp}(s_1)}}{\xleftarrow{\hspace{1cm}}}} \underset{\xrightarrow{\mathsf{Sp}(s_2)}}{} \mathsf{Sp}(S)
$$

with the left hand object being the pullback of $\mathsf{Sp}(s_1)$, $\mathsf{Sp}(s_2)$.

It remains to prove that the axioms for a groupoid are satisfied in Prof. But the functor Sp, like every functor, preserves the commutativity of diagrams, which yields at once all axioms, with the exception maybe of the associativity of the composition, since this axiom involves a pullback. In fact, an argument perfectly analogous to the one we have just developed proves that we also have a three factor pullback in Prof

$$
\mathsf{Sp}\big((S \otimes_R S) \otimes_S (S \otimes_R S) \otimes_S (S \otimes_R S)\big)
$$
$$
\cong \mathsf{Sp}(S \otimes_R S) \times_{\mathsf{Sp}(S)} \mathsf{Sp}(S \otimes_R S) \times_{\mathsf{Sp}(S)} \mathsf{Sp}(S \otimes_R S),
$$

so that the associativity of the composition reduces again to the preservation by the functor Sp of the diagram expressing that associativity for the cokernel pair of σ. $\qquad\qquad\square$

Lemma 4.7.13 thus allows the following definition.

Definition 4.7.14 Let $\sigma: R \longrightarrow S$ be a Galois descent morphism of rings. The Galois groupoid $\mathsf{Gal}\,[\sigma]$ of σ is the following internal groupoid in the category of profinite spaces:

$$\mathsf{Sp}(S \otimes_R S) \times_{\mathsf{Sp}(S)} \mathsf{Sp}(S \otimes_R S) \xrightarrow{\mathsf{Sp}(s_1 \otimes s_2)} \mathsf{Sp}(S \otimes_R S) \underset{\xrightarrow{\mathsf{Sp}(s_2)}}{\overset{\xrightarrow{\mathsf{Sp}(s_1)}}{\underset{\mathsf{Sp}(\mu)}{\longleftarrow}}} \mathsf{Sp}(S).$$

Here now is the Galois theorem for rings (commutative, with a unit).

Theorem 4.7.15 (Galois theorem) *Let $\sigma: R \longrightarrow S$ be a Galois descent morphism of rings and $\mathsf{Gal}\,[\sigma]$ the corresponding Galois groupoid in the category of profinite spaces. There exists an equivalence of categories*

$$\left(\mathsf{Split}_R(\sigma)\right)^{\mathsf{op}} \approx \mathsf{Prof}^{\mathsf{Gal}\,[\sigma]}$$

between the dual of the category of R-algebras split by σ and the category of internal covariant presheaves on $\mathsf{Gal}\,[\sigma]$ in the category of profinite spaces.

Proof The category $\mathsf{Prof}^{\mathsf{Gal}\,[\sigma]}$ is the category of algebras for the monad on $\mathsf{Prof}/\mathsf{Sp}(S)$ described in proposition 4.6.1. We shall prove that the category $\left(\mathsf{Split}_R(\sigma)\right)^{\mathsf{op}}$ is also monadic on $\mathsf{Prof}/\mathsf{Sp}(S)$, for a monad which is isomorphic to that given by proposition 4.6.1. This will yield the expected result.

By corollary 4.7.12, the functor

$$\left(\mathsf{Split}_R(\sigma)\right)^{\mathsf{op}} \longrightarrow \mathsf{Prof}/\mathsf{Sp}(S), \quad A \mapsto \mathsf{Sp}_S(S \otimes_R A)$$

is indeed monadic. Its left adjoint is $U \circ \mathcal{C}_S$. Observe that the functorial part of the corresponding monad is

$$T: \mathsf{Prof}/\mathsf{Sp}(S) \longrightarrow \mathsf{Prof}/\mathsf{Sp}(S), \quad (X, \varphi) \mapsto \mathsf{Sp}_S\left(S \otimes_R \mathcal{C}_S(X, \varphi)\right).$$

Let us now compute explicitly the form of the functorial part of the monad given by proposition 4.6.1. Given $(X, \varphi) \in \mathsf{Prof}/\mathsf{Sp}(S)$, one computes first the pushout in S-Alg

$$S \xrightarrow{\;s_1\;} S \otimes_R^1 S$$

$$\alpha_{C_S(X,\varphi)} \Big\downarrow \qquad\qquad \Big\downarrow \mathrm{id}_S \otimes \alpha_{C_S(X,\varphi)}$$

$$C_S(X,\varphi) \longrightarrow (S \otimes_R^1 S) \otimes_S C_S(X,\varphi) \cong S \otimes_R C_S(X,\varphi)$$

where

$$\alpha_{C_S(X,\varphi)} \colon S \longrightarrow C_S(X,\varphi), \quad s \mapsto s \cdot 1,$$

and the structure of S-module on $S \otimes_R C_S(X,\varphi)$ is given by the action of S on $C_S(X,\varphi)$ (see theorem 4.3.6), that is

$$S \otimes_R C_S(X,\varphi) \longrightarrow C_S(X,\varphi), \quad s \otimes a \mapsto \alpha_{C_S(X,\varphi)}(s) \cdot a.$$

By lemma 4.7.10, this is also a pushout in $\mathsf{Split}_S(S)$. Applying lemma 4.7.5, we get the following pullback in $\mathsf{Prof}/\mathsf{Sp}(S)$, where the isomorphism follows from theorem 4.3.6.

$$\mathsf{Sp}_S\big(S \otimes_R C(X,\varphi)\big) \longrightarrow \mathsf{Sp}_S C_S(X,\varphi) \cong (X,\varphi)$$

$$\mathsf{Sp}_S\big(\mathrm{id}_S \otimes \alpha_{C_S(X,\varphi)}\big) \Big\downarrow \qquad\qquad \Big\downarrow \mathsf{Sp}_S(\alpha_{C_S(X,\varphi)})$$

$$\mathsf{Sp}_S(S \otimes_R S) \xrightarrow[\mathsf{Sp}_S(s_1)]{} \mathsf{Sp}_S(S)$$

Since pullbacks in $\mathsf{Prof}/\mathsf{Sp}(S)$ are computed as in Prof, we get in fact the pullback of diagram 4.7 in Prof. The left vertical composite is precisely $\big(\Sigma_{\mathsf{Sp}(s_2)} \circ \mathsf{Sp}(s_1)^*\big)(X,\varphi)$.

Since the structure of S-module on $S \otimes_R C_S(X,\varphi)$ is given by the action of S on $C_S(X,\varphi)$, it is induced by the corresponding morphism

$$S \xrightarrow{\;s_2\;} S \otimes_R S \xrightarrow{\mathrm{id}_S \otimes \alpha_{C_S(X,\varphi)}} S \otimes_R C_S(X,\varphi).$$

This shows that the left vertical composite in diagram 4.7 is also the object $T(X,\varphi) \in \mathsf{Prof}/\mathsf{Sp}(S)$, for the monad T described above. This shows already that the two monads T and $\Sigma_{\mathsf{Sp}(s_2)} \circ \mathsf{Sp}(s_1)^*$ coincide on objects. It is now routine, left to the reader, to verify that the two monads considered in this proof are in fact isomorphic.   □

Let us conclude this section by observing that theorem 4.7.15 extends the corresponding result for fields, namely, theorem 3.5.8.

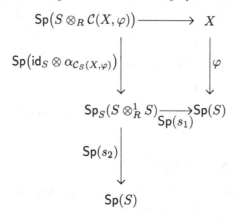

$$\mathsf{Sp}\big(S \otimes_R C(X, \varphi)\big) \longrightarrow X$$

Diagram 4.7

Corollary 4.7.16 *Let* $\sigma \colon K \longrightarrow L$ *be a Galois extension of fields.*

(i) *The Galois groupoid* $\mathsf{Gal}\,[\sigma]$ *of* σ, *viewed as a Galois descent morphism of rings, coincides with the usual profinite Galois group* $\mathsf{Gal}\,[L : K]$ *of the field extension* $[L : K]$.

(ii) *The* K-*algebras split by* L *in the sense of definition 2.3.1 coincide with the* K-*algebras split by* σ.

(iii) *The internal covariant presheaves on the internal groupoid* $\mathsf{Gal}\,[\sigma]$ *coincide with the profinite* $\mathsf{Gal}\,[L : K]$-*spaces.*

(iv) *The equivalence of theorem 4.7.15 reduces to the equivalence of theorem 3.5.8.*

Proof   In the case of the present corollary, $\mathsf{Sp}(K) \cong \{*\} \cong \mathsf{Sp}(L)$, and therefore $\mathsf{Prof}/\mathsf{Sp}(K) \cong \mathsf{Prof} \cong \mathsf{Prof}/\mathsf{Sp}(L)$. In particular the Galois groupoid $\mathsf{Gal}\,[\sigma]$ has a unique object $* \in \mathsf{Sp}(L)$ and is therefore an internal group in Prof, that is, a profinite group G. Obviously, the internal covariant presheaves on $\mathsf{Gal}\,[\sigma]$ are exactly the profinite G-sets. By proposition 4.5.5, this profinite group G is given by

$$G \cong \mathsf{Sp}(L \otimes_K L) \cong \mathsf{Aut}_K(L) \cong \mathsf{Gal}\,[L : K]$$

and the group structure of $\mathsf{Aut}_K(L)$ agrees with that of $\mathsf{Sp}(L \otimes_K L)$.

Theorems 3.5.8 and 4.7.15, together with propositions 4.5.5 and 4.5.4, yield the diagram

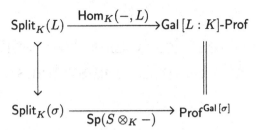

In this commutative diagram of functors, the horizontal morphisms are equivalences. Since the right vertical morphism is an equality, the left vertical morphism is an equivalence as well. □

5

Categorical Galois theorem and factorization systems

This chapter is the core of the book. It could even be given a longer title, namely *Categorical Galois theorem, "non-Grothendieck" examples and factorization systems.*

We show first how the situation for commutative rings, studied in chapter 4, generalizes to develop Galois theory with respect to an axiomatic categorical setting. This setting consists basically in an adjunction with "well-behaved" properties, which mimic the situation of the *Pierce spectrum* functor and its adjoint, as in section 4.3. This categorical setting will contain the situations of the previous chapters: in particular the cases of fields and commutative rings. But our categorical Galois theorem will also apply to many other contexts.

First we apply the categorical Galois theorem to the study of central extensions of groups. This topic is generally not considered as part of Galois theory, but nevertheless the central extensions of groups turn out to be precisely the objects split over extension in a special case of categorical Galois theory.

This chapter also provides a good help for understanding the relationship between the Galois theory and factorization systems, which began in [18]. We focus in particular on the case of semi-left-exact reflections and apply it to the monotone–light factorization of continuous maps between compact Hausdorff spaces. It should be noticed that the situation of chapter 4, the categorical Galois theory of rings, is also a special case of a semi-left-exact reflection.

It might look strange for a topologist to describe light maps of compact Hausdorf spaces as the actions of a compact totally disconnected equivalence relation considered as a topological groupoid. However if we replace the category of compact Hausdorff spaces by the dual category of commutative C^*-algebras, which is equivalent to it, the surprise disappears since general profinite topological groupoids are already used in Galois theory of commutative rings.

5.1 The abstract categorical Galois theorem

This section presents the Galois theorem of Janelidze in its general form. The Galois theory for rings, as developed in section 4.7, is in fact a special case of this more general theory.

Definition 5.1.1 Let $\mathcal{C}$ be a category. A class $\overline{\mathcal{C}}$ of arrows in $\mathcal{C}$ is admissible when

  (i) every isomorphism is in $\overline{\mathcal{C}}$,
  (ii) $\overline{\mathcal{C}}$ is closed under composition,
  (iii) in the pullback

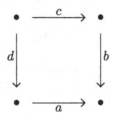

   if $a, b \in \overline{\mathcal{C}}$, then $c, d \in \overline{\mathcal{C}}$.

Definition 5.1.2 Let $\overline{\mathcal{C}}$ be an admissible class of morphisms in a category $\mathcal{C}$. For an object $C \in \mathcal{C}$, we write $\overline{\mathcal{C}}/C$ for the following category:

  (i) the objects are the pairs (X, f) where $f\colon X \longrightarrow C$ is in $\overline{\mathcal{C}}$;
  (ii) the arrows $h\colon (X, f) \longrightarrow (X', f')$ are all arrows $h\colon X \longrightarrow X'$ in $\mathcal{C}$ such that $f' \circ h = f$.

$\overline{\mathcal{C}}/C$ is thus a full subcategory of $\mathcal{C}/C$.

Definition 5.1.3 A relatively admissible adjunction consists in

  (i) an adjunction $A \underset{S}{\overset{C}{\rightleftarrows}} \mathcal{P}$; $S \dashv C$,
  (ii) two admissible classes $\overline{\mathcal{A}} \subseteq \mathcal{A}$, $\overline{\mathcal{P}} \subseteq \mathcal{P}$ of arrows,

such that

  (i) $\forall f \in \mathcal{A} \ \ f \in \overline{\mathcal{A}} \Rightarrow S(f) \in \overline{\mathcal{P}}$,
  (ii) $\forall g \in \mathcal{P} \ \ g \in \overline{\mathcal{P}} \Rightarrow C(g) \in \overline{\mathcal{A}}$,
  (iii) $\forall A \in \mathcal{A} \ \ \eta_A\colon A \longrightarrow CS(A)$, the unit of the adjunction $S \dashv C$, is an arrow of $\overline{\mathcal{A}}$,

(iv) $\forall X \in \mathcal{P}$ $\varepsilon_X \colon SC(X) \longrightarrow X$, the counit of the adjunction $S \dashv C$, is an arrow of $\overline{\mathcal{P}}$.

The various conditions in definitions 5.1.1 and 5.1.3 obviously imply the following two lemmas.

Lemma 5.1.4 *Let $\mathcal{A}$ be a category with pullbacks. In the conditions of definition 5.1.3, lemma 4.3.4 yields an adjunction between $\mathcal{A}/A$ and $\mathcal{P}/S(A)$, which restricts to an adjunction*

$$\mathcal{A}/A \underset{S_A}{\overset{C_A}{\longleftarrow}} \mathcal{P}/S(A), \quad S_A \dashv C_A. \qquad \square$$

Lemma 5.1.5 *Let $\overline{\mathcal{C}}$ be an admissible class of arrows in a category with pullbacks. For every morphism $\sigma \colon S \longrightarrow R$ in $\overline{\mathcal{C}}$, the adjunction $\Sigma_\sigma \dashv \sigma^*$ of section 4.4 restricts to an adjunction*

$$\overline{\mathcal{C}}/R \underset{\sigma^*}{\overset{\Sigma_\sigma}{\longleftarrow}} \overline{\mathcal{C}}/S, \quad \Sigma_\sigma \dashv \sigma^*. \qquad \square$$

Definition 5.1.6 Let $\overline{\mathcal{C}}$ be an admissible class of arrows in a category $\mathcal{C}$ with pullbacks. An arrow $\sigma \colon S \longrightarrow R$ of $\mathcal{C}$ is an effective descent morphism relatively to $\overline{\mathcal{C}}$ when

(i) $\sigma \in \overline{\mathcal{C}}$,
(ii) the functor $\sigma^* \colon \overline{\mathcal{C}}/R \longrightarrow \overline{\mathcal{C}}/S$ is monadic.

Definition 5.1.7 With the notation of definition 5.1.3, consider a relatively admissible adjunction

$$S \dashv C \colon (\mathcal{A}, \overline{\mathcal{A}}) \overset{\longleftarrow}{\longrightarrow} (\mathcal{P}, \overline{\mathcal{P}})$$

where $\mathcal{A}$ is a category with pullbacks. An object $(A, a) \in \overline{\mathcal{A}}/R$ is split by a morphism $\sigma \colon S \longrightarrow R$ of $\overline{\mathcal{A}}$ when the unit

$$\eta^S_{\sigma^*(A,a)} \colon \sigma^*(A, a) \longrightarrow C_S S_S \sigma^*(A, a)$$

of the adjunction $S_S \dashv C_S$ is an isomorphism at the object $\sigma^*(A, a)$.

Definition 5.1.8 With the notation of definition 5.1.3, consider a relatively admissible adjunction

$$S \dashv C \colon (\mathcal{A}, \overline{\mathcal{A}}) \overset{\longleftarrow}{\longrightarrow} (\mathcal{P}, \overline{\mathcal{P}})$$

where $\mathcal{A}$ and $\mathcal{P}$ are categories with pullbacks. A morphism $\sigma \colon S \longrightarrow R$ in $\mathcal{A}$ is of relative Galois descent with respect to these data when

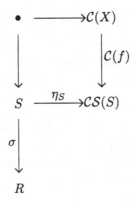

Diagram 5.1

(i) σ is an effective descent morphism relatively to $\overline{\mathcal{A}}$,

(ii) the counit of the adjunction $\mathcal{S}_S \dashv \mathcal{C}_S$

$$\mathcal{S}_S \circ \mathcal{C}_S \Longrightarrow \mathrm{id}_{\overline{\mathcal{P}}/\mathcal{S}(S)}$$

is an isomorphism,

(iii) for every object $(X, f) \in \overline{\mathcal{P}}/\mathcal{S}(S)$, the object $(\Sigma_\sigma \circ \mathcal{C}_S)(X, f) \in \overline{\mathcal{A}}/R$ is split by σ.

Let us recall that the object indicated in condition (iii) is the left vertical composite in diagram 5.1, where the square is a pullback and η is the unit of the adjunction $\mathcal{S} \dashv \mathcal{C}$.

For clarity, let us put on a single diagram the various functors involved in definition 5.1.8.

$$
\begin{array}{ccc}
& \mathcal{C}_S & \\
\overline{\mathcal{A}}/S & \xleftarrow{\hspace{1cm}} & \mathcal{P}/\mathcal{S}(S) \\
& \xrightarrow[\mathcal{S}_S]{} & \\
\Sigma_\sigma \Big\uparrow\Big\downarrow \sigma^* & \quad \Sigma_{\mathcal{S}(\sigma)} \Big\uparrow\Big\downarrow \mathcal{S}(\sigma)^* & \\
& \mathcal{C}_R & \\
\overline{\mathcal{A}}/R & \xleftarrow{\hspace{1cm}} & \mathcal{P}/\mathcal{S}(R) \\
& \xrightarrow[\mathcal{S}_R]{} &
\end{array}
$$

$\Sigma_\sigma \dashv \sigma^*, \quad \mathcal{S}_R \dashv \mathcal{C}_R, \quad \Sigma_{\mathcal{S}(\sigma)} \dashv \mathcal{S}(\sigma)^*, \quad \mathcal{S}_S \dashv \mathcal{C}_S.$

Lemma 5.1.9 *In the conditions of definition 5.1.8,*

$$\Sigma_{\mathcal{S}(\sigma)} \circ \mathcal{S}_S \cong \mathcal{S}_R \circ \Sigma_\sigma, \quad \mathcal{C}_S \circ \mathcal{S}(\sigma)^* = \sigma^* \circ \mathcal{C}_R.$$

Proof The first relation is obvious and the second follows by adjunction.

□

Before going on, it is useful to specify in what sense the situation for rings is a special case of the present situation. In the case of rings,

- $\mathcal{A}$ is the dual of the category of rings,
- $\overline{\mathcal{A}}$ is the class of all morphisms of $\mathcal{A}$, thus $\overline{\mathcal{A}}/S = \mathcal{A}/S$,
- $\mathcal{P}$ is the category of profinite spaces,
- $\overline{\mathcal{P}}$ is the class of all morphisms of $\mathcal{P}$, thus $\overline{\mathcal{P}}/\mathcal{S}(S) = \mathcal{P}/\mathcal{S}(S)$,
- S is the 'Pierce spectrum' functor,
- $\mathcal{C}$ is the functor $\mathcal{C}(-,\mathbb{Z})$.

Indeed, a ring is exactly a $\mathbb{Z}$-algebra and given a ring R, $\mathrm{Ring}^{\mathrm{op}}/R$ is the dual category of R-algebras (see 4.3.1).

The first two of the following results are then obvious, by definition 5.1.7.

Lemma 5.1.10 *In the conditions of definition 5.1.8, the following conditions are equivalent for an object $(A,a) \in \overline{\mathcal{A}}/S$:*

(i) *(A,a) is split by $\mathrm{id}_S\colon S \!=\!=\!= S$;*
(ii) *the morphism $\eta^S_{(A,a)}\colon (A,a) \longrightarrow \mathcal{C}_S\mathcal{S}_S(A,a)$ is an isomorphism.*

□

Corollary 5.1.11 *In the conditions of definition 5.1.8, the following conditions are equivalent for an object $(A,a) \in \overline{\mathcal{A}}/R$:*

(i) *(A,a) is split by $\sigma\colon S \longrightarrow R$;*
(ii) *$\sigma^*(A,a)$ is split by $\mathrm{id}_S\colon S \longrightarrow S$.*

□

Lemma 5.1.12 *In the conditions of definition 5.1.8, the following conditions are equivalent for an object $(A,a) \in \overline{\mathcal{A}}/S$:*

(i) *(A,a) is split by $\mathrm{id}_S\colon S \!=\!=\!= S$;*
(ii) *$(A,a) \cong \mathcal{C}_S(X,\varphi)$ for some $(X,\varphi) \in \overline{\mathcal{P}}/\mathcal{S}(S)$.*

Proof Choosing $(X,\varphi) = \mathcal{S}_S(A,a)$ yields (i) $\Rightarrow$ (ii) by lemma 5.1.10. Conversely, by definition 5.1.8,

$$(A,a) \cong \mathcal{C}_S(X,\varphi) \cong \mathcal{C}_S\mathcal{S}_S\mathcal{C}_S(X,\varphi) \cong \mathcal{C}_S\mathcal{S}_S(A,a)$$

from which we obtain the conclusion by lemma 5.1.10.

□

Corollary 5.1.13 *In the conditions of definition 5.1.8, the following conditions are equivalent for an object $(A, a) \in \overline{A}/R$:*

(i) *(A, a) is split by $\sigma \colon S \longrightarrow R$;*

(ii) *$\sigma^*(A, a) \cong \mathcal{C}_S(X, \varphi)$ for some $(X, \varphi) \in \overline{\mathcal{P}}/\mathcal{S}(S)$.* $\square$

Lemma 5.1.14 *We assume the conditions of definition 5.1.8 and write $\mathsf{Split}_S(S)$ for the full subcategory $\mathsf{Split}_S(S) \subseteq \overline{A}/S$ of those objects which are split by $\mathrm{id}_S \colon S =\!\!=\!\!= S$. The functors*

$$\mathsf{Split}_S(S) \underset{\mathcal{S}_S}{\overset{\mathcal{C}_S}{\longleftrightarrow}} \overline{\mathcal{P}}/\mathcal{S}(S)$$

constitute an equivalence of categories.

Proof By lemma 5.1.12, $\mathcal{C}_S$ takes values in $\mathsf{Split}_S(S)$. One has $\mathcal{S}_S \circ \mathcal{C}_S \cong \mathrm{id}$ by definition 5.1.8 and $\mathrm{id} \cong \mathcal{C}_S \circ \mathcal{S}_S$ by lemma 5.1.10. $\square$

Lemma 5.1.15 *In the conditions of definition 5.1.8, if an object $(A, a) \in \overline{A}/S$ is split by $\mathrm{id}_S \colon S =\!\!=\!\!= S$, the same property holds for the object $(\sigma^* \circ \Sigma_\sigma)(A, a)$.*

Proof By lemma 5.1.12, $(A, a) \cong \mathcal{C}_S(X, \varphi)$ for some $(X, \varphi) \in \overline{\mathcal{P}}/\mathcal{S}(S)$. By definition 5.1.8, $\Sigma_\sigma \mathcal{C}_S(X, \varphi)$ is split by σ. By corollary 5.1.11, this implies that $\sigma^* \Sigma_\sigma \mathcal{C}_S(X, \varphi) \cong \sigma^* \Sigma_\sigma(A, a)$ is split by id_S. $\square$

Lemma 5.1.16 *In the conditions of definition 5.1.8, $(S, \mathrm{id}_S) \in \overline{A}/S$ is split by $\mathrm{id}_S \colon S =\!\!=\!\!= S$.*

Proof The lemma immediately follows from the previous equality $\mathcal{C}_S\big(\mathcal{S}_S(S), \mathrm{id}_{\mathcal{S}(S)}\big) = (S, \mathrm{id}_S)$ and lemma 5.1.12. $\square$

Corollary 5.1.17 *In the conditions and with the notation of definition 5.1.8, the object $(R, \mathrm{id}_R) \in \overline{A}/R$ is split by $\sigma \colon S \longrightarrow R$.* $\square$

Lemma 5.1.18 *In the conditions of definition 5.1.8, for all integers $n \in \mathbb{N}$ and $1 \le i \le n$, the objects $\big(\prod_{i=1}^{n}(S, \sigma), p_i\big) \in \overline{A}/S$ are split by $\mathrm{id}_S \colon S =\!\!=\!\!= S$.*

Proof By iterated application of lemma 5.1.15, starting with $(A, a) = (S, \mathrm{id}_S)$ via lemma 5.1.16. $\square$

Lemma 5.1.19 *In the conditions of definition 5.1.8, if (A, a) and (B, b) are objects of $\overline{A}/S$ split by id_S, then $(A, a) \times (B, b) \in \overline{A}/S$ is split by id_S as well.*

Proof By lemma 5.1.12, we can write $(A, a) \cong C_S(X, \varphi)$ and $(B, b) \cong C_S(Y, \psi)$. We deduce

$$(A, a) \times (B, b) \cong C_S(X, \varphi) \times C_S(Y, \psi) \cong C_S\big((X, \varphi) \times (Y, \psi)\big)$$

because C_S, as a right adjoint, preserves products. One concludes the proof by lemma 5.1.12 again. $\qquad\square$

Lemma 5.1.20 *Assume the conditions of definition 5.1.8 and write $\mathsf{Split}_R(\sigma)$ for the full subcategory $\mathsf{Split}_R(\sigma) \subseteq \overline{A}/R$ of those objects which are split by $\sigma\colon S \longrightarrow R$. The functor*

$$\sigma^*\colon \mathsf{Split}_R(\sigma) \longrightarrow \mathsf{Split}_S(S)$$

is monadic.

Proof By corollary 5.1.11, this functor is correctly defined. Observe that when $(A, a) \in \mathsf{Split}_S(S)$, $\sigma^* \Sigma_\sigma(A, a) \in \mathsf{Split}_S(S)$ by lemma 5.1.15, and thus $\Sigma_\sigma(A, a) \in \mathsf{Split}_R(\sigma)$ by corollary 5.1.11 (which in fact is shown directly in the proof of lemma 5.1.15). This proves that the adjunction

$$\overline{A}/R \underset{\sigma^*}{\overset{\Sigma_\sigma}{\rightleftarrows}} \overline{A}/S, \quad \Sigma_\sigma \dashv \sigma^*$$

of lemma 5.1.5 restricts to an adjunction

$$\mathsf{Split}_R(\sigma) \underset{\sigma^*}{\overset{\Sigma_\sigma}{\rightleftarrows}} \mathsf{Split}_S(S), \quad \Sigma_\sigma \dashv \sigma^*.$$

This is one of the conditions of the Beck criterion for monadicity (see [66]).

Since σ is an effective descent morphism relatively to $\overline{A}$, the functor

$$\sigma^*\colon \overline{A}/R \longrightarrow \overline{A}/S$$

is monadic, thus reflects isomorphisms. The same thus holds for its restriction to the full subcategories of split objects.

Finally consider two morphisms $u, v\colon (B, b) \rightrightarrows (A, a)$ in $\mathsf{Split}_R(\sigma)$ such that the following diagram is a split coequalizer of $\sigma^*(u)$, $\sigma^*(v)$ in $\mathsf{Split}_S(S)$.

$$\sigma^*(B,b) \overset{\overset{\displaystyle t}{\overbrace{\qquad\qquad}}}{\underset{\sigma^*(v)}{\overset{\sigma^*(u)}{\rightrightarrows}}} \sigma^*(A,a) \overset{\overset{\displaystyle r}{\overbrace{\qquad\qquad}}}{\underset{q}{\longrightarrow}} (Q,x)$$

This split coequalizer is preserved by every functor, thus in particular by the inclusion $\mathsf{Split}_S(S) \subseteq \overline{\mathcal{A}}/S$. The Beck criterion in the case of the functor

$$\sigma^* : \overline{\mathcal{A}}/R \longrightarrow \overline{\mathcal{A}}/S$$

thus implies the existence of a coequalizer in $\overline{\mathcal{A}}/R$

$$(B,b) \underset{v}{\overset{u}{\rightrightarrows}} (A,a) \overset{p}{\twoheadrightarrow} (C,c)$$

which is preserved by σ^*. To conclude the proof, it remains to check that $(C,c) \in \mathsf{Split}_R(\sigma)$. By corollary 5.1.11, this reduces to $\sigma^*(C,c) \in \mathsf{Split}_S(S)$. The preservation of the coequalizer $p = \mathsf{Coker}\,(u,v)$ by σ^* implies precisely $\sigma^*(C,c) \cong (Q,x) \in \mathsf{Split}_S(S)$. $\square$

Corollary 5.1.21 *In the conditions of definition 5.1.8, the functor*

$$\mathsf{Split}_R(\sigma) \longrightarrow \overline{\mathcal{P}}/\mathcal{S}(S), \quad (A,a) \mapsto (\mathcal{S}_S \circ \sigma^*)(A,a)$$

is monadic.

Proof   This functor is the composite of those in lemmas 5.1.20 and 5.1.14. $\square$

For clarity, let us consider on a single diagram the various functors involved in our discussion.

$$
\begin{array}{ccccc}
\mathsf{Split}_R(\sigma) & \underset{\sigma^*}{\overset{\Sigma_\sigma}{\leftrightarrows}} & \mathsf{Split}_S(S) & \underset{\mathcal{S}_S}{\overset{\mathcal{C}_S}{\leftrightarrows}} & \overline{\mathcal{P}}/\mathcal{S}(S) \\
\big\downarrow & & \big\downarrow & & \big\| \\
\overline{\mathcal{A}}/R & \underset{\sigma^*}{\overset{\Sigma_\sigma}{\leftrightarrows}} & \overline{\mathcal{A}}/S & \underset{\mathcal{S}_S}{\overset{\mathcal{C}_S}{\leftrightarrows}} & \overline{\mathcal{P}}/\mathcal{S}(S)
\end{array}
$$

Lemma 5.1.22 *In the conditions and with the notation of definition 5.1.8, the functor $\mathcal{S} : \mathcal{A} \longrightarrow \mathcal{P}$ transforms the kernel pair of $\sigma : S \longrightarrow R$, seen as a groupoid in $\mathcal{A}$ (see example 4.6.2), into a groupoid in $\mathcal{P}$.*

Proof   The kernel pair of σ, viewed as a groupoid in $\mathcal{A}$,

$$(S \times_R S) \times_S (S \times_R S) \xrightarrow{\ (p_1, p_4)\ } S \times_R S \underset{\overset{p_2}{\longrightarrow}}{\overset{\overset{p_1}{\longrightarrow}}{\underset{\Delta}{\longleftarrow}}} S,$$

with $\tau \cup$ at $S \times_R S$,

can also be seen as the "object part" of the following groupoid in $\mathcal{A}/R$:

$$\big((S,\sigma) \times (S,\sigma)\big) \underset{(S,\sigma)}{\times} \big((S,\sigma) \times (S,\sigma)\big) \xrightarrow{\ (p_1, p_4)\ } (S,\sigma) \times (S,\sigma) \underset{\overset{p_2}{\longrightarrow}}{\overset{\overset{p_1}{\longrightarrow}}{\underset{\Delta}{\longleftarrow}}} (S,\sigma).$$

with $\tau \cup$.

The pullback defining the left hand object can even be seen as the image by Σ_σ of the pullback

$$
\begin{array}{ccc}
\big((S \times_R S) \times_S (S \times_R S), p\big) & \longrightarrow & (S \times_R S, p_1) \\
\downarrow & & \downarrow{\scriptstyle p_1} \\
(S \times_R S, p_2) & \xrightarrow{\ p_2\ } & (S, \mathrm{id}_S)
\end{array}
$$

in $\overline{\mathcal{A}}/S$. Also observe that

$$\big((S \times_R S) \times_S (S \times_R S), p\big) \cong \big(S \times_R S \times_R S, p_2\big).$$

By lemma 5.1.18, the last pullback in $\overline{\mathcal{A}}/S$ lies in fact in $\mathsf{Split}_S(S)$. By lemma 5.1.14, we get therefore a pullback in $\overline{\mathcal{P}}/\mathcal{S}(S)$

$$
\begin{array}{ccc}
\mathcal{S}_S\big((S \times_R S) \times_S (S \times_R S), p\big) & \longrightarrow & \mathcal{S}_S(S \times_R S, p_1) \\
\downarrow & & \downarrow{\scriptstyle \mathcal{S}_S(p_1)} \\
\mathcal{S}_S(S \times_R S, p_2) & \xrightarrow{\ \mathcal{S}_S(p_2)\ } & \mathcal{S}_S(S, \mathrm{id}_S)
\end{array}
$$

Since pullbacks in $\overline{\mathcal{P}}/\mathcal{S}(S)$ are computed as in $\mathcal{P}$, we get the pullback in $\mathcal{P}$

$$
\begin{array}{ccc}
\mathcal{S}\big((S \times_R S) \times_S (S \times_R S)\big) & \longrightarrow & \mathcal{S}(S \times_R S) \\
\downarrow & & \downarrow{\scriptstyle \mathcal{S}(p_1)} \\
\mathcal{S}(S \times_R S) & \xrightarrow{\ \mathcal{S}(p_2)\ } & \mathcal{S}(S)
\end{array}
$$

which indicates that applying the functor S to the kernel pair of σ yields the following situation in $\mathcal{P}$:

$$\mathcal{S}(S \times_R S) \times_{\mathcal{S}(S)} \mathcal{S}(S \times_R S) \xrightarrow[\mathcal{S}(\tau)]{\overset{(\mathcal{S}(p_1),\,\mathcal{S}(p_4))}{\longrightarrow}} \mathcal{S}(S \times_R S) \begin{array}{c} \xrightarrow{\mathcal{S}(p_1)} \\ \xleftarrow{\mathcal{S}(\Delta)} \\ \xrightarrow{\mathcal{S}(p_2)} \end{array} \mathcal{S}(S).$$

A perfectly analogous argument applies to prove that

$$\mathcal{S}\big((S \times_R S) \times_S (S \times_R S) \times_S (S \times_R S)\big)$$
$$\cong \mathcal{S}(S \times_R S) \times_{\mathcal{S}(S)} \mathcal{S}(S \times_R S) \times_{\mathcal{S}(S)} \mathcal{S}(S \times_R S).$$

This object is useful to describe the associativity of the composition in the expected groupoid.

The axioms for an internal groupoid are expressed by the commutativity of some diagrams involving the previous pullbacks. We have just seen that these pullbacks are the images under S of the corresponding pullbacks defining the kernel pair of σ, seen as a groupoid. Since S, like every functor, preserves the commutativity of diagrams, the proof is done. $\qquad\qquad\square$

Definition 5.1.23 Let $S \dashv C \colon (\mathcal{A}, \overline{\mathcal{A}}) \overset{\longleftarrow}{\longrightarrow} (\mathcal{P}, \overline{\mathcal{P}})$ be a relatively admissible adjunction (notation of definition 5.1.3) and $\sigma \colon S \longrightarrow R$ a morphism of $\mathcal{A}$ which is of relative Galois descent with respect to those data. The Galois groupoid $\mathsf{Gal}\,[\sigma]$ of σ is the internal groupoid in $\mathcal{P}$

$$\mathcal{S}(S \times_R S) \times_{\mathcal{S}(S)} \mathcal{S}(S \times_R S) \xrightarrow[\mathcal{S}(\tau)]{\overset{(\mathcal{S}(p_1),\,\mathcal{S}(p_4))}{\longrightarrow}} \mathcal{S}(S \times_R S) \begin{array}{c} \xrightarrow{\mathcal{S}(p_1)} \\ \xleftarrow{\mathcal{S}(\Delta)} \\ \xrightarrow{\mathcal{S}(p_2)} \end{array} \mathcal{S}(S)$$

given by lemma 5.1.22.

Theorem 5.1.24 (Galois theorem) *Let* $S \dashv C \colon (\mathcal{A}, \overline{\mathcal{A}}) \overset{\longleftarrow}{\longrightarrow} (\mathcal{P}, \overline{\mathcal{P}})$ *be a relatively admissible adjunction (notation of definition 5.1.3) and* $\sigma \colon S \longrightarrow R$ *a morphism of* $\mathcal{A}$ *which is of relative Galois descent with respect to these data. In these conditions, there exists an equivalence of categories*

$$\mathsf{Split}_R(\sigma) \approx \overline{\mathcal{P}}^{\,\mathsf{Gal}\,[\sigma]}$$

between the category of those objects $(A, a) \in \overline{\mathcal{A}}/R$ *which are split by* σ *and the category of internal covariant presheaves* (P, p, π) *on the internal groupoid* $\mathsf{Gal}\,[\sigma]$ *in* $\mathcal{P}$, *in which* $p \in \overline{\mathcal{P}}$.

$$(S \times_R S) \times_S A \xrightarrow{\quad a_2 \quad} A$$

$$
\begin{array}{ccc}
(S \times_R S) \times_S A & \xrightarrow{a_2} & A \\
\Big\downarrow{\scriptstyle a_1} & & \Big\downarrow{\scriptstyle a} \\
S \times_R S & \xrightarrow{p_1} & S \\
\Big\downarrow{\scriptstyle p_2} & & \Big\downarrow{\scriptstyle \sigma} \\
S & \xrightarrow{\quad\sigma\quad} & R
\end{array}
$$

Diagram 5.2

Proof The category $\mathcal{P}^{\mathsf{Gal}\,[\sigma]}$ is the category of algebras for the monad on $\overline{\mathcal{P}}/\mathcal{S}(S)$ described in proposition 4.6.1. We shall prove that $\mathsf{Split}_R(\sigma)$ is also monadic on $\overline{\mathcal{P}}/\mathcal{S}(S)$, for a monad which is isomorphic to that given by proposition 4.6.1. This will yield the expected result.

By corollary 5.1.21, the functor

$$\mathsf{Split}_R(\sigma) \longrightarrow \overline{\mathcal{P}}/\mathcal{S}(S), \quad (A,a) \mapsto \mathcal{S}_S \sigma^*(A,a)$$

is monadic; its left adjoint is $\Sigma_\sigma \circ \mathcal{C}_S$. Observe that the functorial part of the corresponding monad is thus

$$T \colon \overline{\mathcal{P}}/\mathcal{S}(S) \longrightarrow \overline{\mathcal{P}}/\mathcal{S}(S), \quad (X,\varphi) \mapsto \mathcal{S}_S \sigma^* \Sigma_\sigma \mathcal{C}_S(X,\varphi).$$

Let us show that this monad coincides with that given by proposition 4.6.1.

Putting $(A,a) = \mathcal{C}_S(X,\varphi)$, we must first consider pullbacks in $\mathcal{A}$ as in diagram 5.2, which show that

$$\big((S \times_R S) \times_S A, p_2 \circ a_1\big) = \sigma^* \Sigma_\sigma (A,a).$$

The objects $(S \times_R S, p_1)$ and (A,a) of $\overline{\mathcal{A}}/S$ are split by id_S, by lemmas 5.1.18 and 5.1.12 respectively. Thus by lemma 5.1.19, their product

$$(S \times_R S, p_1) \times (A,a) = \big((S \times_R S) \times_S A, p_1 \circ a\big)$$

in $\overline{\mathcal{A}}/S$ is split by id_S as well. By lemma 5.1.14, this product is transformed by $\mathcal{S}_S$ into a product in $\overline{\mathcal{P}}/\mathcal{S}(S)$. Since this product in $\overline{\mathcal{P}}/\mathcal{S}(S)$ corresponds to a pullback in $\overline{\mathcal{P}}$, we get diagram 5.3, where the square is a pullback in $\overline{\mathcal{P}}$. Notice that indeed $\mathcal{S}(A) \cong X$ because $\mathcal{S}_S(A,a) = \mathcal{S}_S \mathcal{C}_S(X,\varphi) \cong (X,\varphi)$. This diagram shows that

$$\mathcal{S}\big((S \times_R S) \times_S A\big) \longrightarrow \mathcal{S}(A) \cong X$$

$$\mathcal{S}(a_1) \Big\downarrow \qquad\qquad \Big\downarrow \mathcal{S}(a) = \varphi$$

$$\mathcal{S}(S \times_R S) \xrightarrow[\;\mathcal{S}(p_1)\;]{} \mathcal{S}(S)$$

$$\mathcal{S}(p_2) \Big\downarrow$$

$$\mathcal{S}(S)$$

Diagram 5.3

$$
\begin{aligned}
\mathcal{S}_S \sigma^* \Sigma_\sigma \mathcal{C}_S(X, \varphi) &\cong \mathcal{S}_S\big((S \times_R S) \times_S A, p_2 \circ a_1\big) \\
&\cong \Big(\mathcal{S}\big((S \times_R S) \times_S A\big), \mathcal{S}(p_2) \circ \mathcal{S}(a_1)\Big) \\
&\cong \big(\Sigma_{\mathcal{S}(p_2)} \circ \mathcal{S}(p_1)^*\big)(X, \varphi).
\end{aligned}
$$

This shows that the functor T of the present proof coincides on the objects with the functorial part of the monad given by proposition 4.6.1.

This is the difficult part of the proof. Completing the proof that the two monads we have indicated are isomorphic is now routine left to the reader.   □

5.2 Central extensions of groups

This section is devoted to applying the general Galois theory of Janelidze (see section 5.1) to recapture various results on central extensions of groups. All groups are written multiplicatively.

We shall apply the results of section 5.1 to the following adjunction, which will be shown to be relatively admissible:

$$(\mathsf{Gr}, \overline{\mathsf{Gr}}) \xleftarrow[\;\mathsf{ab}\;]{\;i\;} (\mathsf{Ab}, \overline{\mathsf{Ab}}), \qquad \mathsf{ab} \dashv i$$

where

- Gr is the category of groups,

- Ab is the category of abelian groups,
- i is the canonical inclusion,
- ab is the abelianization functor,
- $\overline{\text{Gr}} \subseteq \text{Gr}$ is the class of surjective homomorphisms,
- $\overline{\text{Ab}} \subseteq \text{Ab}$ is the class of surjective homomorphisms.

We first recall some elementary results on the abelianization of a group.

Definition 5.2.1 Let $H \subseteq G$ be a subgroup of a group G. The group of H-commutators of G is the subgroup of G generated by the elements of the form

$$xyx^{-1}y^{-1} \quad \text{with} \quad x \in H, \ y \in G,$$

which are called the "elementary commutators" This group of commutators is denoted by $[H, G]$.

Lemma 5.2.2 *When $H \subseteq G$ is a normal subgroup, the subgroup of commutators $[H, G] \subseteq G$ is normal as well. Moreover, in the quotient $G/[H, G]$,*

$$x \in H \quad \text{and} \quad y \in G \Rightarrow [x][y] = [y][x].$$

Proof An element of $[H, G]$ is a finite composite of elements of one of the two forms,

$$xyx^{-1}y^{-1} \quad \text{with} \quad x \in H, \ y \in G,$$

or the inverse

$$xyx^{-1}y^{-1} \quad \text{with} \quad x \in G, \ y \in H$$

of such an element. Observe that for all elements $x, y, z \in G$,

$$z(xyx^{-1}y^{-1})z^{-1} = (zxz^{-1})(zyz^{-1})(zxz^{-1})^{-1}(zyz^{-1})^{-1},$$

which suffices to prove that $[H, G]$ is stable for the operation $u \mapsto zuz^{-1}$. Thus $[H, G]$ is a normal subgroup.

Finally given $x \in H$ and $y \in G$, one has $[x][y][x]^{-1}[y]^{-1} = [1]$ in $G/[H, G]$, from which we obtain the last part of the statement. $\square$

Proposition 5.2.3 *The left adjoint functor of the inclusion functor $i \colon \text{Ab} \longrightarrow \text{Gr}$ is given by*

$$\text{Ab} \colon \text{Gr} \longrightarrow \text{Ab}, \quad G \mapsto G/[G, G].$$

Proof Every group G is normal in itself, from which $G/[G,G]$ is an abelian group by lemma 5.2.2. This yields a surjective group homomorphism

$$\eta_G\colon G \longrightarrow G/[G,G], \quad x \mapsto [x].$$

If $f\colon G \longrightarrow A$ is a group homomorphism, with A an abelian group, for all elements $x, y \in G$

$$f\left(xyx^{-1}y^{-1}\right) = f(x)f(y)f(x)^{-1}f(y)^{-1} = 1$$

since A is abelian. Thus f factors through the quotient $G/[G,G]$ and this factorization is unique, since η_G is surjective.   □

Proposition 5.2.4 *The adjunction* ab ⊣ i: $(\mathsf{Gr}, \overline{\mathsf{Gr}}) \xleftrightarrow{\quad} (\mathsf{Ab}, \overline{\mathsf{Ab}})$, *with* $\overline{\mathsf{Gr}}$ *and* $\overline{\mathsf{Ab}}$ *the classes of surjective homomorphisms, is relatively admissible in the sense of definition 5.1.3.*

Proof In Ab or Gr, a composite or a pullback of surjections is again a surjection. Obviously, the canonical inclusion $i\colon$ Ab $\longrightarrow$ Gr preserves surjections and the unit of the adjunction $\eta_G\colon G \longrightarrow G/[G,G]$ is surjective at each object $G \in$ Gr.

Since i is full and faithful, the counit of the adjunction is an isomorphism, thus certainly a surjection at each object $A \in$ Ab. More directly, when A is abelian, $[A, A] = \{1\}$ and thus $A/[A, A] \cong A$.

It remains to prove that the functor ab respects surjections. Given a group homomorphism $f\colon G_1 \longrightarrow G_2$, the commutative diagram

$$
\begin{array}{ccc}
G_1 & \xrightarrow{\eta_{G_1}} & G_1/[G_1, G_1] \\
{\scriptstyle f}\downarrow & & \downarrow{\scriptstyle \mathsf{ab}(f)} \\
G_2 & \xrightarrow[\eta_{G_2}]{} & G_2/[G_2, G_2]
\end{array}
$$

shows at once that when f is surjective, $\mathsf{ab}(f)$ is surjective as well.   □

Proposition 5.2.5 *For every group G, the counit of the adjunction*

$$\overline{\mathsf{Gr}}/G \xleftrightarrow[\mathsf{ab}_G]{i_G} \overline{\mathsf{Ab}}/\mathsf{ab}(G)$$

is an isomorphism.

Proof Let us consider a surjection $\varphi\colon A \longrightarrow \mathrm{ab}(G)$ in the category of abelian groups and the corresponding pullback

$$
\begin{array}{ccc}
A' & \xrightarrow{\ \varphi''\ } & A \\
{\scriptstyle\varphi'}\big\downarrow & & \big\downarrow{\scriptstyle\varphi} \\
G & \xrightarrow[\ \eta_G\ ]{} & \mathrm{ab}(G)
\end{array}
$$

We thus have $i_G(A, \varphi) = (A', \varphi')$ and it remains to prove that

$$
\left(\mathrm{ab}(A') \xrightarrow{\ \mathrm{ab}(\varphi')\ } \mathrm{ab}(G)\right) \cong \left(A \xrightarrow{\ \varphi\ } \mathrm{ab}(G)\right).
$$

Observe that since η_G and φ are surjective, so are φ'' and φ'. Considering the factorization φ''' obtained from proposition 5.2.4

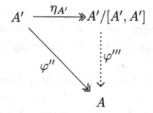

we deduce that φ''' is surjective as well.

In fact, φ''' is also injective. To prove this, consider $a' \in A'$ and the corresponding element $[a'] \in A'/[A', A']$; assume that $\varphi'''([a']) = 1$. The element $a' \in A'$ is thus a pair $a' = (x, a)$ with $x \in G$, $a \in A$ and $\varphi(a) = [x] \in G/[G, G]$. This yields

$$
a = \varphi''(x, a) = (\varphi''' \circ \eta_{A'})(x, a) = \varphi'''([a']) = 1.
$$

We must prove that $a' = (x, a) = (x, 1) \in [A', A']$. Since

$$
\eta_G(x) = (\eta_G \circ \varphi')(x, a) = (\varphi \circ \varphi'')(x, a) = \varphi(1) = 1,
$$

one has $x \in [G, G]$. By definition 5.2.1, this implies

$$
x = x_1 y_1 x_1^{-1} y_1^{-1} \ldots x_n y_n x_n^{-1} y_n^{-1}
$$

with $x_i, y_i \in G$. Since φ is surjective, for each index $i = 1, \ldots, n$ let us choose

$$
a_i \in A \ \text{ with } \ \varphi(a_i) = [x_i], \quad b_i \in A \ \text{ with } \ \varphi(b_i) = [y_i]
$$

so that all (x_i, a_i) and (y_i, b_i) are in A'. Since the group A is abelian

$$a_1 b_1 a_1^{-1} b_1^{-1} \cdots a_n b_n a_n^{-1} b_n^{-1} = 1.$$

It follows that

$$(x, 1) = (x_1, a_1)(y_1, b_1)(x_1^{-1}, a_1^{-1})(y_1^{-1}, b_1^{-1}) \cdots$$
$$\cdots (x_n, a_n)(y_n, b_n)(x_n^{-1}, a_n^{-1})(y_n^{-1}, b_n^{-1})$$

and this last composite is in $[A', A']$ by definition 5.2.1. Thus $a' = (x, 1) \in [A', A']$. This proves that φ''' is injective, and thus an isomorphism.

It remains to notice that $\varphi \circ \varphi''' = \mathsf{ab}(\varphi')$. But

$$\varphi \circ \varphi''' \circ \eta_{A'} = \varphi \circ \varphi'' = \eta_G \circ \varphi' = \mathsf{ab}(\varphi') \circ \eta_{A'}$$

by naturality of η. Thus $\varphi \circ \varphi''' = \mathsf{ab}(\varphi')$ since $\eta_{A'}$ is surjective. $\quad\square$

Proposition 5.2.6 *Every surjection in* Gr *is an effective descent morphism relatively to the class of surjective homomorphisms.*

Proof Let $\sigma\colon S \longrightarrow R$ be a surjective homomorphism of groups. We shall apply the Beck criterion (see [66]).

The functors σ^* and Σ_σ restrict to functors

$$\overline{\mathsf{Gr}} \underset{\sigma^*}{\overset{\Sigma_\sigma}{\rightleftarrows}} \overline{\mathsf{Gr}}/S$$

from which σ^* has a left adjoint.

This functor σ^* reflects isomorphisms since, by lemma 4.4.6, this is the case for the functor

$$\sigma^*\colon \mathsf{Gr}/R \longrightarrow \mathsf{Gr}/S.$$

Finally the categories $\overline{\mathsf{Gr}}/R$ and $\overline{\mathsf{Gr}}/S$ have coequalizers computed as in Gr/R and Gr/S, that is, finally, computed as in Gr. Thus the Beck condition on split coequalizers is valid for $\sigma^*\colon \overline{\mathsf{Gr}}/R \longrightarrow \overline{\mathsf{Gr}}/S$ since it is valid for $\sigma^*\colon \mathsf{Gr}/R \longrightarrow \mathsf{Gr}/S$. $\quad\square$

Lemma 5.2.7 *Let* $\sigma\colon S \longrightarrow R$ *be a surjective homomorphism of groups. Consider the relatively admissible adjunction*

$$(\mathsf{Gr}, \overline{\mathsf{Gr}}) \underset{\mathsf{ab}}{\overset{i}{\rightleftarrows}} (\mathsf{Ab}, \overline{\mathsf{Ab}}).$$

Given $(H, h) \in \overline{\mathsf{Gr}}/R$, *we consider the following pullback in* Gr:

$$S \times_R H \xrightarrow{\ p_2\ } H$$

$$\sigma^*(H,h) = (S \times_R H, p_1)$$

$$S \xrightarrow{\ \sigma\ } R$$

(with vertical maps p_1 and h)

The following conditions are then equivalent:

(i) $(H,h) \in \overline{\mathrm{Gr}}/R$ *is split by* σ;

(ii) *the following diagram is a pullback –*

$$S \times_R H \xrightarrow{\ \eta_{S \times_R H}\ } \mathbf{ab}(S \times_R H)$$

(with vertical maps p_1 and $\mathbf{ab}(p_1)$)

$$S \xrightarrow{\ \eta_S\ } \mathbf{ab}(S)$$

(iii) $p_1 \colon [S \times_R H, S \times_R H] \longrightarrow [S,S]$ *is an isomorphism;*

(iv) $p_1 \colon [S \times_R H, S \times_R H] \longrightarrow [S,S]$ *is injective.*

Proof Let us first compute explicitly condition (i). One considers the pullback

$$H' \longrightarrow \mathbf{ab}(S \times_R H)$$

$$(H',h') = i_S \mathbf{ab}_S \sigma^*(H,h)$$

(with vertical maps h' and $\mathbf{ab}(p_1)$)

$$S \xrightarrow{\ \eta_S\ } \mathbf{ab}(S)$$

Condition (i) means $(H',h') \cong (S \times_R H, p_1)$, that is condition (ii).
  Let us next prove that the map

$$p_1 \colon [S \times_R H, S \times_R H] \longrightarrow [S,S]$$

is always surjective, which will prove (iii) $\Leftrightarrow$ (iv). Notice that obviously, the image of an element of the form $xyx^{-1}y^{-1}$ is again an element of that form, proving that p_1 indeed restricts to the groups of commutators. Let us now choose $s,t \in S$ and prove that $sts^{-1}t^{-1}$ has the form $p_1(u)$ for some commutator u; by definition 5.2.1, this is enough to prove the required surjectivity. By surjectivity of h, we choose $x,y \in H$ such that

$h(x) = \sigma(s)$, $h(y) = \sigma(t)$; this yields $(s, x) \in S \times_R H$ and $(t, y) \in S \times_R H$.
Thus

$$(s, x)(t, y)(s^{-1}, x^{-1})(t^{-1}, y^{-1}) \in [S \times_R H, S \times_R H]$$

that is

$$(sts^{-1}t^{-1}, xyx^{-1}y^{-1}) \in [S \times_R H, S \times_R H].$$

This element is of course mapped by p_1 onto $sts^{-1}t^{-1}$.

It remains to prove (ii) $\Leftrightarrow$ (iv). For this consider the diagram

$$
\begin{array}{ccccccccc}
0 & \longrightarrow & [S \times_R H, S \times_R H] & \rightarrowtail & S \times_R H & \xrightarrow{\eta_{S \times_R H}} & \mathrm{ab}(S \times_R H) & \longrightarrow & 0 \\
& & \downarrow{\scriptstyle p_1} & & \downarrow{\scriptstyle p_1} & & \downarrow{\scriptstyle \mathrm{ab}(p_1)} & & \\
0 & \longrightarrow & [S, S] & \rightarrowtail & S & \xrightarrow{\eta_S} & \mathrm{ab}(S) & \longrightarrow & 0
\end{array}
$$

Since h is surjective, p_1 is surjective as well. Since $\mathrm{ab} \dashv i$, ab preserves epimorphisms and $\mathrm{ab}(p_1)$ is also surjective. The three vertical morphisms in the above diagram are thus surjections.

Suppose first condition (ii), that is, the right hand square is a pullback. We must prove that

$$\forall u \in [S \times_R H, S \times_R H] \quad p_1(u) = 1 \Rightarrow u = 1.$$

We have

$$u = (s, x), \quad s \in S, \quad x \in H, \quad \sigma(s) = h(x).$$

From $p_1(u) = 1$ we get $s = 1$ and from $u \in [S \times_R H, S \times_R H]$ we deduce $\eta_{S \times_R H}(u) = 1$. Since the right hand square is a pullback, we have thus $u = (1, 1)$, which proves the injectivity of p_1 on the groups of commutators.

Conversely assume (iii), or equivalently (iv). Consider diagram 5.4, which is commutative and where the square is a pullback. We must prove that the factorization θ is an isomorphism.

To prove the injectivity of θ, consider $(s, x) \in S \times_R H$, thus $s \in S$, $x \in H$ with $\sigma(s) = h(x)$. Suppose $\theta(s, x) = 1$: we must deduce $(s, x) = (1, 1)$. One immediately gets

$$s = p_1(s, x) = (\pi_1 \circ \theta)(s, x) = \pi_1(1) = 1,$$

$$\eta_{S \times_R H}(s, x) = (\pi_2 \circ \theta)(s, x) = \pi_2(1) = 1.$$

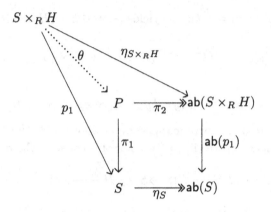

Diagram 5.4

This proves $(s,x) \in [S \times_R H, S \times_R H]$ with $p_1(s,x) = 1$. Thus $(s,x) = (1,1)$ by condition (iv).

To prove the surjectivity of θ, choose $(s,[t,x]) \in P$, thus $s \in S$, $t \in S$, $x \in H$ with $\sigma(t) = h(x)$ and $[s] = [t]$. This yields $[t^{-1}s] = [1]$, that is $t^{-1}s \in [S,S]$. By condition (iii), there exists a unique $y \in H$ such that $(t^{-1}s, y) \in [S \times_R H, S \times_R H]$. In particular, $\sigma(t^{-1}s) = h(y)$ and $[t^{-1}s, y] = [1,1]$. Therefore

$$[s, xy] = [tt^{-1}s, xy] = [t,x][t^{-1}s, y] = [t,x][1,1] = [t,x].$$

Finally,

$$\pi_1\theta(s, xy) = p_1(s, xy) = s, \quad \pi_2\theta(s, xy) = [s, xy] = [t,x],$$

that is $\theta(s, xy) = (s, [t,x])$ and θ is surjective. $\square$

Definition 5.2.8

(i) An extension of a group R is an exact sequence of groups

$$0 \longrightarrow K \overset{k}{\rightarrowtail} S \overset{\sigma}{\twoheadrightarrow} R \longrightarrow 0.$$

(ii) This extension of the group R is central when K is contained in the centre $Z(S)$ of S, where as usual

$$Z(S) = \{x \in S \mid \forall y \in S \ xy = ys\}.$$

(iii) This central extension of the group R is weakly universal when, for every other central extension (k', σ') of R, there exists a factorization φ making the following diagram commutative:

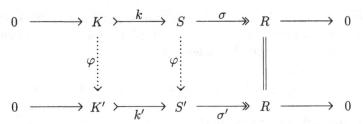

(iv) The central extension of the group R is universal when it is weakly universal and, in the previous condition, the factorization φ is unique.

Proposition 5.2.9 *Consider a weakly universal central extension of the group R;*

$$0 \longrightarrow K \overset{k}{\rightarrowtail} S \overset{\sigma}{\twoheadrightarrow} R \longrightarrow 0.$$

For a surjective homomorphism $h\colon H \twoheadrightarrow R$ of groups, the following conditions are equivalent:

(i) *the object $(H, h) \in \overline{\mathsf{Gr}}/R$ is split by σ;*
(ii) *($\mathsf{Ker}\, h, h$) is a central extension of R.*

Proof (i) $\Rightarrow$ (ii) is a more general fact, which does not require any centrality assumption on σ. We must prove that $\mathsf{Ker}\, h \subseteq Z(H)$, that is

$$\forall x \in H \ \ \forall y \in H \ \ \ h(x) = 1 \Rightarrow xy = yx.$$

Since σ is surjective, choose $s \in S$ such that $\sigma(s) = h(y)$, from which $(s, y) \in S \times_R H$. One also has $h(x) = 1 = \sigma(1)$, thus $(1, x) \in S \times_R H$. Therefore

$$(s, y)(1, x)(s^{-1}, y^{-1})(1, x^{-1}) = (1, yxy^{-1}x^{-1}) \in [S \times_R H, S \times_R H].$$

Condition (iv) of lemma 5.2.7 yields $yxy^{-1}x^{-1} = 1$, that is $yx = xy$.

Let us prove now (ii) $\Rightarrow$ (i). We use condition (iv) of lemma 5.2.7. Since the central extension (k, σ) is weakly universal, we get a factorization φ making commutative the following diagram, where i denotes the canonical inclusion:

$$
\begin{array}{ccccccccc}
0 & \longrightarrow & K & \overset{k}{\rightarrowtail} & S & \overset{\sigma}{\twoheadrightarrow} & R & \longrightarrow & 0 \\
& & \varphi\downarrow & & \varphi\downarrow & & \| & & \\
0 & \longrightarrow & \mathsf{Ker}\, h & \overset{}{\underset{i}{\rightarrowtail}} & H & \overset{}{\underset{h}{\twoheadrightarrow}} & R & \longrightarrow & 0
\end{array}
$$

Given $s \in S$, one has $h\varphi(s) = \sigma(s)$, thus $(s, \varphi(s)) \in S \times_R H$. This yields a group homomorphism $S \longrightarrow S \times_R H$, which thus restricts through the corresponding groups of commutators

$$\begin{pmatrix} \text{id} \\ \varphi \end{pmatrix} : [S, S] \longrightarrow [S \times_R H, S \times_R H], \quad s \mapsto (s, \varphi(s)).$$

We consider now the composite

$$[S \times_R H, S \times_R H] \xrightarrow{\;p_1\;} [S, S] \xrightarrow{\;\begin{pmatrix} \text{id} \\ \varphi \end{pmatrix}\;} [S \times_R H, S \times_R H];$$

proving it is the identity will in particular imply that p_1 is injective as required. It clearly suffices to prove that the composite is the identity on each elementary commutator. Consider thus (s, x) and (t, y) in $S \times_R H$:

$$s, t \in S, \quad x, y \in H, \quad \sigma(s) = h(x), \quad \sigma(t) = h(y).$$

One gets

$$\left(\begin{pmatrix} \text{id} \\ \varphi \end{pmatrix} \circ p_1 \right) \left((s, x)(t, y)(s^{-1}, x^{-1})(t^{-1}, y^{-1}) \right)$$

$$= \left(\begin{pmatrix} \text{id} \\ \varphi \end{pmatrix} \circ p_1 \right) (sts^{-1}t^{-1}, xyx^{-1}y^{-1})$$

$$= \begin{pmatrix} \text{id} \\ \varphi \end{pmatrix} (sts^{-1}t^{-1})$$

$$= \left(sts^{-1}t^{-1}, \varphi(sts^{-1}t^{-1}) \right)$$

$$= \left(sts^{-1}t^{-1}, \varphi(s)\varphi(t)\varphi(s)^{-1}\varphi(t)^{-1} \right).$$

It suffices now to prove that

$$\varphi(s)\varphi(t)\varphi(s)^{-1}\varphi(t)^{-1} = xyx^{-1}y^{-1}.$$

The equalities $h(x) = \sigma(s) = (h \circ \varphi)(s)$ imply $h(\varphi(s) \cdot x^{-1}) = 1$, thus $\varphi(s) \cdot x^{-1} \in \operatorname{Ker} h$. Putting $u = \varphi(s) \cdot x^{-1} \in \operatorname{Ker} h$, we have thus $\varphi(s) = ux$ with $u \in \operatorname{Ker} h$. In the same way, $\varphi(t) = vy$ with $v \in \operatorname{Ker} h$. By assumption, $\operatorname{Ker} h \subseteq Z(H)$; therefore

$$\varphi(s)\varphi(t)\varphi(s)^{-1}\varphi(t)^{-1} = ux \cdot uy \cdot x^{-1}u^{-1} \cdot y^{-1}v^{-1} = xyx^{-1}y^{-1}.$$

This completes the proof that $\begin{pmatrix} \text{id} \\ \varphi \end{pmatrix} \circ p_1$ maps every elementary commutator onto itself.   □

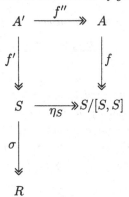

Diagram 5.5

Proposition 5.2.10 *With the notation of 5.2.8, if (k, σ) is a weakly universal central extension of groups, then the morphism $\sigma \colon S \longrightarrow\!\!\!\!\!\longrightarrow R$ is of Galois descent with respect to the relatively admissible adjunction*

$$(\mathsf{Gr}, \overline{\mathsf{Gr}}) \xrightarrow[\text{ab}]{\overset{i}{\longleftarrow}} (\mathsf{Ab}, \overline{\mathsf{Ab}}).$$

Proof By propositions 5.2.4, 5.2.5, 5.2.6, it remains to check condition (iii) of definition 5.1.8. For this consider a surjective homomorphism $f \colon A \longrightarrow\!\!\!\!\!\longrightarrow S/[S, S]$ in the category of abelian groups. The object $(\Sigma_\sigma \circ i_S)(A, f)$ is the left vertical composite in diagram 5.5, where the square is a pullback: We must prove that this composite is split by σ, that is, by proposition 5.2.9, that $(\sigma \circ f', \mathsf{Ker}\,(\sigma \circ f'))$ is a central extension. Observe that the surjectivitiy of f implies that of f'. It remains thus to check that the kernel of $\sigma \circ f'$ is contained in the centre of A'.

Let $a \in A'$ such that $(\sigma \circ f')(a) = 1$. Thus

$$a' = (s, a), \quad s \in S, \quad a \in A, \quad [s] = f(a), \quad \sigma(s) = 1.$$

Consider an arbitrary element $a'' \in A'$, that is

$$a'' = (t, b), \quad t \in S, \quad b \in A, \quad [t] = f(b).$$

We must prove $a'a'' = a''a'$, that is $(st, ab) = (ts, ba)$. Since A is abelian, this reduces to proving $st = ts$. But $s \in \mathsf{Ker}\,\sigma$ by choice of a' and $\mathsf{Ker}\,\sigma \subseteq Z(S)$ by centrality; thus $st = ts$. $\qquad\square$

Theorem 5.2.11 *Given a weakly universal central extension of groups*

$$0 \longrightarrow K \rightarrowtail^{k} S \longrightarrow^{\sigma} \twoheadrightarrow R \longrightarrow 0$$

(i) *the kernel pair of σ, seen as an internal groupoid in the category of groups, is transformed by the functor* Ab: Gr $\longrightarrow$ Ab *into an internal groupoid* Gal $[\sigma]$ *in the category of abelian groups,*

(ii) *the category of central extensions of the group R is equivalent to the category of internal covariant presheaves (P, p, π), with p surjective, on the internal groupoid* Gal $[\sigma]$, *in the category of abelian groups.*

Proof The category of extensions of the group R has for morphisms the morphisms φ as in condition (iii) of definition 5.2.8. With the notation of definition 5.2.8, k is entirely determined by σ, since $k = \text{Ker}\,\sigma$. Thus the category of extensions of R is equivalent to the category $\overline{\text{Gr}}/R$. The result then follows at once from theorem 5.1.24 and proposition 5.2.9.

$\square$

Definition 5.2.12 A group G is perfect when $G = [G, G]$.

A group is thus perfect when it is "perfectly non-abelian": indeed, the abelian group $G/[G, G]$ associated with it is the trivial group.

Let us mention now a standard property of groups, expressed here in more categorical terms. We recall that in a category with binary products and terminal object 1, an internal group consists in giving an object G and three morphisms

$$\mu: G \times G \longrightarrow G,$$
$$\upsilon: 1 \longrightarrow G,$$
$$\iota: G \longrightarrow G$$

called the "multiplication" μ, the "unit" υ and the "inverse" ι, which satisfy diagrammatically the usual group axioms (see [8], volume 2). In other words, an internal group is an internal groupoid whose object of objects is the terminal object.

Lemma 5.2.13 *Consider a group G, written multiplicatively. Suppose G is provided with the structure of an internal group in the category of groups:*

$$\mu: G \times G \longrightarrow G,$$

$$v: 1 \longrightarrow G,$$
$$\iota: G \longrightarrow G$$

with thus μ the multiplication, v the unit and ι the inverse. In these conditions, the two group structures on G coincide and are commutative.

Proof Since ι is a group homomorphism for the original group structure on G, the two units coincide and are thus written as 1. Since μ is a group homomorphism for the original group structure on G, for all elements $a, b, c, d \in G$ one has

$$\mu(a \cdot c, b \cdot d) = \mu\big((a, b) \cdot (c, d)\big) = \mu(a, b) \cdot \mu(c, d).$$

Putting $b = 1 = c$ yields then $\mu(a, d) = a \cdot d$ while putting $a = 1 = d$ yields $\mu(c, b) = b \cdot c$. □

Proposition 5.2.14 *Consider a weakly universal central extension of groups*

$$0 \longrightarrow K \overset{k}{\rightarrowtail} S \overset{\sigma}{\twoheadrightarrow} R \longrightarrow 0.$$

When S is a perfect group

(i) *R is a perfect group,*

(ii) *the Galois groupoid $\mathsf{Gal}\,[\sigma]$ of theorem 5.2.11 is an abelian group isomorphic to K,*

(iii) *the central extension $(\mathrm{Ker}\,\sigma, \sigma)$ is universal.*

Proof Since σ is surjective, every element $r \in R$ can be written $r = \sigma(s)$ with $s \in S$. Since S is perfect, s can be written as a product of elements of the form $xyx^{-1}y^{-1}$. Thus r is a product of elements of the form $\sigma(x)\sigma(y)\sigma(x^{-1})\sigma(y^{-1})$, proving $r \in [R, R]$. Therefore $R = [R, R]$ and R is perfect.

Again since S is perfect, the "object of objects" of the groupoid $\mathsf{Gal}\,[\sigma]$ is the abelian group $\mathrm{ab}(S) = S/[S, S] \cong \{1\}$. Therefore the groupoid $\mathsf{Gal}\,[\sigma]$ is an internal group in Ab, thus by lemma 5.2.13, is an abelian group whose structure coincides with the internal group structure of $\mathsf{Gal}\,[\sigma]$.

By the condition (ii) of lemma 5.2.7, the diagram

$$S \times_R S \xrightarrow{\eta_{S \times_R S}} \mathrm{ab}(S \times_R S) = \mathrm{Gal}\,[\sigma]$$

$$\begin{array}{ccc} & p_1 \downarrow & \mathrm{ab}(p_1) \downarrow \\ S & \xrightarrow{\quad \eta_S \quad} & \mathrm{ab}(S) = \{1\} \end{array}$$

is a pullback, and so $\mathrm{Gal}\,[\sigma] \cong \mathrm{Ker}\,p_1 \cong \mathrm{Ker}\,\sigma$.

It remains to prove the last assertion of the proposition. Consider for this the situation of condition (iii) in definition 5.2.8; we must prove the uniqueness of φ. Let ψ be another factorization. By assumption, $S = [S, S]$ and, via the proofs of proposition 5.2.9 and lemma 5.2.7, we have reciprocal isomorphisms

$$[S \times_R S, S \times_R S] \underset{p_1}{\overset{\binom{\mathrm{id}}{\varphi}}{\rightleftarrows}} [S, S] = S.$$

The same observation applies to ψ, yielding finally

$$\binom{\mathrm{id}}{\varphi} = p_1^{-1} = \binom{\mathrm{id}}{\psi} : S = [S, S] \longrightarrow [S \times_R S, S \times_R S] \subseteq S \times_R S.$$

Composing with the second projection of the pullback yields

$$\varphi = p_2 \circ \binom{\mathrm{id}}{\varphi} = p_2 \circ \binom{\mathrm{id}}{\psi} = \psi. \qquad \square$$

We now need an existence theorem.

Proposition 5.2.15 *Every group has a weakly universal central extension.*

Proof Every group R is a quotient of a free group F (see [61]), which yields an exact sequence

$$0 \longrightarrow K \overset{k}{\rightarrowtail} F \overset{p}{\twoheadrightarrow} R \longrightarrow 0.$$

The group K itself is free, as subgroup of a free group (see [61] again). This last fact will not be needed in the proof. A situation as just exhibited is called a "free resolution of the group" R.

The subgroup K, as a kernel, is a normal subgroup of F. By lemma 5.2.2, $[K, F]$ too is a normal subgroup of F. For every $u \in K = \mathrm{Ker}\,p$

and $v \in F$,

$$p(uvu^{-1}v^{-1}) = p(u)p(v)p(u)^{-1}p(v)^{-1} = 1p(v)1p(v)^{-1} = 1,$$

proving that $[K, F] \subseteq \operatorname{Ker} p$. Therefore we get a factorization σ

$$
\begin{array}{ccccccccc}
0 & \longrightarrow & K & \overset{k}{\rightarrowtail} & F & \overset{p}{\twoheadrightarrow} & R & \longrightarrow & 0 \\
& & & & \Big\downarrow{}^{q} & \diagdown{}^{\sigma} & & & \\
& & & & S = F/[K, F] & & & &
\end{array}
$$

which is still surjective.

Let us prove now that the extension

$$0 \longrightarrow \operatorname{Ker}\sigma \rightarrowtail S \overset{\sigma}{\twoheadrightarrow} R \longrightarrow 0$$

is central. Choose for this $u, v \in F$ and suppose $\sigma([u]) = 1$, that is, $p(u) = 1$. We must prove

$$[u][v] = [v][u] \in F/[K, F].$$

But

$$
\begin{aligned}
u \in \operatorname{Ker} p = K & \Rightarrow uvu^{-1}v^{-1} \in [K, F] \\
& \Rightarrow [u][v][u]^{-1}[v]^{-1} = 1 \\
& \Rightarrow [u][v] = [v][u],
\end{aligned}
$$

which proves the centrality of the extension.

Let us see next that this central extension is weakly universal. Let us choose another central extension

$$0 \longrightarrow \operatorname{Ker} h \rightarrowtail H \overset{h}{\twoheadrightarrow} R \longrightarrow 0$$

of R. We consider diagram 5.6, where $\sigma \circ q \circ k = p \circ k = 0$ induces the existence of the factorization q'. Since F is a free group, it is projective and p factors as $p = h \circ \psi$ through the epimorphism h. Thus $h \circ \psi \circ k = p \circ k = 0$ and we get a factorization ψ' of $\psi \circ k$ through the kernel of h. Given an elementary commutator $uvu^{-1}v^{-1} \in [K, F]$,

$$\psi(uvu^{-1}v^{-1}) = \psi(u)\psi(v)\psi(u^{-1})\psi(v^{-1}) = \psi(u)\psi(u^{-1})\psi(v)\psi(v^{-1}) = 1$$

since $v \in K$ implies $\psi(v) = \psi'(v) \in \operatorname{Ker} h \subseteq Z(H)$, by centrality of the bottom extension. This implies that ψ factors as $\psi = \varphi \circ q$ through the quotient q and, since q is surjective, this yields $h \circ \varphi = \sigma$. Consequently we get the further factorization φ'. $\qquad\square$

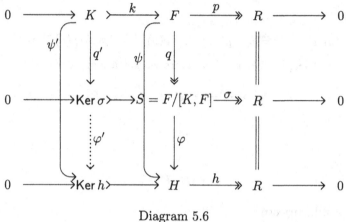

Diagram 5.6

Proposition 5.2.16 *Every perfect group has a universal central extension.*

Proof Let us use freely the notation of the proof of proposition 5.2.15. The surjection σ restricts at once to a surjection $\sigma\colon [S,S] \longrightarrow\!\!\!\!\!\!\rightarrow [R,R] = R$, where $[R,R] = R$ since R is perfect. This yields the diagram of groups

$$
\begin{array}{ccccccccc}
0 & \longrightarrow & L & \overset{l}{\rightarrowtail} & [S,S] & \overset{\sigma}{\longrightarrow\!\!\!\!\rightarrow} & R & \longrightarrow & 0 \\
 & & \downarrow & & \downarrow & & \| & & \\
0 & \longrightarrow & K & \overset{}{\underset{k}{\rightarrowtail}} & S & \overset{}{\underset{\sigma}{\longrightarrow\!\!\!\!\rightarrow}} & R & \longrightarrow & 0
\end{array}
$$

where both lines are exact sequences. The top line is obviously a weakly universal central extension, since so is the bottom line (see definition 5.2.8). To conclude the proof by proposition 5.2.14, it remains to prove that $[S,S]$ is perfect.

For this we observe first that, trivially by definition 5.2.1,

$$[S,S] = \left[\frac{F}{[K,F]}, \frac{F}{[K,F]} \right] = \frac{[F,F]}{[K,F]}.$$

Next, since R is perfect, F is "perfect up to K", i.e. every element $x \in F$ can be written as $x = ck$ with $c \in [F,F]$ and $k \in K$. Indeed, since R is perfect, write

$$[x] = [x_1][y_1][x_1]^{-1}[y_1]^{-1} \cdots [x_n][y_n][x_n]^{-1}[y_n]^{-1}$$

in $F/K \cong R$. This implies

$$x = x_1 y_1 x_1^{-1} y_1^{-1} \cdots x_n y_n x_n^{-1} y_n^{-1} k$$

for some $k \in K$.

Now any $t \in T = [S,S] = [F,F]/[K,F] \subseteq F/[K,F]$ can be written as a product of elements of the form $[xyx^{-1}y^{-1}] = [x][y][x^{-1}][y^{-1}]$ in $F/[K,F]$. Applying the previous observation to each of these elements, we write $x = ck$ and $y = dl$ for some $c,d \in [F,F]$ and $k,l \in K$. Since for every $z \in F$ and every $m \in K$, we have $[z][m] = [m][z]$ in $F/[K,F]$,

$$
\begin{aligned}
[xyx^{-1}y^{-1}] &= [c][k][d][l][k]^{-1}[c]^{-1}[l]^{-1}[d]^{-1} \\
&= [c][d][c]^{-1}[d]^{-1}[k][l][k]^{-1}[l]^{-1} \\
&= [cdc^{-1}d^{-1}][klk^{-1}l^{-1}] \\
&= [cdc^{-1}d^{-1}] \\
&\in [T,T]
\end{aligned}
$$

because $[klk^{-1}l^{-1}] = 1$ in $F/[K,F]$. This holds for each pair x,y, thus finally $t \in [T,T]$, proving that $T = [S,S]$ is perfect. $\qquad\square$

Let us conclude this section with a lemma providing a link with a famous Hopf formula.

Lemma 5.2.17 *In the conditions and with the notation of proposition 5.2.16, the following isomorphism holds:*

$$\operatorname{Ker}\sigma \cong \frac{K \cap [F,F]}{[K,F]}.$$

Proof To construct an isomorphism

$$\beta\colon \frac{K \cap [F,F]}{[K,F]} \longrightarrow \operatorname{Ker}\sigma,$$

we consider diagram 5.7. The morphisms i, u, k, t, v, w are canonical inclusions, thus square (1) is commutative. The commutativity of square (2) holds by definition of σ in the proof of 5.2.16. Moreover the bottom sequence is exact, thus in particular $p \circ k = 1$. Therefore $\sigma q u = p k t = 1$, which implies that qu factors through the kernel w of σ; this yields α such that $w\alpha = qu$. But then $w\alpha v = quv = 1$, proving $\alpha v = 1$ since w is injective. This yields the expected factorization β through the cokernel s of v, completing at the same time the commutative diagram 5.7.

To prove the injectivity of β, choose $a \in K \cap [F,F]$ such that $\beta([a]) =$

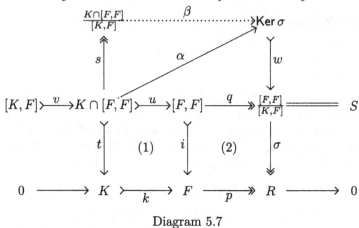

Diagram 5.7

1. This yields

$$qu(a) = w\alpha(a) = w\beta s(a) = w(1) = 1$$

thus $a = u(a) \in [K, F]$ and $[a] = 1$.

To prove the surjectivity of β, let $b \in \mathsf{Ker}\,\sigma$. There is thus an element $a \in [F, F]$ such that $q(a) = b$ and $p(a) = \sigma(b) = 1$. Thus $a \in \mathsf{Ker}\,p = K$ and finally $a \in K \cap [F, F]$. Therefore we can write $b = \alpha(a) = \beta\big(s(a)\big)$. $\quad\square$

The formula of lemma 5.2.17 is precisely the Hopf formula defining $H_2(R, \mathbb{Z})$, starting from a free resolution of R (see [69]). By proposition 5.2.14, in the case where S is a perfect group, the group $H_2(R, \mathbb{Z})$ is thus exactly the Galois group $\mathsf{Gal}\,[\sigma]$.

5.3 Factorization systems

We first review some classical facts about factorization systems; for a more detailed treatment, see [20] or [8], volume 1.

Definition 5.3.1 A factorization system on a category $\mathcal{C}$ consists in two classes $\mathcal{E}$, $\mathcal{M}$ of arrows of $\mathcal{C}$, such that

   (i) every isomorphism is in both $\mathcal{E}$ and $\mathcal{M}$,

   (ii) $\mathcal{E}$ and $\mathcal{M}$ are closed under composition,

   (iii) every morphism f of $\mathcal{C}$ factors as $f = m \circ e$ with $e \in \mathcal{E}$ and $m \in \mathcal{M}$,

(iv) in a commutative square where $e \in \mathcal{E}$ and $m \in \mathcal{M}$, there exists a unique diagonal arrow making both triangles commutative:

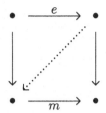

It is important to emphasize the fact that in definition 5.3.1, it is not required at all that the morphisms in $\mathcal{E}$ be epimorphisms or the morphisms in $\mathcal{M}$, monomorphisms.

It is also useful to observe that definition 5.3.1 is selfdual.

Example 5.3.2 In a regular category $\mathcal{C}$, one gets a factorization system $(\mathcal{E}, \mathcal{M})$ by choosing

$$
\begin{array}{lll}
e \in \mathcal{E} & \text{iff} & e \text{ is a regular epimorphism,} \\
m \in \mathcal{M} & \text{iff} & m \text{ is a monomorphism.}
\end{array}
$$

Proof Let us recall that a regular category has enough finite limits (let us say, for short, pullbacks) and coequalizers of kernel pairs; moreover pulling back a regular epimorphism along an arbitrary morphism again yields a regular epimorphism. In such a category, it is well known that conditions (i), (ii), (iii) of 5.3.1 are satisfied by regular epimorphisms and arbitrary monomorphisms (see [4] or [8], volume 2).

To check condition (iv) of 5.3.1, consider diagram 5.8 where e is a regular epimorphism, m is a monomorphism and $m \circ a = b \circ e$. Considering the factorizations $m' \circ e'$ and $m'' \circ e''$ of a and b as a regular epimorphism followed by a monomorphism, we get two factorizations of $m \circ a = b \circ e$ as a regular epimorphism followed by a monomorphism

$$(m \circ m') \circ e' = m \circ a = b \circ e = m'' \circ (e'' \circ e).$$

By uniqueness of such a factorization, we get an isomorphism s making the diagram commutative, and therefrom the expected factorization $m' \circ s^{-1} \circ e''$. $\square$

Proposition 5.3.3 *Let $(\mathcal{E}, \mathcal{M})$ be a factorization system on a category $\mathcal{C}$. The following properties hold:*

(i) *$f \in \mathcal{E} \cap \mathcal{M} \Rightarrow f$ is an isomorphism;*

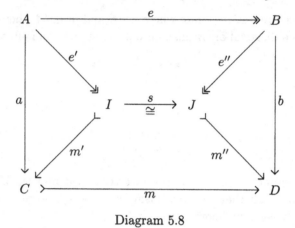

Diagram 5.8

(ii) *the factorization $f = m \circ e$, $m \in \mathcal{M}$, $e \in \mathcal{E}$, of a morphism of $\mathcal{C}$ is unique up to an isomorphism;*

(iii) *$f \in \mathcal{M}$ iff for every commutative square*

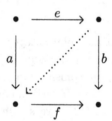

with $e \in \mathcal{E}$, there exists a unique diagonal morphism making both triangles commutative;

(iv) *$f \circ g \in \mathcal{M}$ and $f \in \mathcal{M} \Rightarrow g \in \mathcal{M}$;*

(v) *in a pullback*

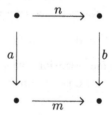

if $m \in \mathcal{M}$, then $n \in \mathcal{M}$.

The dual properties of (iii), (iv), (v) *are valid for the arrows in $\mathcal{E}$.*

Proof (i) It suffices to consider the square

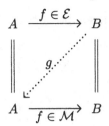

to get g such that $g \circ f = \mathrm{id}_A$, $f \circ g = \mathrm{id}_B$. The converse holds by 5.3.1.

(ii) If $f = m' \circ e'$, with $m' \in \mathcal{M}$, $e' \in \mathcal{E}$, is another factorization, considering both diagrams

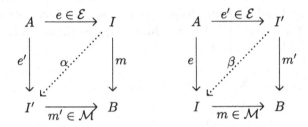

yields the factorizations α and β. Considering both diagrams

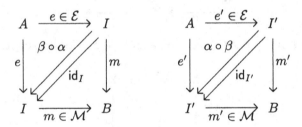

and the uniqueness of the diagonal factorization yields $\beta \circ \alpha = \mathrm{id}_I$ and $\alpha \circ \beta = \mathrm{id}_{I'}$.

(iii) By 5.3.1(iv), the arrows $f \in \mathcal{M}$ have the expected property. Conversely, consider f with property as in condition (iii) of 5.3.1 and write $f = m \circ e$ with $m \in \mathcal{M}$ and $e \in \mathcal{E}$. Considering the square

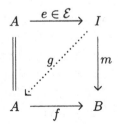

yields a factorization g, by assumption on f. The following diagram is then commutative:

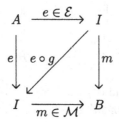

This proves $e \circ g = \mathrm{id}_I$ by the uniqueness condition in 5.3.1(iv). Since on the other hand we have already $g \circ e = \mathrm{id}_A$, g is an isomorphism. Therefore $f = m \circ g^{-1} \in \mathcal{M}$ by conditions (i) and (ii) in 5.3.1.

   (iv) Consider a diagram

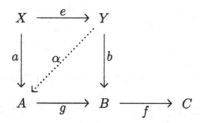

with $f \circ g \in \mathcal{M}$, $f \in \mathcal{M}$, $g \circ a = b \circ e$ and $e \in \mathcal{E}$. Considering the square

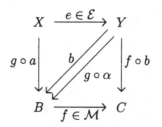

yields a unique factorization α. Considering now the commutative diagram

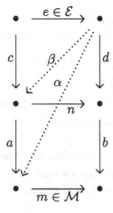

Diagram 5.9

yields $b = g \circ \alpha$, by uniqueness of the factorization. Thus α is already a morphism such that $\alpha \circ e = a$ and $g \circ \alpha = b$. If α' is another such morphism, then $\alpha' \circ e = a$ and $(f \circ g) \circ \alpha' = f \circ b$, thus $\alpha = \alpha'$ by the uniqueness condition in 5.3.1(iv). One concludes the proof by 5.3.3(iii).

(v) Consider the given pullback and a commutative square $n \circ c = d \circ e$, with $e \in \mathcal{E}$, as in diagram 5.9. Since $e \in \mathcal{E}$ and $m \in \mathcal{M}$, we first get a factorization α such that $\alpha \circ e = a \circ c$ and $m \circ \alpha = b \circ d$. This last equality yields a factorization β through the bottom pullback, such that $a \circ \beta = \alpha$ and $n \circ \beta = d$. Therefore

$$a \circ \beta \circ e = \alpha \circ e = a \circ c, \quad n \circ \beta \circ e = d \circ e = n \circ c,$$

from which $\beta \circ e = c$ since the bottom square is a pullback. If β' is another factorization in the upper square,

$$(a \circ \beta') \circ e = a \circ c, \quad m \circ (a \circ \beta') = b \circ n \circ \beta' = b \circ d,$$

from which $a \circ \beta' = \alpha$, by 5.3.1(iv). But then $a \circ \beta' = a \circ \beta$ and $n \circ \beta' = n \circ \beta$, from which $\beta = \beta'$ since the bottom square is a pullback. One concludes the proof again by 5.3.3(iii)). $\qquad \square$

5.4 Reflective factorization systems

We now investigate the relation between factorization systems and reflective subcategories.

Definition 5.4.1 A category $\mathcal{C}$ is finitely well-complete when

(i) C is finitely complete,

(ii) C admits arbitrarily large intersections of subobjects.

Most categories we consider in general are finitely well-complete, for one of the following reasons:

(i) the category is complete and each object has only a set of subobjects, like the categories of sets, groups, topological spaces, and so on;

(ii) the category is finitely complete and each object has only a finite set of subobjects, like the categories of finite sets, finite groups, finite topological spaces, and so on.

Theorem 5.4.2 *Let C be a finitely well-complete category. There exists a bijection between*

(i) *the replete reflective full subcategories of C,*

(ii) *the factorization systems $(\mathcal{E}, \mathcal{M})$ on C which satisfy the additional condition*

$$f \circ g \in \mathcal{E} \text{ and } f \in \mathcal{E} \Rightarrow g \in \mathcal{E}.$$

Moreover, all the morphisms of a given reflective full subcategory belong to the corresponding class $\mathcal{M}$.

We recall that a full subcategory $\mathcal{R} \subseteq C$ is replete when every object $C \in C$ which is isomorphic to an object $R \in \mathcal{R}$ is itself in $\mathcal{R}$. This is just a canonical way to choose one specific subcategory in each class of equivalent full subcategories. This allows us also, without any loss of generality, to assume freely that the counit of a replete reflective subcategory is the identity.

Also observe that in view of axiom 5.3.1(ii) and the dual of property 5.3.3(iv), the additional requirement in 5.4.2(ii) can be stated as

> *If two sides of a commutative triangle are in $\mathcal{E}$,*
> *so is the third side.*

Obviously, example 5.3.2 is not of this type.

Proof of theorem   Let us begin with a replete reflective full subcategory

$$r \dashv i \colon \mathcal{R} \xrightarrow[\longrightarrow]{\longleftarrow} C.$$

Let us put

$$\begin{array}{lll} f \in \mathcal{E} & \text{iff} & r(f) \text{ is an isomorphism,} \\ f \in \mathcal{M} & \text{iff} & f \text{ satisfies condition 5.3.3(iii).} \end{array}$$

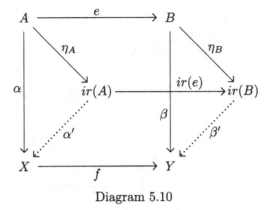

Diagram 5.10

It follows at once that both classes $\mathcal{E}$ and $\mathcal{M}$ contain isomorphisms, that the class $\mathcal{E}$ is closed under composition and that the class $\mathcal{M}$ satisfies condition 5.3.1(iv).

Next, if $m_1, m_2 \in \mathcal{M}$ and the rectangle below is commutative, with $e \in \mathcal{E}$,

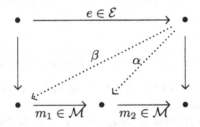

we get at once the unique factorizations α, since $e \in \mathcal{E}$, $m_2 \in \mathcal{M}$, and β, since $e \in \mathcal{E}$, $m_1 \in \mathcal{M}$. By 5.3.3(iii), it follows that $m_2 \circ m_1 \in \mathcal{M}$.

To get a factorization system $(\mathcal{E}, \mathcal{M})$, it remains to prove that every morphism f factors as $f = m \circ e$ with $m \in \mathcal{M}$ and $e \in \mathcal{E}$. To achieve this, we prove first the last assertion, namely that every arrow of $\mathcal{R}$ is in $\mathcal{M}$.

Given $f\colon X \longrightarrow Y$ in $\mathcal{R}$ and and $e\colon A \longrightarrow B$ such that $r(e)$ is an isomorphism, consider diagram 5.10 where $f \circ \alpha = \beta \circ e$ and $\eta\colon \mathsf{id} \Rightarrow ir$ is the unit of the adjunction $r \dashv i$. The factorizations α' and β' exist by the adjunction property, since X and Y are in $\mathcal{R}$. This yields the "diagonal" morphism

$$B \xrightarrow{\quad \eta_B \quad} ir(B) \xrightarrow{\ (ir(e))^{-1}\ } ir(A) \xrightarrow{\quad \alpha' \quad} X$$

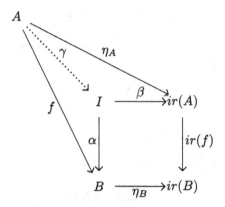

Diagram 5.11

which makes the diagram commutative. This factorization is unique because given $g: B \longrightarrow X$ with $g \circ e = \alpha$, $f \circ g = \beta$, we get at once

$$g = \eta_X \circ g = ir(g) \circ \eta_B = ir(\alpha) \circ \left(ir(e)\right)^{-1} \circ \eta_B = \alpha' \circ \left(ir(e)\right)^{-1} \circ \eta_B$$

because $X \in \mathcal{R}$, thus η_X can be chosen to be id_X, as already observed. This concludes the proof that $f \in \mathcal{M}$.

Next observe that

$$f \circ g \in \mathcal{E} \text{ and } f \in \mathcal{E} \Rightarrow g \in \mathcal{E}$$

since when two sides of the triangle $r(f \circ g)$, $r(f)$, $r(g)$ are isomorphisms, so certainly is the third side.

It remains to prove that a morphism $f \in \mathcal{C}$ factors as $f = m \circ e$ with $m \in \mathcal{M}$ and $e \in \mathcal{E}$. To do this, consider diagram 5.11, where the square is a pullback and γ is the factorization through it of the external commutative quadrilateral. Since $ir(f) \in \mathcal{R}$, we know already that $ir(f) \in \mathcal{M}$; it is easy to see that the proof of 5.3.3(v) applies to show that $\alpha \in \mathcal{M}$. But in general, γ has no reason to be a morphism in $\mathcal{E}$. Thus we must factor γ itself as a morphism in $\mathcal{E}$ followed by a morphism in $\mathcal{M}$.

Let us consider all the possible factorizations

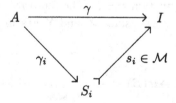

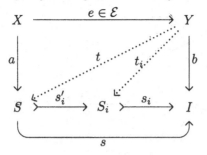

Diagram 5.12

where s_i is a monomorphism belonging to $\mathcal{M}$; there is of course at least one such factorization, namely, $\gamma = \mathrm{id}_I \circ \gamma$. Since $\mathcal{C}$ is finitely well-complete, we can consider $S = \bigcap S_i$ and we get a factorization

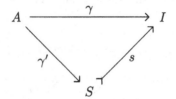

We shall prove that $s \in \mathcal{M}$ and $\gamma' \in \mathcal{E}$. This will yield the factorization $f = (\alpha \circ s) \circ \gamma'$ with $\gamma' \in \mathcal{E}$ and $\alpha \circ s \in \mathcal{M}$, since $\alpha \in \mathcal{M}$ and $s \in \mathcal{M}$.

It is easy to observe that $s \in \mathcal{M}$. Consider diagram 5.12, where the outer part is commutative and $e \in \mathcal{E}$. The monomorphism s factors through each s_i via a monomorphism s_i'. Since $s_i \in \mathcal{M}$ and $e \in \mathcal{E}$, we get a unique factorization t_i. Since $s_i \circ t_i = b = s_j \circ t_j$ for all indices i, j, this yields a further unique factorization t through the intersection S. This proves precisely $s \in \mathcal{M}$.

To prove that $\gamma' \in \mathcal{E}$, that is, $r(\gamma')$ is an isomorphism, we consider diagram 5.13 where the left hand square is a pullback and w is the corresponding factorization arising from $\eta_S \circ \gamma' = ir(\gamma') \circ \eta_A$. The right hand square is commutative since

$$ir(\beta) \circ ir(s) \circ ir(\gamma') = ir(\beta \circ s \circ \gamma') = ir(\beta \circ \gamma) = ir(\eta_A) = \mathrm{id}_{ir(A)}.$$

This last equality implies already that $ir(\gamma')$ is a monomorphism; on the other hand $ir(\gamma') \in \mathcal{R}$, thus $ir(\gamma') \in \mathcal{M}$ by a previous part of the proof. By pulling back, we get that u as well is a monomorphism belonging to $\mathcal{M}$. But the definition of S as intersection of the subobjects of I

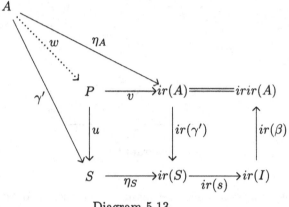

Diagram 5.13

in $\mathcal{M}$ then forces u to be an isomorphism. This allows considering the following situation:

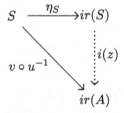

where the unique morphism $z\colon r(S)\longrightarrow r(A)$ making the diagram commutative exists by the adjointness property. This implies

$$i\big(r(\gamma')\circ z\big)\circ\eta_S = ir(\gamma')\circ i(z)\circ\eta_S = ir(\gamma')\circ v\circ u^{-1} = \eta_S\circ u\circ u^{-1} = \eta_S$$

from which $r(\gamma')\circ z = \mathrm{id}_S$ by uniqueness of the factorization through the unit η_S of the adjunction. Therefore $ir(\gamma')\circ i(z) = \mathrm{id}_{i(S)}$ and $ir(\gamma')$ is a retraction in $\mathcal{C}$; since we know already it is also a monomorphism, it is an isomorphism. Thus $r(\gamma')$ is an isomorphism as well, because $\mathcal{R}$ is full in $\mathcal{C}$. This means $\gamma' \in \mathcal{E}$. This ends the proof that $(\mathcal{E},\mathcal{M})$ is a factorization system satisfying all the requirements of the statement.

Let us now start with a factorization system $(\mathcal{E},\mathcal{M})$ satisfying the additional condition

$$f\circ g\in\mathcal{E} \text{ and } f\in\mathcal{E} \Rightarrow g\in\mathcal{E}.$$

Let us define $\mathcal{R}$ as the full subcategory of $\mathcal{C}$ generated by those objects $C\in\mathcal{C}$ such that the unique morphism $\xi_C\colon C\longrightarrow \mathbf{1}$ to the terminal

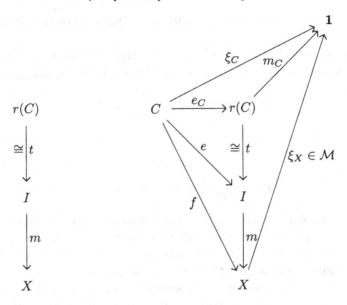

Diagram 5.14

object is in $\mathcal{M}$. Let us prove that the inclusion $i\colon \mathcal{R} \hookrightarrow \mathcal{C}$ has a left adjoint functor r. For this, we consider diagram 5.14 in the category $\mathcal{C}$, with the left hand part of this diagram in the subcategory $\mathcal{R}$.

Given $C \in \mathcal{C}$, we consider the unique morphism ξ_C to the terminal object and we factor it as $\xi_C = m_C \circ e_C$, with $m_C \in \mathcal{M}$ and $e_C \in \mathcal{E}$. By definition of $\mathcal{R}$, the corresponding object $r(C)$ belongs to $\mathcal{R}$. To prove the adjointness property, choose $X \in \mathcal{R}$ and $f\colon C \longrightarrow X$. By definition of $\mathcal{R}$, $\xi_X \in \mathcal{M}$. Factorizing f as $f = m \circ e$ with $m \in \mathcal{M}$ and $e \in \mathcal{E}$, we get two $(\mathcal{E}, \mathcal{M})$-factorizations of ξ_C, namely, $m_C \circ e_C$ and $(\xi_X \circ m) \circ e$. The uniqueness of such a factorization implies the existence of an isomorphism $t\colon r(C) \xrightarrow{\cong} I$ making diagram 5.14 commutative. Since the full subcategory $\mathcal{R}$ is replete, $I \in \mathcal{R}$ and thus $t \in \mathcal{R}$ as well. This yields the expected factorization $m \circ t$. If $t'\colon r(C) \longrightarrow X$ is another morphism such that $t' \circ e_C = f$, the trivial equality $\xi_X \circ t' = m_C$ implies $t' \in \mathcal{M}$ by 5.3.3(iv). Thus $f = t' \circ e_C$ is an $(\mathcal{E}, \mathcal{M})$-factorization of f and, by uniqueness, $t' = m \circ t$. This concludes the proof that the inclusion functor i admits a left adjoint functor taking values $r(C)$ on the objects.

It remains to check that both constructions that we have just described define mutually inverse bijections. First, in the construction of the reflection $\mathcal{R}$, starting from a factorization system $(\mathcal{E}, \mathcal{M})$, we must prove that the morphisms inverted by the reflexion r are exactly those of $\mathcal{E}$.

Let $g\colon C \longrightarrow D$ be a morphism of $\mathcal{C}$; $r(g)\colon r(C) \longrightarrow r(D)$ is the unique morphism of $\mathcal{R}$ making commutative the following diagram:

$$
\begin{array}{ccccc}
C & \xrightarrow{\ e_C\ } & r(C) & \xrightarrow{\ m_C\ } & 1 \\
\downarrow{\scriptstyle g} & & \downarrow{\scriptstyle r(g)} & & \| \\
D & \xrightarrow{\ e_D\ } & r(D) & \xrightarrow{\ m_D\ } & 1
\end{array}
$$

where $e_C, e_D \in \mathcal{E}$ are the units of the adjunction and $m_C, m_D \in \mathcal{M}$. If $r(g)$ is an isomorphism, then $r(g) \circ e_C$ and e_D are in $\mathcal{E}$, thus $g \in \mathcal{E}$. Conversely if $g \in \mathcal{E}$, then $m_C \circ e_C$ and $m_D \circ (e_D \circ g)$ are two $(\mathcal{E}, \mathcal{M})$-factorizations of the same arrow, thus $r(g)$ is an isomorphism.

On the other hand starting from a reflective subcategory $\mathcal{R}$ and constructing the corresponding factorization system $(\mathcal{E}, \mathcal{M})$ as at the beginning of the proof, we must check that for every object $C \in \mathcal{C}$, the composite

$$
C \xrightarrow{\ \eta_C\ } ir(C) \xrightarrow{\ \xi_{ir(C)}\ } 1
$$

is such that $\eta_C \in \mathcal{E}$ and $\xi_{ir(C)} \in \mathcal{M}$.

Indeed, $r(\eta_C) = \mathrm{id}_{r(C)}$ by adjunction, thus $\eta_C \in \mathcal{E}$. Moreover $ir(C) \in \mathcal{R}$ and $1 \in \mathcal{R}$, thus $\xi_{ir(C)} \in \mathcal{R}$ since $\mathcal{R}$ is full in $\mathcal{C}$; we know already that this implies $\xi_{ir(C)} \in \mathcal{M}$. $\qquad\square$

5.5 Semi-exact reflections

First we recall without proof a result of Gabriel and Ulmer (see [32] or [8], volume 2). This result will not be used in this book, but will at least justify the terminology "semi-left-exact" used in this section. The notion of "semi-left-exact reflection" appears also under the name "fibred reflection" in [11].

Proposition 5.5.1 *Let $\mathcal{C}$ be a category with finite limits. For a full reflective subcategory $r \dashv i\colon \mathcal{R} \underset{\longrightarrow}{\longleftarrow} \mathcal{C}$, the following conditions are equivalent:*

  (i) *the reflection $r\colon \mathcal{C} \longrightarrow \mathcal{R}$ preserves finite limits, that is, is left exact;*

(ii) *the class of those arrows which are inverted by the reflection r is stable under change of base; this means, given a pullback*

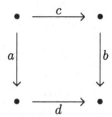

if $r(d)$ is an isomorphism, then $r(c)$ is an isomorphism as well.

□

In other words, the reflection is left exact iff the class $\mathcal{E}$ of the corresponding factorization system (see theorem 5.4.2) is stable under change of base. The next proposition presents a weakening of this situation.

Proposition 5.5.2 *Let C be a finitely well-complete category. Let $r \dashv i \colon \mathcal{R} \xrightarrow{\longleftarrow} C$ be a full reflective subcategory of C and $(\mathcal{E}, \mathcal{M})$ the corresponding factorization system, as in theorem 5.4.2. The following conditions are equivalent:*

(i) *in the pullback in C*

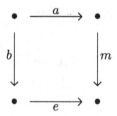

if $e \in \mathcal{E}$ and $m \in \mathcal{M}$, then $a \in \mathcal{E}$ (and, of course, $b \in \mathcal{M}$);

(ii) *in the pullback in C*

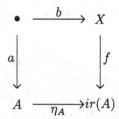

where $A \in C$, η_A is the unit of the adjunction $r \dashv i$ and $X \in \mathcal{R}$, one has $b \in \mathcal{E}$;

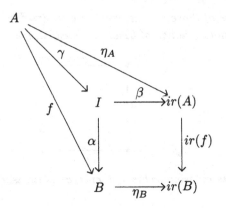

Diagram 5.15

(iii) *for every object $C \in \mathcal{C}$, the adjunction*

$$\mathcal{R}/r(C) \underset{i_C}{\overset{r_C}{\rightleftarrows}} \mathcal{C}/C, \quad r_C \dashv i_C$$

given by lemma 4.3.4 still presents $\mathcal{R}/r(C)$ as a full reflective subcategory of $\mathcal{C}/C$.

Moreover, in these conditions,

- *a morphism f is in the class $\mathcal{M}$ iff the following square is a pullback –*

$$
\begin{array}{ccc}
A & \xrightarrow{\;\eta_A\;} & ir(A) \\
{\scriptstyle f}\downarrow & & \downarrow{\scriptstyle ir(f)} \\
B & \xrightarrow[\;\eta_B\;]{} & ir(B)
\end{array}
$$

- *the reflection $r: \mathcal{C} \longrightarrow \mathcal{R}$ preserves those pullbacks mentioned in condition (i).*

Proof Condition (ii) is just a special case of condition (i) since $\eta_A \in \mathcal{E}$ and $f \in \mathcal{R} \subseteq \mathcal{M}$ (see theorem 5.4.2). Moreover in condition (i), the statement $b \in \mathcal{M}$ holds by 5.3.3(v). To prove (ii) $\Rightarrow$ (i), we show first that condition (ii) implies the characterization of the arrows $f \in \mathcal{M}$ as at the end of the statement.

   If the square

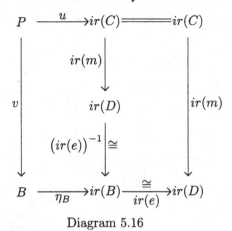

Diagram 5.16

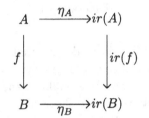

is a pullback, then $ir(f) \in \mathcal{R} \subseteq \mathcal{M}$ (see 5.4.2) gives $f \in \mathcal{M}$ (see 5.3.3(v)). Conversely if $f \in \mathcal{M}$, consider diagram 5.15, already used in the proof of 5.4.2 and where the square is a pullback. By assumption (ii), $\beta \in \mathcal{E}$; but since $\eta_A \in \mathcal{E}$, it follows from 5.4.2(ii) that $\gamma \in \mathcal{E}$. On the other hand $ir(f) \in \mathcal{R} \subseteq \mathcal{M}$ and thus $\alpha \in \mathcal{M}$ by 5.3.3(v). This yields $f = \alpha \circ \gamma$ with $\alpha \in \mathcal{M}$ and $\gamma \in \mathcal{E}$. But when $f \in \mathcal{M}$, the uniqueness of the $(\mathcal{E}, \mathcal{M})$-factorization implies that γ is an isomorphism. Thus the outer part of the diagram is a pullback, since the inner square is.

We are now able to prove (ii) $\Rightarrow$ (i). Let us consider the situation

$$
\begin{array}{ccccc}
A & \xrightarrow{\ a\ } & C & \xrightarrow{\ \eta_C\ } & ir(C) \\
\downarrow{\scriptstyle b} & & \downarrow{\scriptstyle m} & & \downarrow{\scriptstyle ir(m)} \\
B & \xrightarrow{\ e\ } & D & \xrightarrow{\ \eta_D\ } & ir(D)
\end{array}
$$

where the left hand square is a pullback as in condition (i), thus with $e \in \mathcal{E}$ and $m \in \mathcal{M}$. From $m \in \mathcal{M}$ and the characterization of arrows

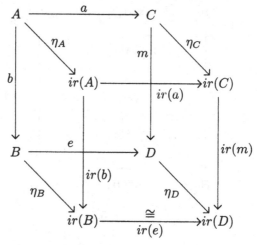

Diagram 5.17

in $\mathcal{M}$ we have just exhibited, we deduce that the right hand square is a pullback as well. Let us compare this with diagram 5.16, where the right hand rectangle is obviously a pullback and the left hand rectangle is a pullback by definition. Notice that $ir(e)$ is indeed an isomorphism since $e \in \mathcal{E}$. Since $ir(C) \in \mathcal{R}$, by assumption (ii), $u \in \mathcal{E}$. By naturality, $\eta_D \circ e = ir(e) \circ \eta_B$, proving that the two diagrams have the same bottom arrow. Therefore $\eta_C \circ a \colon A \longrightarrow ir(C)$ coincides with $u \colon P \longrightarrow ir(C)$ up to an isomorphism. Since $\eta_C \in \mathcal{E}$ and $u \in \mathcal{E}$, it follows that $a \in \mathcal{E}$ by 5.4.2(ii). This concludes the proof of the equivalence (i) $\Leftrightarrow$ (ii).

Let us prove now (i) $\Rightarrow$ (iii). It is useful to go back to lemma 4.3.4 to observe how pullbacks occur at once in the definition of the functor i_C.

Thus choose $f \colon X \longrightarrow r(C)$ in $\mathcal{R}$ and consider the following pullback in $\mathcal{C}$:

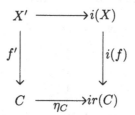

We must prove that the counit of the adjunction $r_C \dashv i_C$ is an isomorphism (see [8], volume 1), that is the existence of an isomorphism in $\mathcal{R}/r(C)$

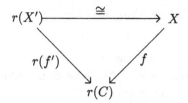

Observe first that given the pullback of condition (i), one gets the "cubical" diagram 5.17, about which we use freely a "three-dimensional" terminology. Since $e \in \mathcal{E}$ implies that $ir(e)$ is an isomorphism, for the face containing $ir(a)$, $ir(e)$ to be a pullback it is necessary and sufficient that $ir(a)$ be an isomorphism as well. This is further equivalent to $a \in \mathcal{E}$, which is precisely condition (i). In particular, condition (i) implies that the pullbacks mentioned in this condition are preserved by ir, thus by r since pullbacks in $\mathcal{R}$ are computed as in $\mathcal{C}$.

Observe next that the pullback defining f' is such that $\eta_C \in \mathcal{E}$ and $i(f) \in \mathcal{R} \subseteq \mathcal{M}$, thus it is preserved by the reflection r, as we have already seen. This yields the following pullback in $\mathcal{R}$, where the bottom line is an identity:

$$
\begin{array}{ccc}
r(X') & \xrightarrow{\;s\;} & ri(X) = X \\
{\scriptstyle r(f')}\Big\downarrow & & \Big\downarrow{\scriptstyle f} \\
r(C) & =\!=\!=\!= & r(C)
\end{array}
$$

The arrow s is thus the expected isomorphism.

Finally we prove (iii) $\Rightarrow$ (ii). Thus assume condition (iii) and consider the situation of condition (ii):

$$
\begin{array}{ccccc}
X' & \xrightarrow{\;f''\;} & i(X) & X & X \in \mathcal{R} \\
{\scriptstyle f'}\Big\downarrow & & \Big\downarrow{\scriptstyle i(f)} & \Big\downarrow{\scriptstyle f} & \\
C & \xrightarrow[\;\eta_C\;]{} & ir(C) & r(C) &
\end{array}
$$

We must prove that $f'' \in \mathcal{E}$, that is, $r(f'')$ is an isomorphism. But $(X', f') = i_C(X, f)$ and, by assumption, the counit of the adjunction

$$
r_C i_C(X, f) = \big(r(X'), r(f')\big) \longrightarrow (X, f) \cong \big(ri(X), f\big)
$$

is an isomorphism. This counit is precisely $r(f'')$. □

Definition 5.5.3 Let C be a category with pullbacks and $r \dashv i \colon \mathcal{R} \overset{\longleftarrow}{\longrightarrow} C$ a reflective full subcategory of C. The reflection is semi-left-exact when, for every object $C \in C$, the adjunction

$$\mathcal{R}/r(C) \overset{r_C}{\underset{i_C}{\overset{\longleftarrow}{\longrightarrow}}} C/C, \quad r_C \dashv i_C$$

given by lemma 4.3.4 still presents $\mathcal{R}/r(C)$ as a reflective full subcategory of C/C.

Definition 5.5.4 Let us consider the following data:

- a category C with pullbacks;
- a semi-left-exact reflection $r \dashv i \colon \mathcal{R} \overset{\longleftarrow}{\longrightarrow} C$;
- a morphism $\sigma \colon S \longrightarrow R$ in C.

We particularize definitions 5.1.7 and 5.1.8 by saying that

(i) a pair $(A, f) \in C/R$ is split by σ with respect to the semi-left-exact reflection $r \dashv i$, when it is so in the sense of definition 5.1.7, relatively to the classes $\overline{\mathcal{R}}, \overline{C}$ of all arrows of $\mathcal{R}$ and C,

(ii) σ is a morphism of Galois descent with respect to the semi-left-exact reflection $r \dashv i$, when it is so in the sense of definition 5.1.8, relativeley to the classes $\overline{\mathcal{R}}, \overline{C}$ of all arrows of $\mathcal{R}$ and C.

In the special case of semi-left-exact reflections, we shall now investigate simpler descriptions of split objects and Galois descent morphisms. First, we observe that restricting one's attention to semi-left-exact reflections allows working with finitely complete categories instead of finitely well-complete ones (see definition 5.4.1).

Proposition 5.5.5 *Let C be a category with finite limits. There exists a bijection between*

(i) *the semi-left-exact reflective replete full subcategories of C,*

(ii) *the factorization systems $(\mathcal{E}, \mathcal{M})$ on C which satisfy the two aditional conditions*

  (a) *$f \circ g \in \mathcal{E}$ and $f \in \mathcal{E} \Rightarrow g \in \mathcal{E}$,*
  (b) *given a pullback*

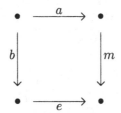

with $e \in \mathcal{E}$ and $m \in \mathcal{M}$, one has $a \in \mathcal{E}$ (and, of course, $b \in \mathcal{M}$).

In these conditions, an arrow $f \in C$ is in $\mathcal{M}$ precisely when the square

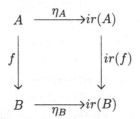

is a pullback, with η the unit of the adjunction $r \dashv i$. Moreover the reflection $r\colon C \longrightarrow \mathcal{R}$ preserves the pullbacks mentioned in condition (ii).(b) and each arrow of $\mathcal{R}$ is in $\mathcal{M}$.

Proof   In the case of a finitely well-complete category C, this is an immediate consequence of 5.4.2 and 5.5.2. So it suffices to observe that when restricting one's attention to semi-left-exact reflections, one can get rid of the existence of arbitrary intersections. This assumption has been used at a single place, in the proof of 5.4.2, to prove the existence of the $(\mathcal{E}, \mathcal{M})$-factorization of a morphism f, starting with a given reflection.

Observe first that the class $\mathcal{M}$ constructed in 5.4.2 from an arbitrary reflection is stable under change of base. The proof of this fact is exactly the same as that of point (v) in proposition 5.3.3.

Observe next that the following facts remain valid here, since they have been proved without any reference to the existence of arbitrary intersections in C:

(i) the inclusion $\mathcal{R} \subseteq \mathcal{M}$, in the proof of 5.4.2;
(ii) the implication $f \circ g \in \mathcal{E}$ and $f \in \mathcal{E} \Rightarrow g \in \mathcal{E}$, in the proof of 5.4.2;
(iii) the implication (iii) $\Rightarrow$ (ii), in the proof of 5.5.2.

When the reflection $r \dashv i\colon \mathcal{R} \longleftrightarrow C$ is semi-left-exact and $\mathcal{E}$, $\mathcal{M}$ are defined as in the proof of 5.4.2, the $(\mathcal{E}, \mathcal{M})$-factorization of an arbitrary

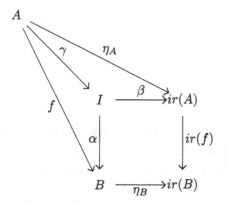

Diagram 5.18

morphism $f\colon A \longrightarrow B$ can be constructed at once, without any intersection. Indeed, as in the proof of 5.5.2, we consider diagram 5.18 where the square is a pullback and γ is the factorization through it of the external quadrilateral. The previous observations show that $ir(f) \in \mathcal{R} \subseteq \mathcal{M}$ thus, by pulling back, $\alpha \in \mathcal{M}$. On the other hand, $\beta \in \mathcal{E}$ and $\eta_A \in \mathcal{E}$ imply $\gamma \in \mathcal{E}$. So $f = \alpha \circ \gamma$ is the expected $(\mathcal{E}, \mathcal{M})$-factorization.   □

Avoiding the existence of arbitrary intersections in 5.5.5 is useful if one intends to apply the result to, for example, elementary toposes.

We now investigate the form of split pairs.

Proposition 5.5.6 *Let us consider the following data:*

- *a finitely complete category $\mathcal{C}$;*
- *a semi-left-exact reflection $r \dashv i \colon \mathcal{R} \overset{\longleftarrow}{\longrightarrow} \mathcal{C}$;*
- *the corresponding factorization system $(\mathcal{E}, \mathcal{M})$;*
- *a morphism $\sigma \colon S \longrightarrow R$ in $\mathcal{C}$.*

For an object $(A, f) \in \mathcal{C}/R$, the following conditions are equivalent:

(i) *the pair (A, f) is split by σ with respect to the semi-left-exact reflection $r \dashv i$;*

(ii) *the canonical morphism $\sigma^*(A, f) \longrightarrow i_S r_S \sigma^*(A, f)$ is an isomorphism;*

(iii) *in the pullback square*

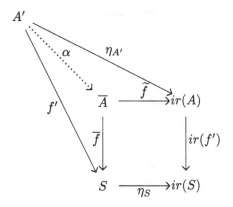

Diagram 5.19

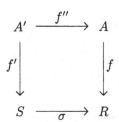

one has $f' \in \mathcal{M}$.

Proof With the notation of the statement, one has $(A', f') = \sigma^*(A, f)$. Next computing the pullback

$$\overline{A} \xrightarrow{\tilde{f}} ir(A')$$
$$\overline{f} \downarrow \qquad \downarrow ir(f')$$
$$S \xrightarrow{\eta_S} ir(S)$$

we find $i_S r_S(A', f') = (\overline{A}, \overline{f})$. The canonical morphism involved in the statement is then the factorization α in diagram 5.19. This factorization is an isomorphism precisely when the outer part of the diagram is a pullback, that is, when $f' \in \mathcal{M}$ (see 5.5.5). But α being an isomorphism is also the definition of (A, f) being split by σ relatively to $r \dashv i$. □

Next we investigate, in the present context, condition (iii) in definition 5.1.8.

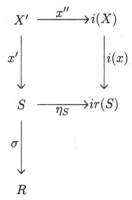

Diagram 5.20

Lemma 5.5.7 *Let us consider the following data:*

- *a finitely complete category $\mathcal{C}$;*
- *a semi-left-exact reflection $r \dashv i \colon \mathcal{R} \rightleftharpoons \mathcal{C}$;*
- *the corresponding factorization system $(\mathcal{E}, \mathcal{M})$;*
- *a morphism $\sigma \colon S \longrightarrow R$ in $\mathcal{C}$.*

The following conditions are equivalent:

(i) *for every object $(X, x) \in \mathcal{R}/r(S)$, the object $(\Sigma_\sigma \circ i_S)(X, x) \in \mathcal{C}/R$ is split by σ with respect to $r \dashv i$;*

(ii) *in the pullback*

$$
\begin{array}{ccc}
S \times_R S & \xrightarrow{\;p_2\;} & S \\
{\scriptstyle p_1}\downarrow & & \downarrow{\scriptstyle \sigma} \\
S & \xrightarrow[\;\sigma\;]{} & R
\end{array}
$$

one has $p_1 \in \mathcal{M}$.

Proof   Considering diagram 5.20, where the square is a pullback, we get $(X', \sigma \circ x') = (\Sigma_\sigma \circ i_S)(X, x)$. We consider next diagram 5.21, where both squares are pullbacks: this yields $(\overline{X}, p_1 \circ \overline{x}) = \sigma^*(X', \sigma \circ x')$. By proposition 5.5.6, $(\Sigma_\sigma \circ i_S)(X, x)$ is split by σ when $p_1 \circ \overline{x} \in \mathcal{M}$. But $i(x) \in \mathcal{R} \subseteq \mathcal{M}$ (see 5.5.5), which implies $x' \in \mathcal{M}$ and $\overline{x} \in \mathcal{M}$ (see 5.3.3(v)). Thus when $p_1 \in \mathcal{M}$, one has $p_1 \circ \overline{x} \in \mathcal{M}$ (see definition 5.3.1(ii)). Conversely if condition (i) of the statement holds, putting

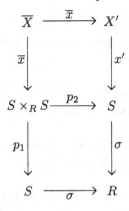

Diagram 5.21

$(X, x) = \bigl(r(S), \mathrm{id}_{r(S)}\bigr)$ yields $x' = \mathrm{id}_S$ and $\overline{x} = \mathrm{id}_{S \times_R S}$; therefore $p_1 \circ \overline{x} \in \mathcal{M}$ implies $p_1 \in \mathcal{M}$.   □

Observe that in condition 5.5.7(ii), we can equivalently write $p_1 \in \mathcal{M}$, or $p_2 \in \mathcal{M}$, or $p_1, p_2 \in \mathcal{M}$ since the twisting isomorphism

$$\tau \colon S \times_R S \longrightarrow S \times_R S,$$

transforms p_1 into p_2. In other words $p_1 \in \mathcal{M}$ implies $p_2 = p_1 \circ \tau \in \mathcal{M}$ by definition 5.3.1.

Finally we are able to characterize morphisms of Galois descent in the present situation:

Corollary 5.5.8 *Let us consider the following data:*

- *a finitely complete category $\mathcal{C}$;*
- *a semi-left-exact reflection $r \dashv i \colon \mathcal{R} \overset{\longleftarrow}{\underset{\longrightarrow}{}} \mathcal{C}$;*
- *the corresponding factorization system $(\mathcal{E}, \mathcal{M})$;*
- *a morphism $\sigma \colon S \longrightarrow R$ in $\mathcal{C}$.*

The following conditions are equivalent for an effective descent morphism σ:

(i) *σ is a morphism of Galois descent with respect to the semi-left-exact reflection $r \dashv i$;*

(ii) *$p_1 \colon S \times_R S \longrightarrow S$ is in the class $\mathcal{M}$.*

These conditions are in particular satisfied when $\sigma \in \mathcal{M}$.

Proof The equivalence (i) ⇔ (ii) follows from lemma 5.5.7 and condition (v) of proposition 5.3.3.   □

5.6 Connected components of a space

First, we recall some basic facts about connected components in an arbitrary topological space.

Definition 5.6.1 A topological space X is connected when it is non-empty and cannot be written as the union of two disjoint non-trivial open subsets.

A subset $A \subseteq X$ of a topological space is connected when, provided with the induced topology, it is a connected space.

Lemma 5.6.2 *Let X be a topological space. Let $(A_i)_{i \in I}$ be a family of connected subsets of X. When the intersection $\bigcap_{i \in I} A_i$ is non-empty, the union $\bigcup_{i \in I} A_i$ is connected.*

Proof Let $x \in \bigcap_{i \in I} A_i$. We prove the statement by a reduction *ad absurdum*. Assume $\bigcup_{i \in I} A_i = M \cup N$ where M and N are non trivial disjoint open subsets of $\bigcup_{i \in I} A_i$. For each A_i we have

$$A_i = (A_i \cap M) \cup (A_i \cap N)$$

with $A_i \cap M$ and $A_i \cap N$ open in A_i. Since A_i is connected, this implies $A_i \cap M = A_i$ or $A_i \cap N = A_i$, that is, $A_i \subseteq M$ or $A_i \subseteq N$. On the other hand, $x \in M \cup N$, thus $x \in M$ or $x \in N$. Let us handle the case $x \in M$. If $x \in M$, since $x \in A_i$ for all $i \in I$, then $A_i \subseteq M$ for all $i \in I$. This proves $\bigcup_{i \in I} A_i = M$, which is a contradiction. $\qquad\square$

Definition 5.6.3 The connected component of a point x in a topological space X is the union of all connected subsets containing x, that is, the largest connected subset containing x. We shall write Γ_x for the connected component of the point x.

By lemma 5.6.2 and the fact that the singleton $\{x\}$ is obviously connected, definition 5.6.3 indeed makes sense.

Lemma 5.6.4 *In a topological space, the closure of a connected subset is again connected.*

Proof Let $A \subseteq X$ be a connected subset of the space X. Assume $\overline{A} = M \cup N$ where M, N are non trivial disjoint open subsets of $\overline{A}$. One has in particular

$$A = (A \cap M) \cup (A \cap N)$$

with $A \cap M$ and $A \cap N$ open subsets of A. This forces $A \cap M = \emptyset$ or $A \cap N = \emptyset$. Let us handle the case $A \cap N = \emptyset$. Since N is non-empty, choose $x \in N$. Since $x \in \overline{A}$ and N is open in $\overline{A}$, one has $A \cap N \neq \emptyset$, which is a contradiction. □

Corollary 5.6.5 *In a topological space, the connected component of a point is closed.*

Proof The closure of the connected component of a point x is still connected by lemma 5.6.4, thus is equal to Γ_x by maximality of Γ_x (see definition 5.6.3). □

Lemma 5.6.6 *The image of a connected subset under a continuous map is still connected.*

Proof Let $f: X \longrightarrow Y$ be a continuous map and $A \subseteq X$ a connected subset. If $f(A) = M \cup N$ with M, N disjoint non-empty open subsets in $f(A)$, then

$$A \subseteq f^{-1}f(A) = f^{-1}(M) \cup f^{-1}(N)$$

with $f^{-1}(M)$ and $f^{-1}(N)$ disjoint non-empty open subsets of $f^{-1}f(A)$. Thus

$$A = \left(A \cap f^{-1}(M)\right) \cup \left(A \cap f^{-1}(N)\right)$$

and since A is connected, one of those open subsets of A is empty while the other one equals A. Let us handle the case $A = A \cap f^{-1}(M)$. In this case $A \subseteq f^{-1}(M)$ and thus $f(A) \subseteq ff^{-1}(M) \subseteq M$. Since $M \subseteq f(A)$, we get $M = f(A)$, which is a contradiction. □

Corollary 5.6.7 *If $f: X \longrightarrow Y$ is a continuous map and $x \in X$, then $f(\Gamma_x) \subseteq \Gamma_{f(x)}$.* □

Lemma 5.6.8 *If x is a point of a topological space X, then*

$$\Gamma_x \subseteq \bigcap \{U | U \text{ is a clopen of } X, \ x \in U\}.$$

Proof If U is a clopen containing x, then

$$\Gamma_x = (\Gamma_x \cap U) \cup (\Gamma_x \cap \complement U)$$

and by connectedness of Γ_x, one of those two terms of the union must be empty. Since $x \in \Gamma_x \cap U$, this forces $\Gamma_x \cap \complement U = \emptyset$, that is $\Gamma_x \subseteq U$. □

5.7 Connected components of a compact Hausdorff space

We shall provide the set of connected components of a compact Hausdorff space with a profinite topology. To achieve this, we shall first show that in a compact Hausdorff space, the inclusion of lemma 5.6.8 is in fact an equality. The proof uses the uniform structure of a compact space.

Let us recall that a relation R on a set X is a subset $R \subseteq X \times X$. Given relations R, S on X we use the classical notation

$$
\begin{aligned}
R^{-1} &= \{(x,y)|(y,x) \in R\}, \\
R \circ S &= \{(x,z)|\exists y \in X \ (x,y) \in R, \ (y,z) \in S\}, \\
\Delta_X &= \{(x,x)|x \in X\}.
\end{aligned}
$$

When no confusion can occur, we shall just write Δ instead of Δ_X for the diagonal of X. With the previous notation, it follows at once that

$$
\begin{aligned}
R \text{ is reflexive} &\quad \text{iff} \quad \Delta \subseteq R, \\
R \text{ is symmetric} &\quad \text{iff} \quad R = R^{-1}, \\
R \text{ is transitive} &\quad \text{iff} \quad R \circ R \subseteq R.
\end{aligned}
$$

The notion of neighbourhood of a point x tries to recapture the idea of "points which are sufficiently close to the fixed point x"; for example in a metric space X, given $\varepsilon > 0$,

$$
V = \{y \in X \mid d(x,y) < \varepsilon\}.
$$

The notion of entourage tries to recapture an analogous idea, but "uniformly", for all the points of the space at the same time; for example on the metric space X:

$$
V = \{(y,z)|y \in X, \ z \in X, \ d(y,z) < \varepsilon\}.
$$

An "entourage" in a set X is thus a relation V on X expressing the "proximity" of the points $(y,z) \in V$.

There is an abstract notion of *uniform space*, which is a set X provided with a family of subsets $V \subseteq X \times X$, called "entourages" and satisfying precisely conditions (i) to (iii) of 5.7.2 below . The reader interested in these questions will find an explicit treatment of them in [10]. In this book, we shall only be interested in the uniform structure which is naturally present on a compact Hausdorff space, not in the abstract theory of uniform spaces.

Definition 5.7.1 In a compact Hausdorff space X, a neighbourhood of the diagonal $\Delta \subset X \times X$ is called an entourage of X.

A neighbourhood of the diagonal is of course a subset containing an open subset containing the diagonal.

Proposition 5.7.2 *Let X be a compact Hausdorff space and $\mathcal{V}$ the set of its entourages. The following properties hold:*

(i) $\forall V \in \mathcal{V} \quad \Delta \subseteq V$;

(ii) $\forall V \in \mathcal{V} \quad V^{-1} \in \mathcal{V}$;

(iii) $\forall V \in \mathcal{V} \quad \exists W \in \mathcal{V} \quad with \quad W \circ W \subseteq V$.

Proof Property (i) holds by definition of an entourage.

To prove (ii), it suffices to consider the twisting morphism

$$\tau \colon X \times X \longrightarrow X \times X, \quad (x,y) \mapsto (y,x)$$

and observe that given $V \in \mathcal{V}$, one has $V^{-1} = \tau^{-1}(V)$. Since τ is continuous, V^{-1} is again a neighbourhood of Δ.

We prove (iii) by reduction *ad absurdum*. Assume thus the existence of $V \in \mathcal{V}$ such that for every $W \in \mathcal{V}$, one has $W \circ W \not\subseteq V$, that is, $(W \circ W) \cap \complement V \neq \emptyset$. Trivially, for all $W_1, W_2 \in \mathcal{V}$, one has

$$(W_1 \cap W_2) \circ (W_1 \cap W_2) \subseteq (W_1 \circ W_1) \cap (W_2 \circ W_2)$$

which proves that the subsets $(W \circ W) \cap \complement V$, with $W \in \mathcal{V}$, constitute a filter base in $X \times X$. Since X and thus $X \times X$ are compact Hausdorff, there exists a point (x,y) in the adherence of this filter base, that is, every neighbourhood of (x,y) meets every subset in the filter base (see [58]). Since V is in particular a neighbourhood of (x,x) and does not meet any $(W \circ W) \cap \complement V$, it follows that $(x,x) \neq (x,y)$, thus $x \neq y$. By normality of the compact Hausdorff space X (see [58]), we can find

$$x \in A' \subseteq A, \quad y \in B' \subseteq B$$

where A, B are open and disjoint and A', B' are, respectively, closed neighbourhoods of x and y. We consider $C = \complement(A' \cup B')$ which is thus an open subset of X and finally

$$W_0 = (A \times A) \cup (B \times B) \cup (C \times C)$$

which is open in $X \times X$ and contains the diagonal, since $A \cup B \cup C = X$. Thus $W_0 \in \mathcal{V}$. Since $A' \times B'$ is a neighbourhood of (x,y), we have thus

$$(A' \times B') \cap (W_0 \circ W_0) \cap \complement V \neq \emptyset;$$

we choose (u,v) in this intersection. Since $(u,v) \in W_0 \circ W_0$, there exists $w \in X$ such that $(u,w) \in W_0$ and $(w,v) \in W_0$. But $u \in A'$ implies

$w \in A$ by definition of W_0, just because A' is disjoint from B and C. Analogously, $v \in B'$ implies $w \notin B$. But then $w \in A \cap B$, which is a contradiction. $\qquad \square$

Proposition 5.7.3 *In a compact Hausdorff space, every entourage contains an open symmetric entourage.*

Proof With the notation of 5.7.2, $V \in \mathcal{V}$ contains an open entourage $W \subsetneq V$ and $W^{-1} = \tau^{-1}(W)$ is thus another open entourage. It follows that $W \cap W^{-1} \subseteq W \subseteq V$ is still an open neighbourhood of the diagonal, obviously symmetric. $\qquad \square$

In view of proposition 5.7.3, we shall concentrate our attention on open symmetric entourages, since these "generate" the uniform structure of X.

Lemma 5.7.4 *Consider a subset $B \subset X$ of a compact Hausdorff space X. For every open symmetric entourage V, we define*

$$V(B) = \{ x \in X \mid \exists b \in B \ \ (x, b) \in V \}.$$

This subset $V(B)$ is open in X.

Proof Let us consider, for each element $y \in X$, the continuous map

$$(-, y) \colon X \longrightarrow X \times X, \quad x \mapsto (x, y).$$

One gets at once

$$
\begin{aligned}
V(B) &= \{ x \in X \mid \exists b \in B \ \ (x, b) \in V \} \\
&= \bigcup_{b \in B} \{ x \in X \mid (x, b) \in V \} \\
&= \bigcup_{b \in B} (-, b)^{-1}(V)
\end{aligned}
$$

which proves that $V(B)$ is a union of open subsets. $\qquad \square$

Lemma 5.7.5 *Let B, C be disjoint closed subspaces of a compact Hausdorff space X. There exists an open symmetric entourage V such that*

$$\big(V(B) \times V(C) \big) \cap V = \emptyset.$$

In particular, $V(B) \cap V(C) = \emptyset$.

Proof In the space $X \times X$, the subset $B \times C$ is closed and disjoint from the diagonal Δ, which is closed as well, by Hausdorffness of X (see [58]). By normality of $X \times X$, we choose open subsets M, N of $X \times X$ such that

$$B \times C \subseteq M, \quad \Delta \subseteq N, \quad M \cap N = \emptyset.$$

By 5.7.2(iii) and 5.7.3, we choose an open symmetric entourage V such that $V \circ V \circ V \subseteq N$. We proceed by a reduction *ad absurdum*. If there exists a point

$$(x, y) \in \big(V(B) \times V(C)\big) \cap V,$$

one gets

$$x \in V(B) \quad \Rightarrow \quad \exists b \in B \ (x, b) \in V,$$
$$y \in V(C) \quad \Rightarrow \quad \exists c \in C \ (y, c) \in V.$$

Together with $(x, y) \in V$, the symmetry of V and the inclusion $V \circ V \circ V \subseteq N$ then imply $(b, c) \in N$. This is a contradiction since $(b, c) \in B \times C \subseteq M$ and M is disjoint from N. This concludes the proof of

$$\big(V(B) \times V(C)\big) \cap V = \emptyset$$

which implies a fortiori

$$\big(V(B) \times V(C)\big) \cap \Delta = \emptyset.$$

This last equality means precisely $V(B) \cap V(C) = \emptyset$. $\square$

We shall now focus on the nearness relation associated with an entourage.

Definition 5.7.6 Let X be a compact Hausdorff space.

(i) For an entourage V, the relation of V-nearness is the equivalence relation on X generated by the pairs $(x, y) \in V$.

(ii) The nearness relation on the space X is the intersection of all the V-nearness relations, for all entourages V.

A pair (x, y) is thus in the relation of V-nearness when there exists a finite sequence of elements of X

$$x = z_1, \ldots, z_i, \ldots, z_n = y$$

with each pair (z_i, z_{i+1}) in V or V^{-1}. When V is symmetric, this reduces to each pair (z_i, z_{i+1}) being in V.

Observe that in the example of a metric space already mentioned, if

$(z_i, z_{i+1}) \in V$ means $d(z_i, z_{i+1}) < \varepsilon$, two points x, y are in the relation of V-nearness when "one can travel from x to y by steps of length less than ε". The case of the nearness relation can be interpreted analogously, but using "arbitrarily small steps". Thus the V-nearness relations do not tell anything about the distance between points, but about the "width of the cracks" which can prevent you from travelling from one point to another one. Theorem 5.7.9 will emphasize the mathematical relevance of this intuition.

Lemma 5.7.7 *The nearness relation on a compact Hausdorff space X is the equivalence relation obtained as intersection of all the V-nearness relations, for all the open symmetric entourages V.*

Proof By proposition 5.7.3 and the fact that an intersection of equivalence relations is still an equivalence relation.   □

Lemma 5.7.8 *Let X be a compact Hausdorff space. For every open symmetric entourage V, the equivalence classes in X for the relation of V-nearness are clopens of X.*

Proof Refering to lemma 5.7.4, let us consider

$$V(z) = V(\{z\}) = \{x \in X \,|\, (x, z) \in V\}$$

which is thus open in X. Writing further $[z]_V$ for the equivalence class of z for the relation of V-nearness, we have trivially $z \in V(z) \subseteq [z]_V$. Since $V(z)$ is open, $[z]_V$ is a neighbourhood of each of its points, thus is open. But since distinct equivalence classes are disjoint, an equivalence class is the complement of the union of the other classes, which are open; thus it is closed as well.   □

Theorem 5.7.9 *Let $x \in X$ be a point of a compact Hausdordff space. Write Γ_x for the connected component of x and $[x]_\sim$ for the equivalence class of x for the nearness relation. The following equalities hold:*

$$\Gamma_x = [x]_\sim = \bigcap\{U \subseteq X \mid U \text{ clopen, } x \in U\}.$$

Proof By lemma 5.7.7, one has

$$[x]_\sim = \bigcap\{[x]_V \mid V \text{ open symmetric entourage}\}.$$

By lemma 5.7.8, $[x]_\sim$ is thus an intersection of clopens containing x, which proves already, together with lemma 5.6.8, that

$$\Gamma_x \subseteq \bigcap \{U \subseteq X \mid U \text{ clopen, } x \in U\} \subseteq [x]_\sim.$$

To conclude the proof, it remains to prove that $[x]_\sim$ is connected, since this will imply $[x]_\sim \subseteq \Gamma_x$. We do this by a reduction *ad absurdum*.

If $[x]_\sim$ is not connected, let us write $[x]_\sim = B \cup C$ where B, C are non trivial disjoint open subsets of $[x]_\sim$. Notice that since B, C are mutual complements in $[x]_\sim$, they are also closed in $[x]_\sim$. But $[x]_\sim$ itself is closed as intersection of the clopens $[x]_V$, as in lemma 5.7.8. Thus B and C are also closed in X. Applying lemma 5.7.5, let us choose an open symmetric entourage V such that

$$\big(V(B) \times V(C)\big) \cap V = \emptyset, \quad V(B) \cap V(C) = \emptyset.$$

Further, applying propositions 5.7.2 and 5.7.3, we choose an open symmetric entourage W such that $W \circ W \subseteq V$. Clearly $W = W \circ \Delta \subseteq W \circ W \subseteq V$, thus $W(B) \subseteq V(B)$ and $W(C) \subseteq V(C)$. In particular,

$$\big(W(B) \times W(C)\big) \cap W = \emptyset, \quad W(B) \cap W(C) = \emptyset.$$

Moreover, by lemma 5.7.4, $W(B)$ and $W(C)$ are open. We shall consider $H = \complement\big(W(B) \cup W(C)\big)$ which is thus closed in X.

Since $x \in B \cup C$, one has $x \in B$ or $x \in C$; let us handle the case $x \in B$. Since C is non-empty, choose $y \in C$. Since $x, y \in [x]_\sim$, the pair (x, y) is in the E-nearness relation for every open symmetric entourage $E \subseteq W$; thus there exists a chain

$$x = z_1, z_2, \ldots, z_i, \ldots, z_{n-1}, z_n = y$$

with each pair (z_i, z_{i+1}) in the E-nearness relation. Let us prove that at least one z_i must be in H. Otherwise one would have

$$\forall i = 1, \ldots, n \quad z_i \in \complement H = W(B) \cup W(C) \subseteq V(B) \cup V(C)$$

where $V(B) \cap V(C) = \emptyset$ by lemma 5.7.5. Notice that

$$
\begin{aligned}
z_i \in W(B) \quad &\Rightarrow \quad \exists b_i \in B \ (z_i, b_i) \in W \\
&\Rightarrow \quad \exists b_i \in B \ (z_i, b_i) \in W \text{ and } (z_i, z_{i+1}) \in E \subseteq W \\
&\Rightarrow \quad \exists b_i \in B \ (z_{i+1}, b_i) \in W \circ W \subseteq V \\
&\Rightarrow \quad z_{i+1} \in V(B) \\
&\Rightarrow \quad z_{i+1} \in W(B)
\end{aligned}
$$

where the last implication holds because $z_{i+1} \in W(B) \cup W(C)$ and $W(C)$ is disjoint from $V(B)$. Starting from $x = z_1 \in B \subseteq W(B)$ and iterating the previous argument would yield $z_n = y \in W(B)$. But this is impossible because $y \in W(C)$ and $W(C)$ is disjoint from $W(B)$. Thus indeed, at least one of the elements z_i belongs to H. On the other hand, by definition of the E-nearness relation, all elements z_i are in $[x]_E$. This proves that for each open symmetric entourage $E \subseteq W$, one has $[x]_E \cap H \neq \emptyset$.

Now when E runs through all the open symmetric subentourages of W, the corresponding subsets $[x]_E \cap H$ constitute a filter basis constituted of closed subsets of X. Since X is a compact Hausdorff space, there exists a point x_0 in the intersection of the filter basis, that is a point $x_0 \in [x]_\sim \cap H$. This is a contradiction since

$$[x]_\sim = B \cup C \subseteq W(B) \cup W(C) = \complement H. \qquad \square$$

Corollary 5.7.10 *For a compact Hausdorff space X, the following conditions are equivalent:*

(i) *X is totally disconnected;*

(ii) *the connected component of any point $x \in X$ is reduced to that point.*

Proof (i) $\Rightarrow$ (ii) If X is totally disconnected and $x \in X$, for every $y \in X$, $y \neq x$, choose a clopen U_y such that $x \in U_y$ and $y \notin U_y$. This implies $\{x\} = \bigcap_{y \neq x} U_y$ and one concludes the proof by theorem 5.7.9.

Conversely, if $x \neq y$, then $y \notin \{x\} = \Gamma_x$ and by theorem 5.7.9, there exists a clopen U such that $x \in U$ and $y \notin U$. $\qquad \square$

Corollary 5.7.11 *Let X be a compact Hausdorff space. The quotient of X by the nearness relation is a profinite space. The equivalence classes for this quotient are exactly the connected components of X.*

Proof By theorem 5.7.9, the equivalence classes for the nearness relation are exactly the connected components. The quotient is obviously compact, as continuous image of a compact. It remains to prove that it is totally disconnected (see theorem 3.4.7).

Let $\Gamma_x \neq \Gamma_y$ be distinct connected components. One has

$$\Gamma_x = \bigcap \{U \mid U \text{ clopen}, \ x \in U\}$$

by theorem 5.7.9. Since $y \notin \Gamma_x$, there exists a clopen U such that $x \in U$ and $y \notin U$. Observe that this clopen U is saturated for the nearness relation, that is, by theorem 5.7.9, the relation "having the same connected component": indeed, if $z \in U$, then $\Gamma_z \subseteq U$ by theorem

5.7.9. Since U is a saturated clopen in X which separates x and y, the image of U in the quotient is still a clopen which saturates the classes of x and y. □

The next result makes more precise a part of the statement of 3.4.6.

Proposition 5.7.12 *The category of profinite spaces is a reflective full subcategory of the category of compact Hausdorff spaces. The reflection of a compact Hausdorff space is its space of connected components.*

Proof Write $\Gamma(X)$ for the space of connected components of the compact Hausdorff space X and $\gamma_X\colon X \longrightarrow\!\!\!\!\!\rightarrow \Gamma(X)$ for the canonical quotient, sending a point $x \in X$ onto its connected component Γ_x. By corollary 5.7.11, the space $\Gamma(X)$ is profinite and γ_X is continuous.

Consider now a profinite space Y together with a continuous map $f\colon X \longrightarrow Y$. We must prove the existence of a continous factorization g,

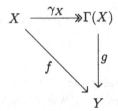

necessarily unique since γ_X is surjective.

Given $x, y \in X$ such that $\Gamma_x = \Gamma_y$, let us prove that $f(x) = f(y)$, from which the definition $g(\Gamma_x) = f(x)$ will make sense. Indeed if $f(x) \neq f(y)$, there is a clopen $U \subseteq Y$ such that $f(x) \in U$ and $f(y) \notin U$. This yields a clopen $f^{-1}(U) \subseteq X$ such that $x \in f^{-1}(U)$ and $y \notin f^{-1}(U)$. But then $\Gamma_x \neq \Gamma_y$ by theorem 5.7.9, which contradicts our assumption. This allows us thus to define $g(\Gamma_x) = f(x)$, yielding a map g such that $g \circ \gamma_X = f$. Since f is continuous and $\Gamma(X)$ is provided with the quotient topology, g is continuous as well. □

5.8 The monotone–light factorization

We shall now describe categorically, in the category of compact Hausdorff spaces, the *monotone* and the *light* maps introduced by Eilenberg (in the metric case) and Whyburn (see [28] and [75]). Let us mention – we shall not prove it and shall not need it – that every continuous map between compact Hausdorff spaces factors as a monotone map followed by a light one.

Definition 5.8.1 Let $f: X \longrightarrow Y$ be a continuous map between compact Hausdorff spaces.

(i) The map f is monotone when the inverse image of each point is connected.

(ii) The map f is light when the inverse image of each point is totally disconnected.

The "opposition" between both notions of monotone and light map is emphasized by corollary 5.7.10. In the situation of definition 5.8.1, for each point $y \in Y$, $f^{-1}(y)$ has exactly one connected component in the monotone case, while in the light case, it has as many connected components as points.

Lemma 5.8.2 *Every monotone map between compact Hausdorff spaces is surjective.*   □

Proposition 5.8.3 *Consider the reflection* $r \dashv i\colon \mathsf{Prof} \xrightleftharpoons{\quad} \mathsf{Comp}$ *of the category of profinite spaces in that of compact Hausdorff spaces (see 5.7.12) and the corresponding factorization system* $(\mathcal{E}, \mathcal{M})$ *(see theorem 5.4.2). For a continuous map* $f\colon X \longrightarrow Y$ *between compact Hausdorff spaces, the following conditions are equivalent:*

(i) f *is monotone;*

(ii) *for every pullback*

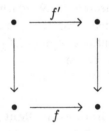

in Comp, *one has* $f' \in \mathcal{E}$, *that is,* $r(f')$ *is an isomorphism.*

Proof In the presence of condition (ii), for every element $y \in Y$, the pullback

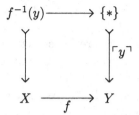

shows that $f^{-1}(y)$ and $\{*\}$ are identified by r, that is, in view of proposition 5.7.12, $f^{-1}(y)$ has exactly one connected component. Thus $f^{-1}(y)$ is connected and non-empty and f is monotone.

Conversely, let f be monotone and thus surjective (lemma 5.8.2). Consider the pullback

$$
\begin{array}{ccc}
P & \xrightarrow{\ h\ } & Z \\
{\scriptstyle k}\downarrow & & \downarrow{\scriptstyle g} \\
X & \xrightarrow[\ f\ ]{} & Y
\end{array}
$$

By construction of a pullback in Comp, if $z \in Z$,

$$h^{-1}(z) = \{(z,x) \mid g(z) = f(x)\} \cong f^{-1}(g(z)).$$

Thus $h^{-1}(z)$ is connected and non-empty since, by assumption, this is the case for $f^{-1}(g(z))$. This proves already that h is monotone. It remains to prove that $r(h)$ is an isomorphism.

Since $h : P \longrightarrow Z$ is monotone, it is surjective. With the notation of the proof of proposition 5.7.12, in the commutative square

$$
\begin{array}{ccc}
P & \xrightarrow{\ h\ } & Z \\
{\scriptstyle \gamma_P}\downarrow & & \downarrow{\scriptstyle \gamma_Z} \\
r(P) & \xrightarrow[\ r(h)\ ]{} & r(Z)
\end{array}
$$

the arrows h, γ_P and γ_Z are surjective, thus $r(h)$ is surjective as well. It remains to prove that $r(h)$ is injective, since a continuous bijection between profinite, thus compact Hausdorff, spaces is a homeomorphism.

Consider thus $a, b \in P$ such that $\Gamma_{h(a)} = \Gamma_{h(b)}$; we must prove that $\Gamma_a = \Gamma_b$. For this observe first that $h^{-1}(\Gamma_{h(a)})$ is connected. Indeed, otherwise one would have $h^{-1}(\Gamma_{h(a)}) = F_1 \cup F_2$ where F_1, F_2 are disjoint non-empty clopens of $h^{-1}(\Gamma_{h(a)})$. Since $h^{-1}(\Gamma_{h(a)})$ is itself closed in P by corollary 5.6.5, F_1 and F_2 are in fact closed in P as well. Let us prove that these closed subsets are saturated for the equivalence relation $\sim$ generated by h, i.e. defined by $c \sim d$ when $h(c) = h(d)$. Indeed given $c \in F_1$, one has $h(c) \in \Gamma_{h(a)}$. We know that $h^{-1}(h(c))$ is connected, because h is monotone. Since F_1, F_2 are clopens in $h^{-1}(\Gamma_{h(a)})$ and $h^{-1}(h(c)) \subseteq h^{-1}(\Gamma_{h(a)})$, the connectedness of $h^{-1}(h(c))$ forces $h^{-1}(h(c)) \subseteq F_1$ or $h^{-1}(h(c)) \subseteq F_2$. We thus have $h^{-1}(h(c)) \subseteq F_1$ because $c \in h^{-1}(h(c)) \cap F_1$. Thus F_1 is saturated for the equivalence relation induced by h, and analogously for F_2. But the continuous surjection $h \colon P \longrightarrow\!\!\!\!\!\rightarrow Z$ between compact Hausdorff spaces necessarily has the quotient topology; indeed, writing $h(P)$ for the image of h provided with the quotient topology, $h(P)$ is compact as continuous image of a compact and thus the identity $h(P) =\!=\!= Z$ is a continuous bijection between compact Hausdorff spaces, thus a homeomorphism. Therefore $h(F_1)$ and $h(F_2)$ remain disjoint non-empty closed subsets of Z and, by surjectivity of h, they cover $hh^{-1}(\Gamma_{h(a)}) = \Gamma_{h(a)}$. But this is a contradiction, because $\Gamma_{h(a)}$ is connected. This completes the proof that $h^{-1}(\Gamma_{h(a)})$ is connected. Analogously, $h^{-1}(\Gamma_{h(b)})$ is connected.

Since $a \in h^{-1}(\Gamma_{h(a)})$ and $b \in h^{-1}(\Gamma_{h(b)})$ and these subsets are connected, we have $h^{-1}(\Gamma_{h(a)}) \subseteq \Gamma_a$ and $h^{-1}(\Gamma_{h(b)}) \subseteq \Gamma_b$. On the other hand, by corollary 5.6.7, $h(\Gamma_a) \subseteq \Gamma_{h(a)}$ and $h(\Gamma_b) \subseteq \Gamma_{h(b)}$. This implies

$$\Gamma_a \subseteq h^{-1}h(\Gamma_a) \subseteq h^{-1}(\Gamma_{h(a)}), \quad \Gamma_b \subseteq h^{-1}h(\Gamma_b) \subseteq h^{-1}(\Gamma_{h(b)})$$

and thus finally, since by assumption $\Gamma_{h(a)} = \Gamma_{h(b)}$,

$$\Gamma_a = h^{-1}(\Gamma_{h(a)}) = h^{-1}(\Gamma_{h(b)}) = \Gamma_b. \qquad \Box$$

Corollary 5.8.4 *The reflection $r \dashv i \colon \mathsf{Prof} \xleftarrow{\quad} \mathsf{Comp}$ of profinite spaces in compact Hausdorff spaces is semi-left-exact.*

Proof If X is a compact Hausdorff space, by definition, the unit of the adjunction $\gamma_X \colon X \longrightarrow \Gamma X$ is such that the inverse image of a point is a connected component of X. Thus γ_X is monotone. Condition (ii) of proposition 5.8.3 implies at once condition (ii) of proposition 5.5.2. $\Box$

We recall now a very classical result; we refer to [58] for a detailed treatment of the Stone–Čech compactifications.

Theorem 5.8.5 (Stone–Čech compactification) *The forgetful functor $|-| \colon \mathsf{Comp} \longrightarrow \mathsf{Set}$ of the category of compact Hausdorff spaces to* *that of sets has a left adjoint which takes values in the category of profinite spaces.*

Proof It is well known that $|-|$ has a left adjoint functor β, namely, the Stone–Čech compactification functor. We recall the construction of β. Given a set X, we consider the map

$$\eta_X : X \longrightarrow 2^X, \quad x \mapsto \{x\}$$

where 2^X is identified with the power set of X, that is, $\{x\}$ is the family $(\varepsilon_y)_{y \in X}$ with $\varepsilon_x = 1$ and $\varepsilon_y = 0$ for $y \neq x$. Viewing 2 as a discrete space, we put on 2^X the product topology and define $\beta(X)$ as the closure of the image of η_X in this product space. Since 2 is finite and discrete, 2^X is profinite by corollary 3.4.8. Thus $\beta(X)$ is closed in a profinite, that is compact, totally disconnected space (see theorem 3.4.7); thus it is still compact and totally disconnected, that is, profinite. $\square$

Lemma 5.8.6 *The unit η of the Stone–Čech adjunction $\beta \dashv |-|$ (see theorem 5.8.5) is injective in each component. The counit σ of the same adjunction is surjective in each component.*

Proof With the notation of theorem 5.8.5, $|\beta(X)|$ is a subset of 2^X in which η_X takes values. The unit of the adjunction is given, for each set X, by the corestriction

$$\eta_X : X \longrightarrow |\beta(X)|, \quad x \mapsto \{x\}$$

of this map η_X; it is indeed injective.

On the other hand, given a compact Hausdorff space Y and writing σ for the counit of the adjunction $\beta \dashv |-|$, the following diagram is one of the triangular identities of this adjunction:

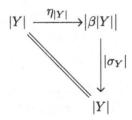

This proves the surjectivity of the map $|\sigma_Y|$, that is, of the continuous map σ_Y. $\square$

Proposition 5.8.7 *The forgetful functor* $|-|\colon \mathsf{Comp}\longrightarrow\mathsf{Set}$ *from the category of compact Hausdorff spaces to that of sets is monadic.*

Proof We know already that this functor has a left adjoint (see 5.8.5). It reflects isomorphisms, since a continuous bijection between compact Hausdorff spaces is necessarily a homeomorphism (see [58]). Since the category of compact Hausdorff spaces has coequalizers, it remains, by the Beck criterion (see [8], volume 2), to prove that the functor $|-|$ preserves the coequalizer of those pairs (u,v) such that $(|u|,|v|)$ has a split coequalizer in the category of sets.

Thus we consider two continuous maps $u,v\colon X\Longrightarrow Y$, with X, Y compact Hausdorff spaces. We suppose they have a split coequalizer in the category of sets; this means the existence of a set Q and maps q, r, s such that

$$|X| \overset{r}{\underset{\substack{|u|\\|v|}}{\Longrightarrow}} |Y| \overset{s}{\underset{q}{\longrightarrow}} Q.$$

$$q\circ s = 1_Q,\quad |u|\circ r = 1_Y,\quad q\circ|u| = q\circ|v|,\quad |v|\circ r = s\circ q.$$

Let us provide Q with the quotient topology, induced by the topology of Y. If we prove that Q is Hausdorff, Q will be compact Hausdorff as image of the compact Hausdorff space Y by the continuous surjection q. It will follow at once that q is the coequalizer of (u,v) in the category of compact Hausdorff spaces, since it is in the category of sets and Q is provided with the quotient topology. But proving the Hausdorffness of Q reduces to proving that its kernel pair is a closed equivalence relation in $Y\times Y$ (see [58]).

First we consider the relation

$$R = \Big\{\big(u(x),v(x)\big)\,\Big|\,x\in X\Big\}$$

which is a closed subspace of $Y\times Y$. Indeed, R is the image of the compact Hausdorff space X by the continuous map

$$X\longrightarrow Y\times Y,\quad x\mapsto \big(u(x),v(x)\big);$$

thus it is a compact, and therefore closed, subset of the compact Hausdorff space $Y\times Y$. The kernel pair of q is the equivalence relation R^+ generated by R, and we must prove it remains closed.

Let us prove that the fact of having a split coequalizer in the category of sets forces R^+ to be $R^{-1}\circ R$, the composite of the relation R with its opposite relation R^{-1}. If $y \in Y$, one has immediately

$$(y, sq(y)) = (ur(y), vr(y)) \in R.$$

Therefore, if $y, y' \in Y$ with $q(y) = q(y')$

$$(y, sq(y)) \in R \text{ and } (sq(y'), y') \in R^{-1} \Rightarrow (y', y) \in R^{-1} \circ R.$$

Thus the kernel pair of q is contained in $R^{-1} \circ R$, while the converse inclusion is obvious.

Now R^{-1} is the image of R by the twisting homeomorphism $\tau: Y \times Y \longrightarrow Y \times Y$, thus it is compact and therefore closed. Next $R^{-1} \circ R$ is the image of the pullback $R^{-1} \times_Y R$ by the projection $p_{1,3}: Y \times Y \times Y \longrightarrow Y \times Y$, thus it is again compact and therefore closed. $\qquad\square$

Proposition 5.8.8 *Consider the reflection* $r \dashv i: \mathsf{Prof} \underset{\longrightarrow}{\overset{\longleftarrow}{}} \mathsf{Comp}$ *of profinite spaces in compact Hausdorff spaces and the corresponding factorization system* $(\mathcal{E}, \mathcal{M})$. *For a continuous map* $f: X \longrightarrow Y$ *between compact Hausdorff spaces, the following conditions are equivalent:*

(i) f *is light;*

(ii) *in the following pullback, where* $|Y|$ *denotes the underlying set of* Y *and* σ *is the counit of the Stone–Čech adjunction (see theorem 5.8.5) the space* P *is profinite:*

$$
\begin{array}{ccc}
P & \overset{g}{\longrightarrow} & X \\
{\scriptstyle h}\downarrow & & \downarrow{\scriptstyle f} \\
\beta|Y| & \underset{\sigma_Y}{\longrightarrow\!\!\!\rightarrow} & Y
\end{array}
$$

(iii) *in the above pullback,* $h \in \mathcal{M}$;

(iv) $(X, f) \in \mathsf{Comp}/Y$ *is split by* σ_Y.

Proof (i) $\Rightarrow$ (ii) By corollary 5.7.10, it suffices to prove that given $(z, x) \in P$, its connected component $\Gamma_{(z,x)}$ is reduced to the point (z, x). By corollary 5.6.7, $h(\Gamma_{(z,x)})$ is connected, thus

$$h(\Gamma_{(z,x)}) \subseteq \Gamma_{h(z,x)} = \Gamma_z = \{z\},$$

since $\beta|Y|$ is profinite (see theorem 5.8.5 and corollary 5.7.10). It follows that $\Gamma_{(z,x)} = \{z\} \times g(\Gamma_{(z,x)})$. By lemma 5.6.6, $g(\Gamma_{(z,x)})$ is connected and contained in $f^{-1}(\sigma_Y(z))$. But since f is light, $f^{-1}(\sigma_Y(z))$ is totally disconnected (see definition 5.8.1), from which $g(\Gamma_{(z,x)})$ has only one element. This proves that $\Gamma_{(z,x)} = \{(z,x)\}$ and P is totally disconnected by corollary 5.7.10.

(ii) $\Leftrightarrow$ (iii) Consider the following commutative diagram:

$$
\begin{array}{ccc}
P & \xrightarrow{\;\gamma_P\;} & \Gamma(P) \\
{\scriptstyle h}\downarrow & & \downarrow{\scriptstyle r(h)} \\
\beta|Y| & \xrightarrow[\gamma_{\beta|Y|}]{} & \Gamma(\beta|Y|)
\end{array}
$$

By theorem 5.8.5, $\beta|Y|$ is profinite; via corollary 5.8.4 and proposition 5.5.2, we get

$$
\begin{aligned}
h \in \mathcal{M} \quad &\Leftrightarrow \quad \text{the square above is a pullback} \\
&\Leftrightarrow \quad \gamma_P \text{ is an isomorphism} \\
&\Leftrightarrow \quad P \text{ is totally disconnected.}
\end{aligned}
$$

Indeed, since $\beta|Y|$ is totally disconnected, $\gamma_{\beta|Y|}$ is an isomorphism by corollary 5.7.10.

(iii) $\Leftrightarrow$ (iv) This holds by definition 5.5.4 and proposition 5.5.6.

(ii) $\Rightarrow$ (i) For each element $y \in Y$, applying lemma 5.8.6, let us choose $z \in \beta|Y|$ such that $\sigma_Y(z) = y$. This yields

$$
h^{-1}(z) = \{(z,x)\,|\,\sigma_Y(z) = f(x)\} \cong f^{-1}\big(\sigma_Y(z)\big) = f^{-1}(y).
$$

But $h^{-1}(z)$ is totally disconnected as a subspace of the totally disconnected space P. This proves that $f^{-1}(y)$ is totally disconnected and thus f is light. $\square$

Theorem 5.8.9 *For every compact Hausdorff space Y,*

(i) *the morphism $\sigma_Y\colon \beta|Y| \longrightarrow Y$, which is the counit of the Stone–Čech adjunction (see theorem 5.8.5), is a morphism of Galois descent with respect to the adjunction $r \dashv i\colon \mathsf{Prof} \overset{\longrightarrow}{\underset{\longleftarrow}{}} \mathsf{Comp}$ between compact Hausdorff spaces and profinite spaces,*

(ii) *the objects $(X,f) \in \mathsf{Comp}/Y$ which are split by σ_Y are precisely those for which $f\colon X \longrightarrow Y$ is light.*

Proof   The arrow $\sigma_Y\colon \beta|Y| \longrightarrow\!\!\!\!\!\longrightarrow Y$ is a surjection between compact Hausdorff spaces, thus is a topological quotient. By proposition 5.8.7

and lemma 4.4.6, σ_Y is thus an effective descent morphism. We also know, by corollary 5.8.4, that the reflection $r \dashv i$ is semi-left-exact. Referring to corollary 5.5.8, it remains to verify that in the pullback

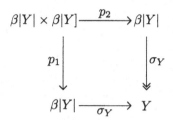

one has $p_1 \in \mathcal{M}$. By proposition 5.8.8, this is equivalent to proving that $\sigma_Y \colon \beta|Y| \longrightarrow Y$ is light, which is indeed the case since each $\sigma_Y^{-1}(y)$ is totally disconnected as a subspace of the totally disconnected space $\beta|Y|$ (see theorem 5.8.5). An alternative argument consists in observing that p_1 is in **Prof**, and applying the last statement of theorem 5.4.2. This proves that σ_Y is a morphism of Galois descent.

The second assertion follows from proposition 5.8.8. $\square$

To conclude this chapter, let us recall the Gelfand duality theorem between the category of commutative $\mathbb{C}^*$-algebras and that of compact Hausdorff spaces. With every commutative $\mathbb{C}^*$-algebra, one associates the spectrum of its locale of closed ideals, which is a compact Hausdorff space. And with every compact Hausdorff space X, one associates the $\mathbb{C}^*$-algebra $\mathcal{C}(X, \mathbb{C})$ of continuous maps. This yields a contravariant equivalence of categories.

Theorem 5.8.9, via the Gelfand duality, can thus also be seen as a Galois theorem for commutative $\mathbb{C}^*$-algebras, the corresponding Galois equivalence being this time contravariant, as in the case of rings.

6

Covering maps

Those Galois theories which involve adjunctions with the category of sets are especially simple and natural; in particular their Galois groups are really the automorphism groups of extensions. The main purpose of this chapter is to describe the classical theory of covering maps of locally connected topological spaces as such a Galois theory.

6.1 Categories of abstract families

For a family $(A_i)_{i \in I}$ of objects of a category $\mathcal{A}$ we will write

$$(A_i)_{i \in I} = A = (A_i)_{i \in I(A)}$$

considering I as a functor

$$I \colon \mathsf{Fam}(\mathcal{A}) \longrightarrow \mathsf{Set}$$

from the category of all families of objects in $\mathcal{A}$ to the category of sets.

According to that notation, a morphism $\alpha \colon A \longrightarrow B$ in $\mathsf{Fam}(\mathcal{A})$ consists of a map $I(\alpha) \colon I(A) \longrightarrow I(B)$ and an $I(A)$-indexed family of morphisms $\alpha_i \colon A_i \longrightarrow B_{I(\alpha)(i)}$. That is, in fact the category $\mathsf{Fam}(\mathcal{A})$ and the functor I are constructed simultaneously.

The categories of families often occur in geometry:

Proposition 6.1.1 *The category* $\mathsf{Fam}(\mathsf{CTop})$, *where* CTop *is the category of connected topological spaces, is equivalent to the category of topological spaces with open connected components.*

Proof Let $A = \coprod_{i \in I} A_i$ be a coproduct in the category Top of topological spaces, that is, A is a topological space which is the disjoint union of the open subsets A_i. Assume that each A_i is connected. To give a continuous map from A to a topological space is the same as to give a continuous map from each A_i to that space. On the other hand every continuous map from a connected space to A has its image contained in one of the A_i. Therefore the proof is straightforward: any topological space with open connected components corresponds to the family of its connected components. □

This and other similar examples of $\mathsf{Fam}(\mathcal{A})$, some of which will be mentioned at the end of this section, suggest various categorical properties for the general $\mathsf{Fam}(\mathcal{A})$, related to the notion of connectedness. As we will see, they easily follow from the fact that each $A \in \mathsf{Fam}(\mathcal{A})$ can be considered as the coproduct

$$A = \coprod_{i \in I(A)} A_i$$

where A_i $\big(i \in I(A)\big)$ are considered as one member families. In particular this implies that $\mathsf{Fam}(\mathcal{A})$ admits all (small) coproducts, and gives the following.

Proposition 6.1.2 *For an object C in* $\mathsf{Fam}(\mathcal{A})$*, the following conditions are equivalent:*

  (i) *the functor*

$$\mathsf{Hom}(C, -)\colon \mathsf{Fam}(\mathcal{A}) \longrightarrow \mathsf{Set}$$

    preserves coproducts;

  (ii) $C \cong X \amalg Y$ *implies either X or Y is the empty family (= the initial object 0 in* $\mathsf{Fam}(\mathcal{A})$*), and $C \neq 0$;*

  (iii) $C \cong X \amalg Y$ *implies either X or Y is canonically isomorphic to C, and $C \neq 0$;*

  (iv) $I(C) = 1$ *(i.e. C is a one member family).* □

The objects C satisfying the equivalent conditions of proposition 6.1.2 are called *connected*. If we replace $\mathsf{Fam}(\mathcal{A})$ by an arbitrary category $\mathcal{C}$, then the connectedness is to be defined via a condition similar to 6.1.2(i), i.e. we introduce

Definition 6.1.3 An object C in a category $\mathcal{C}$ with coproducts is said to be connected if the functor $\mathsf{Hom}(C, -)\colon \mathcal{C} \longrightarrow \mathsf{Set}$ preserves coproducts.

And we obtain

Proposition 6.1.4 *A topological space is connected in the usual sense if and only if it is a connected object in* Top. □

Proposition 6.1.5 *The following conditions on a category C are equivalent:*

(i) C *admits coproducts and every object in C can be presented as a coproduct of connected objects;*

(ii) *the same condition, requiring in addition that the presentation is unique up to an isomorphism;*

(iii) C *is equivalent to a category of the form* Fam(A);

(iv) C *is canonically equivalent (up to choice of coproducts) to the category* Fam(A), *where A is the category of connected objects in C.* □

Since not every topological space has open connected components, the category Top does not satisfy these equivalent conditions. For example the subspace $\{1, \frac{1}{2}, \frac{1}{3}, \ldots\} \cup \{0\}$ of $\mathbb{R}$ is not the coproduct of its connected components (all of which are one element sets).

Readers are invited to deduce proposition 6.1.1 from proposition 6.1.5 and to check that the following categories also satisfy the equivalent conditions of 6.1.5:

- the category of small categories and various "similar" categories, such as the category of n-categories (strict or weak, whatever the term "weak" means), n-groupoids or multiple categories or groupoids, also with ω- instead of n-, etc.; indeed, coproducts in these cases are disjoint unions;

- categories of graphs, oriented or not (and again n-, ω-, multiple, etc.);

- various categories of relational systems like preorders, posets, sets provided with a reflexive or arbitrary relation, etc.;

- any category of the form Set^K, where K is a small category; in particular the category of sets itself, or of simplicial sets, or of presheaves over a topological space or a locale, or of sets provided with a monoid action (to investigate these examples, observe that a functor $F: K \longrightarrow \mathsf{Set}$ is connected in Set^K when its category of elements is connected in Cat);

- any category of the form C/B where C satisfies the equivalent conditions of 6.1.5, and B is an arbitrary ("base") object in C.

6.2 Some limits in Fam($\mathcal{A}$)

We begin with

Proposition 6.2.1 *For a category $\mathcal{A}$ the following conditions are equivalent:*

  (i) $\mathcal{A}$ *has a terminal object;*

  (ii) Fam($\mathcal{A}$) *has a connected terminal object;*

  (iii) Fam($\mathcal{A}$) *has a terminal object which is a terminal object in $\mathcal{A}$ (regarding the objects of $\mathcal{A}$ as one member families).*   □

Assuming from now on that these equivalent conditions hold, we obtain the functors

$$\mathsf{Fam}(\mathcal{A}) \underset{\underset{\Gamma}{\overset{H}{\longleftarrow}}}{\overset{I}{\longrightarrow}} \mathsf{Set},$$

where H is the right adjoint of I and Γ is the right adjoint of H. Explicitly

$$H(S) = S \cdot 1 = \coprod_S 1,$$

i.e. H sends a set S to the coproduct of "S copies" of the terminal object 1 ($=$ to the S-indexed family of terminal objects) and

$$\Gamma(A) = \mathsf{Hom}(1, A)$$

for every $A \in \mathsf{Fam}(\mathcal{A})$.

Remark 6.2.2 The adjunctions $I \dashv H \dashv \Gamma$ can of course be established and all details checked by a straightforward calculation. However for readers familiar with 2-categories, it would also be nice to look at them from the 2-categorical viewpoint.

  (i) Fam($\mathcal{A}$) can be described as a "free coproduct completion" of $\mathcal{A}$, and this makes Fam: Cat $\longrightarrow$ Cat a 2-functor (it is strict due to the explicit construction of Fam we are using).

  (ii) $\mathcal{A}$ has a terminal object if and only if there exists an adjunction

$$\mathcal{A} \underset{\longleftarrow}{\overset{\longrightarrow}{\rule{0pt}{0pt}}} \mathbf{1}$$

where $\mathbf{1}$ is the terminal object in CAT, that is, the category $\mathcal{9}$.

(iii) Fam, being a 2-functor, preserves adjunctions and therefore gives the induced adjunction

$$\mathsf{Fam}(A) \overset{\longrightarrow}{\longleftarrow} \mathsf{Fam}(1)$$

which is just $I \dashv H$ (note that $\mathsf{Fam}(1)$ is nothing but Set).

(iv) For every category C with coproducts (in fact it is sufficient to have copowers) and every object $C \in C$ there exists a unique (up to an isomorphism) adjunction

$$C \overset{F}{\underset{G}{\longleftarrow}} \mathsf{Set}, \quad F \dashv G$$

with $F(1) = C$; this can be described as $F(S) = S \cdot 1$ and $G = \mathrm{Hom}(C, -)$, and so $H \dashv \Gamma$ is a special case of it.

(v) Moreover identifying (up to an equivalence) $\mathsf{Fam}(A)$ with a full subcategory of $\mathsf{Set}^{A^{\mathrm{op}}}$, one could present $I \dashv H$ and $H \dashv \Gamma$ as special cases of the same construction. $\qquad\square$

Now we are going to describe certain pullbacks in $\mathsf{Fam}(A)$. The first lemma however looks better when it is formulated for general limits first:

Lemma 6.2.3 *Let K be a small category and $D \colon K \longrightarrow \mathsf{Fam}(A)$ a functor. For every $x = (x_K)_{K \in K} \in \lim ID$ let D_x be the functor $K \longrightarrow A$ defined by*

$$
\begin{array}{ccc}
K & & D(K)_{x_K} \\
{\scriptstyle f}\Big\downarrow & \longmapsto & \Big\downarrow{\scriptstyle D(f)_{x_K}} \\
K' & & D(K')_{x_{K'}}
\end{array}
$$

If each D_x has a limit

$$\lim D_x = \Big(L_x, \big(\pi_{x,K} \colon L_x \longrightarrow D_x(K)\big)_{K \in K} \Big),$$

then D also has a limit, which can be described as

$$\lim D = \Big(L, \big(\pi_K \colon L \longrightarrow D(K)\big)_{K \in K} \Big)$$

where

- $I(L) = \lim ID$,
- *for $x \in I(L)$, L_x is as above,*

- $I(\pi_K)$ *is the projection* $\lim ID \longrightarrow ID(K)$, *i.e.* $I(\pi_K)(x_K)_{K \in \mathcal{K}} = x_K$,
- $(\pi_K)_x = \pi_{x,K} : L_x \longrightarrow D(K)_{x_K}$.

Proof is again straightforward, but since so many letters and indices are involved, it is better to sketch it.

The fact that the collection of π_K, $K \in \mathcal{K}$, forms a cone reduces to the commutativity of the diagrams

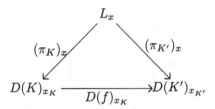

(for all f and x), which are the same as the diagrams

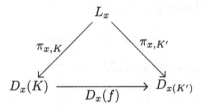

and these commute because each of the collections of $\pi_{x,K}$ forms a cone for the corresponding D_x.

Now, for proving the universal property of the limit, we take another cone

$$\left(L', (\pi'_K : L' \longrightarrow D(K))_{K \in \mathcal{K}}\right)$$

and it is easy to check that the existence and uniqueness of a morphism from it to the cone described above reduce to the existence and uniqueness of morphisms from the cone

$$\left(L'_i, ((\pi'_K)_i : L'_i \longrightarrow D(K)_{I(\pi'_K)(i)})_{K \in \mathcal{K}}\right)$$

(over D_x, where $x_K = I(\pi'_K)(i)$) to the limiting cone $\lim D_x$. $\square$

Corollary 6.2.4 *Let*

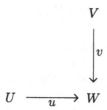

be a diagram in $\mathsf{Fam}(\mathcal{A})$ *such that for each* $(s,t) \in I(U) \times_{I(W)} I(V)$ *there exists a pullback*

$$
\begin{array}{ccc}
P_{(s,t)} & \xrightarrow{\ q_{s,t}\ } & V_t \\
{\scriptstyle p_{s,t}}\big\downarrow & & \big\downarrow{\scriptstyle v_t} \\
U_s & \xrightarrow[\ u_s\ ]{} & W_{I(u)(s)}\ \big(= W_{I(v)(t)}\big)
\end{array}
$$

in $\mathcal{A}$. *Then the family* P *of all* $P_{(s,t)}$ *together with the morphisms* $p_{s,t}$ *and* $q_{s,t}$ *forms a pullback of* u *and* v. $\square$

In particular consider a pullback of the form

$$
\begin{array}{ccc}
B \times_{HI(B)} H(X) & \xrightarrow{\ \pi_2\ } & H(X) \\
{\scriptstyle \pi_1}\big\downarrow & & \big\downarrow{\scriptstyle H(\varphi)} \\
B & \xrightarrow[\ \eta_B\ ]{} & HI(B),
\end{array}
$$

where

- $\eta \colon 1 \longrightarrow HI$ is the unit of the adjunction $I \dashv H$,
- B any object in $\mathsf{Fam}(\mathcal{A})$,
- X any set and φ a map from X to $I(B)$.

Identifying HI with the identity functor of Set, we can describe the object $B \times_{HI(B)} H(X)$ as follows.

Corollary 6.2.5 $B \times_{HI(B)} H(X)$ *is the family* $\big(B_{\varphi(x)}\big)_{x \in X}$ *where*

- *the projection* $\pi_1 \colon \big(B_{\varphi(x)}\big)_{x \in X} \longrightarrow (B_i)_{i \in I(B)}$ *is determined by the map* $\varphi \colon X \longrightarrow I(B)$ *and the family* $B_{\varphi(x)} \longrightarrow B_{\varphi(x)}$, $x \in X$, *of identity morphisms,*

- *the projection* $\pi_2 \colon \left(B_{\varphi(x)}\right)_{x \in X} \longrightarrow (1)_{x \in X}$ *is determined by the identity map* $X \longrightarrow X$ *and the family* $B_{\varphi(x)} \longrightarrow 1$, $x \in X$, *of morphisms.*

In particular that pullback always exists. □

If $X = 1$, i.e. X is a one element set, then φ is determined by its image, which is a one element subset $\{i\}$ in $I(B)$; we will write $\varphi = \tilde{i}$. The inclusion $\{i\} \longrightarrow I(B)$ together with the identity morphism $B_i \longrightarrow B_i$ determines a morphism $[i] \colon B_i \longrightarrow B$, where B_i is considered as a one member family. Applying corollary 6.2.5 (up to an isomorphism) we obtain

Corollary 6.2.6 *The diagram*

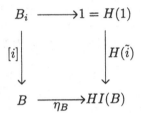

is a pullback for each $i \in I(B)$. *That is, the "connected components" of B can be described as the pullbacks of η_B along (H-images of) all possible maps from a (fixed) one element set $I(B)$.* □

Finally, let us mention the "most straightforward" consequence of lemma 6.2.3:

Corollary 6.2.7 *The canonical embedding* $\mathcal{A} \longrightarrow \mathsf{Fam}(\mathcal{A})$ *preserves all limits which exist in* $\mathcal{A}$. □

Readers familiar with fibrations of categories will of course recognize that all these observations on limits are special cases of simple and more elegant results (on general fibrations). Indeed, the functor $I \colon \mathsf{Fam}(\mathcal{A}) \longrightarrow \mathsf{Set}$ of 6.1 is the basic example of a fibration.

6.3 Involving extensivity

Having in mind the adjunction $I \dashv H$ between $\mathsf{Fam}(\mathcal{A})$ and Set, let us recall (with an adapted notation) from chapter 5:

For an abstract adjunction

$$C \underset{H}{\overset{I}{\rightleftarrows}} \mathcal{X}, \quad \eta\colon \mathrm{id}_C \Rightarrow HI, \quad \varepsilon\colon IH \Rightarrow \mathrm{id}_{\mathcal{X}}$$

between categories C and $\mathcal{X}$ with pullbacks, and an object B in C, there is an induced adjunction

$$C/B \underset{H^B}{\overset{I^B}{\rightleftarrows}} \mathcal{X}/I(B), \quad \eta^B\colon \mathrm{id}_{C/B} \Rightarrow H^B I^B, \quad \varepsilon^B\colon I^B H^B \Rightarrow \mathrm{id}_{\mathcal{X}/I(B)}$$

in which

- $I^B(A,\alpha) = \big(I(A), I(\alpha)\big)$, i.e. I^B sends a morphism $\alpha\colon A \longrightarrow B$, considered as an object (A,α) in C/B, to $I(\alpha)\colon I(A) \longrightarrow I(B)$ considered as $\big(I(A), I(\alpha)\big) \in \mathcal{X}/I(B)$,
- $H^B(X,\varphi) = \big(B \times_{HI(B)} H(X), \pi_1\big)$ constructed as the pullback

$$
\begin{array}{ccc}
B \times_{HI(B)} H(X) & \xrightarrow{\ \pi_2\ } & H(X) \\
{\scriptstyle \pi_1}\downarrow & & \downarrow{\scriptstyle H(\varphi)} \\
B & \xrightarrow[\ \eta_B\ ]{} & HI(B)
\end{array}
$$

- $\eta^B_{(A,\alpha)} = \langle \alpha, \eta_A \rangle\colon A \longrightarrow B \times_{HI(B)} HI(A)$, i.e. $\eta^B_{(A,\alpha)}$ is the morphism making diagram 6.1 commute,
- $\varepsilon^B_{(X,\varphi)}$ is the composite

$$I\big(B \times_{HI(B)} H(X)\big) \xrightarrow{\ I(\pi_2)\ } IH(X) \xrightarrow{\ \varepsilon_X\ } X$$

in the notation above.

Corollary 6.2.5 indicates very clearly that in our special case $(I \dashv H)\colon \mathrm{Fam}(\mathcal{A}) \longrightarrow \mathrm{Set}$, this description is to be simplified. In order to do this nicely, we need to recall

Definition 6.3.1 A category C with coproducts is said to be (infinitary) extensive if for every family $(C_\lambda)_{\lambda \in \Lambda}$ of objects of C the coproduct functor

$$\amalg\colon \prod_{\lambda \in \Lambda} C/C_\lambda \longrightarrow C \Big/ \coprod_{\lambda \in \Lambda} C_\lambda,$$

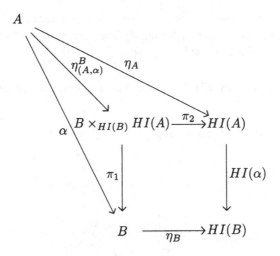

Diagram 6.1

$$(f_\lambda \colon A_\lambda \to C_\lambda)_{\lambda \in \Lambda} \mapsto \left(\coprod_{\lambda \in \Lambda} f_\lambda \colon \coprod_{\lambda \in \Lambda} A_\lambda \to \coprod_{\lambda \in \Lambda} C_\lambda \right)$$

is a category equivalence.

Proposition 6.3.2 *Every category of the form* Fam($\mathcal{A}$) *is extensive.*

Proof Let us just recall the obvious construction of the inverse functor:
Since $I\left(\coprod_{\lambda \in \Lambda} C_\lambda \right)$ is the disjoint union of the sets $I(C_\lambda)$ $(\lambda \in \Lambda)$, for a given object

$$f \colon A \longrightarrow \coprod_{\lambda \in \Lambda} C_\lambda$$

in $\mathcal{C}/ \coprod_{\lambda \in \Lambda} C_\lambda$, we define A_λ as the "subfamily"

$$A_\lambda = (A_i)_{i \in I(f)^{-1}(I(C_\lambda))}$$

of A, and define

$$f_\lambda \colon A_\lambda \longrightarrow C_\lambda$$

as the restriction of f to that subfamily. The correspondence

$$(A, f) \mapsto ((A_\lambda, f_\lambda))_{\lambda \in \Lambda}$$

determines the desired inverse functor from $\mathcal{C}/ \coprod_{\lambda \in \Lambda} C_\lambda$ to $\prod_{\lambda \in \Lambda} \mathcal{C}/C_\lambda$.
It is a routine calculation to check that both of the composites are identity functors (up to isomorphism of course). □

In the next proposition we will write $I_A \dashv H_A$ instead of $I \dashv H$, which we must do because not just $\mathsf{Fam}(\mathcal{A})$ but also $\mathsf{Fam}(\mathcal{A}/B_i)$ (see below) is involved. However, we will keep using $I^B \dashv H^B$.

Proposition 6.3.3 *There is a canonical equivalence of adjunctions*

$$
\begin{array}{ccc}
\mathsf{Fam}(\mathcal{A})/B & \underset{H^B}{\overset{I^B}{\rightleftarrows}} & \mathsf{Set}/I(B) \\[2mm]
\wr \updownarrow & & \updownarrow \wr \\[2mm]
\prod_{i\in I(B)} \mathsf{Fam}(\mathcal{A}/B_i) & \underset{\prod_{i\in I(B)} H_{\mathcal{A}/B_i}}{\overset{\prod_{i\in I(B)} I_{\mathcal{A}/B_i}}{\rightleftarrows}} & \prod_{i\in I(B)} \mathsf{Set}.
\end{array}
$$

Proof   Since $\mathsf{Fam}(\mathcal{A})$ is extensive we have

$$
\mathsf{Fam}(\mathcal{A})/B \sim \mathsf{Fam}(\mathcal{A}) \Big/ \coprod_{i\in I(B)} B_i
$$

$$
\sim \prod_{i\in I(B)} \mathsf{Fam}(\mathcal{A})/B_i \sim \prod_{i\in I(B)} \mathsf{Fam}(\mathcal{A}/B_i)
$$

and since $\mathcal{A}/B_i$ has a terminal object, $I_{\mathcal{A}/B_i}$ and $H_{\mathcal{A}/B_i}$ are well defined. The comutativity of the square involving I^B and $\prod_{i\in I(B)} I_{\mathcal{A}/B_i}$ is obvious.   □

There are various conclusions and remarks to be made. Let us put them as

Conclusions 6.3.4

(i) The alternative description of the adjunction $I^B \dashv H^B$ given by proposition 6.3.3 should be considered as the "external version" of the description given above. This is a good example of a situation where the external description is simpler than the internal, although the internal one also works in more general situations (namely, for arbitrary adjunctions).

(ii) Proposition 6.3.3 appropriately reformulated tells us that $\mathsf{Fam}(\mathcal{A})$ always has those pullbacks which are needed to construct $I^B \dashv H^B$ for any $B \in \mathsf{Fam}(\mathcal{A})$, which also follows from corollary 6.2.5.

On the other hand since the pullback constructed in 6.2.5 is involved in the construction of H^B, one can deduce 6.2.5 from 6.3.3 without using lemma 6.2.3 (or corollary 6.2.4).

(iii) Proposition 6.3.3 and corollary 6.2.5 also independently tell us that the counit $\varepsilon^B : I^B H^B \Rightarrow \mathrm{id}$ is an isomorphism for any $\mathcal{A}$ and any $B \in \mathsf{Fam}(\mathcal{A})$.

(iv) The fact that "the extensivity helped us to remove pullbacks from the construction of $I^B \dashv H^B$" is not at all surprising: just note that the right adjoint of the coproduct functor used in definition 6.3.1 is given by the family of the pullback functors along all coproduct injections $C_\lambda \longrightarrow \coprod_{\lambda \in \Lambda} C_\lambda$ (and those pullbacks therefore exist in any extensive category).

(v) The whole story can of course be repeated in the more general context where the existence of a terminal object in $\mathcal{A}$ is not required. Although the functor H itself would not exist, we will still have the adjunctions $I^B \dashv H^B$, $B \in \mathcal{C}$, defined now via the equivalence of proposition 6.3.3, since each $\mathcal{A}/B_i$ has a terminal object and therefore we can use $I_{\mathcal{A}/B_i} \dashv H_{\mathcal{A}/B_i}$, $i \in I(B)$. $\qquad \square$

6.4 Local connectedness and étale maps

The category Top of topological spaces is extensive, but not of the form $\mathsf{Fam}(\mathcal{A})$. On the other hand the category of topological spaces with open connected components (see proposition 6.1.1), which is the "obvious best replacement" of Top by a category of the form $\mathsf{Fam}(\mathcal{A})$, does not have pullbacks. In this section we consider the category LoCo with objects all locally connected topological spaces (see 6.4.3 below) and morphisms all étale maps (= local homeomorphisms) between them. This category has "all nice properties" and is still large enough to contain all covering maps which categorical Galois theory is to be applied for.

The étale maps were already considered in the previous chapters (see definition 4.2.13); let us begin here with

Proposition 6.4.1 *A continuous map $\alpha \colon A \longrightarrow B$ is étale if and only if it is open, and locally injective, i.e. every $a \in A$ has an open neighbourhood U such that the restriction $\alpha|_U \colon U \longrightarrow B$ is injective.* $\qquad \square$

Also note that for any family $(U_\lambda)_{\lambda \in \Lambda}$ of open subsets of a topological space B, the canonical map

$$\coprod_{\lambda \in \Lambda} U_\lambda \longrightarrow B$$

is étale; we call this map the map associated with the family $(U_\lambda)_{\lambda \in \Lambda}$. In fact it is more convenient to make this construction "up to a homeomorphism", i.e. to replace a family of open subsets by a family of spaces equipped with injective homeomorphisms into the space B.

Proposition 6.4.2 *For every étale map $p \colon E \longrightarrow B$ there exist a family $(U_\lambda)_{\lambda \in \Lambda}$ of open subsets in B and a factorization*

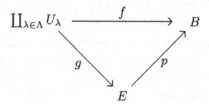

in which f is the map associated with the family $(U_\lambda)_{\lambda \in \Lambda}$ and g is a surjective étale map. Moreover, g also is the map associated with $(U_\lambda)_{\lambda \in \Lambda}$ (up to a homeomorphism).

Proof   Just take $(V_\lambda)_{\lambda \in \Lambda}$ to be the family of all open subsets in E for which the restriction p is injective, and take $U_\lambda = p(V_\lambda)$ for all $\lambda \in \Lambda$. $\qquad \square$

The next proposition explains the reason why the category LoCo is better for our purposes than the category Top:

Proposition 6.4.3 *For a topological space B the following conditions are equivalent:*

- (i) *B is locally connected, i.e. every open subset in B has open connected components;*
- (ii) *for every étale map $\alpha \colon A \longrightarrow B$, the space A has open connected components;*
- (iii) *for every étale map $\alpha \colon A \longrightarrow B$, the space A is locally connected.*

Proof   (iii) $\Rightarrow$ (ii) $\Rightarrow$ (i) is trivial; (i) $\Rightarrow$ (ii) follows from the obvious fact that a topological space which is a union of open subsets with open connected components itself has the same property; (ii) $\Rightarrow$ (iii) follows from (i) $\Leftrightarrow$ (ii) and the (easy) fact that the class of étale maps is closed under composition. $\qquad \square$

We will also need

Lemma 6.4.4 *Let*

$$X \underset{g}{\overset{f}{\rightrightarrows}} Y \overset{h}{\longrightarrow} Z$$

be a coequalizer diagram in Top, *in which f and g are open maps. Then the map h is also open.*

Proof Just note that since h is a quotient map, a subset Z' in Z is open if and only if its inverse image $h^{-1}(Z')$ is open. Whenever Y' is open in Y, so are also $fg^{-1}(Y')$ and $gf^{-1}(Y')$ and $hh^{-1}(Y')$ is obtained from Y' by taking the union of all possible iterations $fg^{-1}(-)$ and $gf^{-1}(-)$. □

Lemma 6.4.5 *Let*

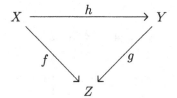

be a commutative diagram in Top *with f étale. Then*

   (i) *if g is étale, so is h,*
   (ii) *if h is open and surjective, then g and h are étale.*

Proof (i) For a point x in X choose open subsets $X' \subseteq X$ and $Y' \subseteq Y$ such that x is in X', $h(x)$ in Y', and the restrictions $f|_{X'}$ and $g|_{Y'}$ are injective. Then take $U = X' \cap h^{-1}(Y')$ and consider the commutative diagram

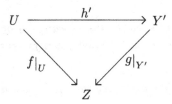

in which h' is induced by h.

Since U is a subset of X', the map $f|_U$ is injective, and therefore so is h'. On the other hand since $g|_{Y'}$ is also injective, for every $V \subseteq U$ we have

$$h'(V) = (g|_{Y'})^{-1} f|_U(V),$$

and so since $f|_U$ is open, so is h'.

Since U is open (which uses the fact that h is continuous!), and this can be done for any x in X, we conclude that h is étale.

(ii) Of course we only need to prove that g is étale. For any $y \in Y$ take $x \in X$ with $h(x) = y$ and an open neighbourhood X' of x for which f induces a homeomorphism $X' \longrightarrow f(X')$. Then $h(X')$ is an open neighbourhood of y for which $g|_{h(X')}$ is injective. That is g is locally injective. On the other hand since f is open and h surjective, we know that g is open: indeed, just note that for every $Y' \subseteq Y$ we have $g(Y') = f(h^{-1}(Y'))$. $\qquad\square$

Lemma 6.4.6 *If*

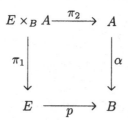

is a pullback diagram in Top *with α étale, then π_1 also is étale.*

Proof For any $(e, a) \in E \times_B A$ choose an open neighbourhood U of a for which α induces a homemorphism $U \longrightarrow \alpha(U)$; then $\pi_2^{-1}(U) \approx E \times_B U$ is open in $E \times_B A$ and π_1 induces a homeomorphism

$$\pi_2^{-1}(U) \longrightarrow p^{-1}(\alpha(U)) \approx E \times_B \alpha(U). \qquad\square$$

Proposition 6.4.7 *Let* Étale *be the subcategory of* Top *with the same objects and morphisms all étale maps. Then for every topological space B we have*

(i) Étale$/B$ *is a full subcategory in* Top$/B$ *closed under colimits and finite limits,*

(ii) *the forgetful functor* Étale$/B \longrightarrow$ Set *reflects isomorphisms and preserves colimits and pullbacks.*

Proof (i) The fact that Étale/B is a full subcategory of Top/B is just a reformulation of 6.4.5(i). It is obviously closed under coproducts, closed under coequalizers by 6.4.4 and 6.4.5(ii), and therefore closed under all colimits. It also contains the terminal object $(B, 1_B)$ and is closed under pullbacks by 6.4.6 – and so it is closed under all finite limits.

(ii) The reflection of isomorphisms is obvious, and the preservation of colimits and finite limits follows from (i). □

Finally, let us list "all the nice properties" of LoCo, which easily follow from the previous results:

Proposition 6.4.8

(i) LoCo *is canonically equivalent to* Fam(CLoCo), *where* CLoCo *is the full subcategory in* LoCo *with objects all connected locally connected spaces.*

(ii) *For every locally connected space B,* LoCo/B *is a full subcategory in* Top/B *closed under colimits and finite limits, and the forgetful functor* LoCo/B ⟶ Set *reflects isomorphisms and preserves colimits and pullbacks.*

(iii) *All pullback functors in* LoCo *(and also in* Étale*) preserve colimits.* □

A fundamental property definitely missed in this list is the fact that each LoCo/B is a topos. However, we prefer not to use it in order to make clear that the theory of covering spaces can be developed categorically without any use of topos theory.

6.5 Localization and covering morphisms

Let $\mathcal{M}$ be a class of continuous maps between topological spaces. Classically one says that a map $\alpha\colon A \longrightarrow B$ is locally in $\mathcal{M}$ if every point b in B has an open neighbouhood U such that the map $\alpha^{-1}(U) \longrightarrow U$ induced by α belongs to $\mathcal{M}$. Equivalently, there exists a family $(U_\lambda)_{\lambda \in \Lambda}$ of open subsets in B such that $B = \bigcup_{\lambda \in \Lambda} U_\lambda$ and each $\alpha^{-1}(U_\lambda) \longrightarrow U_\lambda$ belongs to $\mathcal{M}$.

Let us make some simple remarks on that notion of "being locally in $\mathcal{M}$", always assuming

Convention *Throughout this section, $\mathcal{M}$ denotes a pullback stable class of morphisms in the category* Top *of topological spaces and continuous maps.*

Proposition 6.5.1 *Let*

$$E \times_B A \xrightarrow{\ \pi_2\ } A$$

$$\pi_1 \downarrow \qquad\qquad \downarrow \alpha$$

$$E \xrightarrow{\ p\ } B$$

be a pullback diagram in **Top**. *Then*

(i) *if α is locally in $\mathcal{M}$, then so is π_1,*

(ii) *if p is surjective and étale, and π_1 is locally in $\mathcal{M}$, then α also is locally in $\mathcal{M}$.*

Proof (i) Just take an appropriate family $(U_\lambda)_{\lambda \in \Lambda}$ of open subsets in B and note that the diagram

$$\pi_1^{-1}\left(p^{-1}(U_\lambda)\right) \longrightarrow \alpha^{-1}(U_\lambda)$$

$$\downarrow \qquad\qquad\qquad \downarrow$$

$$p^{-1}(U_\lambda) \longrightarrow U_\lambda$$

is a pullback for each $\lambda \in \Lambda$.

(ii) For each point b in B we choose

- e in E with $p(e) = b$,
- an open neighbourhood U of e for which p induces a homemorphism $U \longrightarrow p(U)$,
- an open neighbourhood V of e for which $\pi_1^{-1}(V) \longrightarrow V$ is in $\mathcal{M}$,
- $W = p(U \cap V)$, which is open and contains b.

Since the diagram

$$\pi_1^{-1}(U \cap V) \xrightarrow{\ \subseteq\ } \pi_1^{-1}(V)$$

$$\downarrow \qquad\qquad\qquad \downarrow$$

$$U \cap V \xrightarrow{\ \subseteq\ } V$$

is a pullback, $\pi_1^{-1}(U \cap V) \longrightarrow U \cap V$ is in $\mathcal{M}$. The homeomorphism $U \longrightarrow p(U)$ induces a homeomorphism $U \cap V \longrightarrow p(U \cap V) = W$ and hence also a homeomorphism $\pi_1^{-1}(U \cap V) \longrightarrow \alpha^{-1}(W)$. The "trivial pullback"

$$
\begin{array}{ccc}
\alpha^{-1}(W) & \longrightarrow & \pi_1^{-1}(U \cap V) \\
\downarrow & & \downarrow \\
W & \longrightarrow & U \cap V
\end{array}
$$

then shows that $\alpha^{-1}(W) \longrightarrow W$ is in $\mathcal{M}$. $\qquad \square$

Now let $\Sigma\mathcal{M}$ be the class of all continuous maps (isomorphic to the maps) of the form

$$
\coprod_{\lambda \in \Lambda} f_\lambda : \coprod_{\lambda \in \Lambda} X_\lambda \longrightarrow \coprod_{\lambda \in \Lambda} Y_\lambda,
$$

where each $f_\lambda : X_\lambda \longrightarrow Y_\lambda$ is in $\mathcal{M}$; this new class obviously contains $\mathcal{M}$ and is pullback stable.

Proposition 6.5.2 *For a continuous map* $\alpha : A \longrightarrow B$ *the following conditions are equivalent:*

(i) *α is locally in $\mathcal{M}$;*

(ii) *α is locally in $\Sigma\mathcal{M}$;*

(iii) *there exists a surjective étale map $p : E \longrightarrow B$ such that the pullback $\pi_1 : E \times_B A \longrightarrow E$ of α along p is in $\Sigma\mathcal{M}$.*

Proof Since (i) $\Rightarrow$ (ii) is trivial and (iii) $\Rightarrow$ (ii) follows from 6.5.1(ii), it suffices to prove (ii) $\Rightarrow$ (i) and (i) $\Rightarrow$ (iii).

(ii) $\Rightarrow$ (i) The condition (ii) tells us that every $b \in B$ has an open neighbourhood U such that $\alpha^{-1}(U) \longrightarrow U$ can be identified with some

$$
\coprod_{\lambda \in \Lambda} f_\lambda : \coprod_{\lambda \in \Lambda} X_\lambda \longrightarrow \coprod_{\lambda \in \Lambda} Y_\lambda,
$$

with each f_λ in $\mathcal{M}$. Then b belongs to one of the Y_λ, and since the diagram

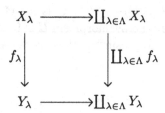

is a pullback and therefore $\alpha^{-1}(Y_\lambda) \longrightarrow Y_\lambda$ can be identified with f_λ, we conclude that $\alpha^{-1}(Y_\lambda) \longrightarrow Y_\lambda$ is in $\mathcal{M}$.

(i) $\Rightarrow$ (iii) Let $(U_\lambda)_{\lambda \in \Lambda}$ be a family of open subsets in B such that each $\alpha^{-1}(U_\lambda) \longrightarrow U_\lambda$ is in $\mathcal{M}$. We take $p: E \longrightarrow B$ to be the map $\coprod_{\lambda \in \Lambda} U_\lambda \longrightarrow B$ considered in the previous section, and note that the diagram

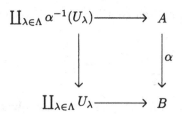

is a pullback, i.e. $\pi_1: E \times_B A \longrightarrow E$ can be identified with

$$\coprod_{\lambda \in \Lambda} \alpha^{-1}(U_\lambda) \longrightarrow \coprod_{\lambda \in \Lambda} U_\lambda.$$

Therefore π_1 is in $\Sigma\mathcal{M}$.   □

That is, the notion "is locally in", which is an instance of the general idea of "localization", can be described using the pullbacks along surjective étale maps instead of the inverse images of open neighbourhoods. In order to make this language of étale maps purely categorical (in LoCo) let us prove

Proposition 6.5.3 *An étale map* $p: E \longrightarrow B$ *between locally connected topological spaces is an effective descent morphism in LoCo if and only if it is surjective.*

Proof   As follows from the "general theory" (see section 4.4) and 6.4.8(iii), p is an effective descent morphism if and only if the functor $p^*: \mathsf{LoCo}/B \longrightarrow \mathsf{LoCo}/E$ reflects isomorphisms.

The image under p^* of the morphism

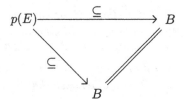

is an isomorphism in LoCo/E, and so if p^* reflects isomorphisms, then $p(E) = B$, i.e. p is surjective. The converse follows from 6.4.8(ii) and the fact that any pullback functor along a surjective map in Set reflects isomorphisms.   □

The next step is to arrive at the categorical notion of a covering morphism. Let us begin by recalling the classical definition:

Definition 6.5.4 A continuous map $\alpha: A \longrightarrow B$ of topological spaces is said to be

  (i) a trivial covering map if A is a disjoint union of open subsets each of which is mapped homeomorphically onto B by α,
 (ii) a covering map if every point in B has an open neighbourhood whose inverse image is a disjoint union of open subsets each of which is mapped homeomorphically onto it by α.

One might also look again at the simplest standard example of a non trivial covering map, which is the canonical map $\mathbb{R} \longrightarrow \mathbb{R}/\mathbb{Z}$, the "covering of a circle by a line".

Lemma 6.5.5 *In* Top, *every covering map is étale.*   □

Proposition 6.5.6 *We have the following:*

  (i) $\alpha: A \longrightarrow B$ *is a trivial covering map if and only if it is (up to an isomorphism) the projection* $B \times X \longrightarrow B$ *for some discrete* X;
 (ii) α *is a covering map if and only if it is locally a trivial covering map (i.e. it is locally in the class of trivial covering maps).*

Together with 6.5.2 and 6.5.3 this gives

Proposition 6.5.7 *Let* $\mathcal{M}$ *be the class of trivial covering maps between locally connected topological spaces, and* $\alpha: A \longrightarrow B$ *is a morphism in* LoCo. *The following hold:*

(i) α is in the class $\Sigma\mathcal{M}$ if and only if for every connected component B_i (with $i \in I(B)$ in the notation of section 6.1) the induced map $\alpha^{-1}(B_i) \longrightarrow B_i$ is of the form $B_i \times X_i \longrightarrow B_i$ as in 6.5.6(i);

(ii) α is a covering map if and only if there exists an effective descent morphism $p \colon E \longrightarrow B$ (in LoCo) such that the pullback of α along p is in $\Sigma\mathcal{M}$. $\square$

Here 6.5.7(i) tells us that for a given $B \in$ LoCo, in order to build up a morphism $\alpha \colon A \longrightarrow B$ that belongs to $\Sigma\mathcal{M}$, we should

- choose a family $(X_i)_{i \in I(B)}$ of sets indexed by the set $I(B)$ of connected components of B,
- define $\alpha_i \colon A_i \longrightarrow B_i$ for $i \in I(B)$ as the projection $B_i \times X_i \longrightarrow B_i$ (or, equivalently as the canonical map

$$\coprod_{x \in X_i} B_i \longrightarrow B_i$$

from the coproduct of "X_i copies" of B_i to B_i),
- and then define α as

$$\alpha = \coprod_{i \in I(B)} \alpha_i \colon \coprod_{i \in I(B)} A_i \longrightarrow \coprod_{i \in I(B)} B_i = B.$$

However, as we see from proposition 6.3.3 and conclusion 6.3.4(v), this is exactly how the functor H^B from Set$/I(B)$ (which is equivalent to the category $\prod_{i \in I(B)}$ Set of $I(B)$-indexed families of sets) to Fam$(\mathcal{A})/B$ (which is LoCo$/B$ here) is defined. That is, since H^B is full and faithful, we have

Proposition 6.5.8 *For a morphism $\alpha \colon A \longrightarrow B$ in LoCo the following conditions are equivalent:*

(i) α *is in $\Sigma\mathcal{M}$ (where $\mathcal{M}$ is as in 6.5.7);*

(ii) (A, α) *is (up to an isomorphism) in the image of*

$$H^B \colon \mathsf{Set}/I(B) \longrightarrow \mathsf{LoCo}/B;$$

(iii) *the canonical morphism $(A, \alpha) \longrightarrow H^B I^B(A, \alpha)$ is an isomorphism.*

If we replace LoCo by LoCo$/B$, or more generally by Fam$(\mathcal{A})$ with a terminal object in $\mathcal{A}$, then this condition (iii) becomes the assertion that the diagram

$$A \xrightarrow{\;\eta_A\;} HI(A)$$

$$\alpha \Big\downarrow \qquad\qquad \Big\downarrow HI(\alpha)$$

$$B \xrightarrow[\;\eta_B\;]{} HI(B)$$

(where η is the unit of $I \dashv H$) is a pullback. $\qquad\qquad$ □

And in this context we introduce

Definition 6.5.9 A morphism $\alpha \colon A \longrightarrow B$ in $\mathcal{C} = \mathsf{Fam}(\mathcal{A})$ is said to be a covering morphism if there exists an effective descent morphism $p \colon E \longrightarrow B$ such that (A, α) is split by p in the sense of definition 5.1.7, i.e. the diagram

$$E \times_B A \xrightarrow{\;\eta_{E \times_B A}\;} HI(E \times_B A)$$

$$\pi_1 \Big\downarrow \qquad\qquad \Big\downarrow HI(\pi_1)$$

$$E \xrightarrow[\;\eta_E\;]{} HI(E)$$

is a pullback.

And from 6.5.7 and 6.5.8 we obtain

Theorem 6.5.10 *An étale map $\alpha \colon A \longrightarrow B$ of locally connected topological spaces is a covering map (in the classical sense) if and only if it is a covering morphism in LoCo/B in the sense of definition 6.5.9.* $\quad$ □

6.6 Classification of coverings

We are going to describe the category $\mathsf{Cov}(B)$ of pairs (A, α), where $\alpha \colon A \longrightarrow B$ is a covering morphism with fixed $B \in \mathcal{C}$. The context we have in mind is as in definition 6.5.9 although all arguments of this section can be repeated in the context of an abstract adjunction as at the beginning of section 6.3, provided we assume H^C to be full and faithful (or, equivalently, $\varepsilon^C \colon I^C H^C \longrightarrow 1$ to be an isomorphism for every $C \in \mathcal{C}$). For simplicity we assume that $\mathcal{C}$ has pullbacks.

The full subcategory in $\mathsf{Cov}(B)$ with objects all (A, α) split by a fixed effective descent morphism $p \colon E \longrightarrow B$ will be denoted by $\mathsf{Split}_B(p)$ in accordance with the notation of section 5.1. We can simply write

$$\mathsf{Cov}(B) = \bigcup_p \mathsf{Split}_B(p) \qquad\qquad (*)$$

and say that each $\mathsf{Split}_B(p)$ can be described via the categorical Galois theory; but it is important to know that $(*)$ is in fact a "good union", and that under the existence of what we call a universal covering it reduces to its largest member. Furthermore we will show that $\mathsf{Split}_B(p)$, which is thus the largest member in which p is actually a universal covering of B, also admits p as a morphism of Galois descent (the definition will be recalled below), and therefore can be described as in the Galois theorem 5.1.24 (with $\overline{\mathcal{P}} = \mathcal{P}$ in the notation of section 5.1). As we will show in the next sections, this extends the classical classification theorem of covering spaces over a "good space".

First we need

Lemma 6.6.1 *The class of effective descent morphisms in $\mathcal{C}$ is closed under composition, pullback stable, and has the right cancellation property, i.e. if it contains $p_1 p_2$, then it contains p_1.*

Proof Let us restrict ourselves to the situation where the pullback functors in $\mathcal{C}$ preserve coequalizers (whose existence we also assume in this proof). In particular this condition holds in LoCo (see 6.4.8(iii)) and in many other important examples. As we already mentioned in the proof of proposition 6.5.3, under this condition, p is an effective descent morphism if and only if $p^* \colon \mathcal{C}/B \longrightarrow \mathcal{C}/E$ reflects isomorphisms.

The closedness under composition and the right cancellation property are now obvious since $(p_1 p_2)^* = p_2^* p_1^*$, and the pullback stability easily follows from the so-called Beck–Chevalley property ("condition"): if the first square below is a pullback diagram, the second square commutes up to isomorphism.

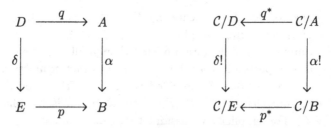

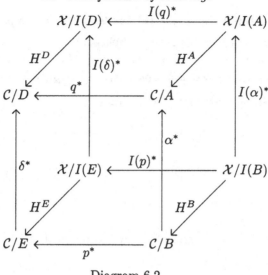

Diagram 6.2

Here $\alpha!$ is the composition with α, which obviously reflects isomorphisms. □

Corollary 6.6.2 *The class of covering morphisms of C is pullback stable.*

Proof Although lemma 6.6.1 makes this straightforward, let us explain the details.

For a given pullback diagram as in the proof of lemma 6.6.1, consider diagram 6.2. "Up to isomorphism" we can say

- the diagram commutes,
- the image of H^B coincides with the category of "trivial coverings" of B, i.e. the category of those objects in C/B which are split by the identity morphism of B,
- since the diagram commutes, p^*, q^*, α^* and δ^* send trivial coverings to trivial coverings,
- $\mathsf{Split}_B(p)$ is the category of all those objects in C/B which are sent to trivial coverings by p^*, and of course the same is true for $\mathsf{Split}_A(q)$ (with $q^*: C/A \longrightarrow C/D$ instead of $p^*: C/B \longrightarrow C/E$),
- therefore α^* restricts to a functor $\mathsf{Split}_B(p) \longrightarrow \mathsf{Split}_A(q)$.

Together with (∗) this tells us that the class of covering morphisms is pullback stable as desired. □

Proposition 6.6.3 *The union* $(*)$ *has the following properties:*

(i) $p = p'p''$ *implies* $\mathsf{Split}_B(p') \subseteq \mathsf{Split}_B(p)$;

(ii) *the union is directed, i.e. for any* $p_i\colon E_i \longrightarrow B$, $i = 1, 2$, *there exists* $p\colon E \longrightarrow B$ *with* $\mathsf{Split}_B(p_i) \subseteq \mathsf{Split}_B(p)$, $i = 1, 2$.

Proof   (i) Given a commutative diagram

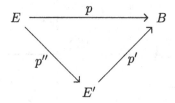

we have ("up to isomorphism" as in the proof of corollary 6.6.2) the following:

- if (A, α) is in $\mathsf{Split}_B(p')$, then $p'^*(A, \alpha)$ is in the image of $H^{E'}$ and therefore $p^*(A, \alpha) = p''^* p'^*(A, \alpha)$ is in the image of H^E;
- if $p^*(A, \alpha)$ is in the image of H^E, then (A, α) is in $\mathsf{Split}_B(p)$ (by definition).

Now (ii) follows from (i), since we can form the pullback

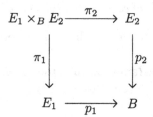

and use lemma 6.6.1 to show that $p_1 \pi_1 = p_2 \pi_2$ is an effective descent morphism (when so are p_1 and p_2). $\square$

Remark 6.6.4 The morphism p'' in 6.6.3(i) may not be an effective descent morphism, which is important for various applications.

Now we introduce

Definition 6.6.5

(i) An object E in $\mathcal{C}$ is said to be Galois closed if it has no non trivial coverings, i.e. every covering morphism $E' \longrightarrow E$ is split by the identity morphism of E.

(ii) A covering morphism $p: E \longrightarrow B$ is said to be a universal covering (of B) if it is an effective descent morphism, and E is Galois closed.

Proposition 6.6.6 *Every universal covering morphism* $p: E \longrightarrow B$ *has the following properties:*

(i) *it is a morphism of Galois descent, i.e.* $(E, p) \in \mathsf{Split}_B(p)$;

(ii) $\mathsf{Cov}(B) = \mathsf{Split}_B(p)$.

Proof Consider the pullback

$$
\begin{array}{ccc}
E \times_B A & \xrightarrow{\;\pi_2\;} & A \\
\Big\downarrow{\scriptstyle \pi_1} & & \Big\downarrow{\scriptstyle \alpha} \\
E & \xrightarrow{\;p\;} & B
\end{array}
$$

where $\alpha: A \longrightarrow B$ is a covering morphism. Corollary 6.6.2 tells us that $\pi_1: E \times_B A \longrightarrow E$ is a covering morphism, and since E is Galois closed we conclude that (A, α) is in $\mathsf{Split}_B(p)$. This proves (ii) and therefore also (i). $\qquad\square$

Now let us describe more precisely the relationship between the (terminology and) notation of this section and of section 5.1.

The data

$$
\overline{\mathcal{A}} \subseteq \mathcal{A} \underset{S}{\overset{\mathcal{C}}{\longleftrightarrow}} \mathcal{P} \supseteq \overline{\mathcal{P}}
$$

of section 5.1 have now $\overline{\mathcal{A}} = \mathcal{A}$ and $\overline{\mathcal{P}} = \mathcal{P}$, i.e. $\overline{\mathcal{A}}$ and $\overline{\mathcal{P}}$ are the classes of all morphisms in $\mathcal{A}$ and $\mathcal{P}$ respectively ($\overline{\mathcal{P}} = \mathcal{P}$ was already mentioned before lemma 6.6.1) with the following correspondence for the notation:

Section 5.1	This section
$\mathcal{A}$	$\mathcal{C} = \mathsf{Fam}(\mathcal{A})$
$\mathcal{P}$	$\mathcal{X} = \mathsf{Set}$
$\mathcal{C}$	H
S	I

Accordingly, definition 5.1.8 is to be translated as follows: a morphism $p: E \longrightarrow B$ is of Galois descent (we do not say "relative" since now $\overline{\mathcal{A}} = \mathcal{A}$ and $\overline{\mathcal{P}} = \mathcal{P}$, just as in chapter 5) if

- p is an effective descent morphism,
- $\varepsilon^E \colon I^E H^E \longrightarrow 1$ is an isomorphism – however, we can omit this condition since now ε^C is an isomorphism for every object $C \in \mathcal{C}$ –
- $\Sigma_p H^E(X, \varphi) \in \mathsf{Split}_B(p)$ for every object (X, φ) in $\mathcal{C}/I(E)$ – however, this is equivalent to $(E, p) \in \mathsf{Split}_B(p)$ required in 6.6.6(i) (the equivalence follows from proposition 5.5.6 and lemma 5.5.7).

For such a $p \colon E \longrightarrow B$ the Galois groupoid $\mathsf{Gal}\,[p]$ is displayed as

$$I(E \times_B E) \times_{I(E)} I(E \times_B E) \longrightarrow I(E \times_B E) \overset{\longrightarrow}{\underset{\longrightarrow}{\longleftarrow}} I(E),$$

or, equivalently, as

$$I(E \times_B E \times_B E) \longrightarrow I(E \times_B E) \overset{\longrightarrow}{\underset{\longrightarrow}{\longleftarrow}} I(E),$$

where

- the domain and codomain morphisms $I(E \times_B E) \longrightarrow I(E)$ are the I-images of the two projections $E \times_B E \longrightarrow E$,
- the composition morphism $I(E \times_B E \times_B E) \longrightarrow I(E \times_B E)$ is the I-image of the morphism $E \times_B E \times_B E \longrightarrow E \times_B E$ induced by the first and the third projection $E \times_B E \times_B E \longrightarrow E$,
- the identity morphism $I(E) \longrightarrow I(E \times_B E)$ is the I-image of the diagonal $E \longrightarrow E \times_B E$,
- the inverse $I(E \times_B E) \longrightarrow I(E \times_B E)$ is the I-image of the "interchange morphism" $E \times_B E \longrightarrow E \times_B E$.

After this, from the Galois theorem 5.1.24 and proposition 6.6.6, we obtain the following classification theorem for coverings.

Theorem 6.6.7 *If* $p \colon E \longrightarrow B$ *is a universal covering morphism, then there exists a category equivalence*

$$\mathsf{Cov}(B) \approx \mathsf{Set}^{\mathsf{Gal}\,[p]}. \qquad \qquad \square$$

6.7 The Chevalley fundamental group

There are two classical definitions of the fundamental group of a topological space which give isomorphic groups for certain "good" spaces.

- The Poincaré fundamental group $\pi_1(B, b)$ is the group of homotopy classes of paths in B from b to b. Recall that although this group depends on the choice of the base point b, it is determined uniquely up to isomorphism when B is path connected.
- The Chevalley fundamental group $\mathsf{Aut}(p) = \mathsf{Aut}_B(E, p)$ is defined only for connected spaces B which admit a universal covering map $p\colon E \longrightarrow B$ with connected E, and of course depends on it, but again, different p produce isomorphic groups. It is the group of all automorphisms u of E with $pu = p$.

The Chevalley definition is less useful for calculations, but it can be literally repeated in our general context, and in this section we are going to show that its generalized form "follows" from theorem 6.6.7, which means

- if B is connected and admits a universal covering, then there exists a universal covering $p\colon E \longrightarrow B$ with connected E,
- if p is as above and therefore $\mathsf{Gal}\,[p]$ is a group (since E is connected), then the groups $\mathsf{Gal}\,[p]$ and $\mathsf{Aut}(p)$ are isomorphic, and moreover, the isomorphism formally follows from theorem 6.6.7.

Accordingly:

Convention *In this section, the ground category C is supposed to be a category of the form $\mathsf{Fam}(\mathcal{A})$, and with pullbacks.*

We will need a number of simple observations listed in

Lemma 6.7.1 *Let $p\colon E \longrightarrow B$ be a universal covering morphism with connected B. Then the following hold:*

(i) $\mathsf{Gal}\,[p]$ *is a connected groupoid, i.e. for any two objects x and y in it, there exists a morphism $x \longrightarrow y$;*

(ii) *there exists a group G such that the category $\mathsf{Cov}(B)$ is equivalent to Set^G;*

(iii) *if*

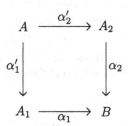

is a pullback diagram in C/B in which α_1 and α_2 are covering morphisms and A_1 and A_2 are not initial objects (i.e. are non-empty families), then A also is not initial;

(iv) *for every covering morphism $\alpha\colon A \longrightarrow B$, in which A is not an initial object in C, there exists a morphism $(E,p) \longrightarrow (A,\alpha)$ in C/B, and therefore α is an effective descent morphism.*

Proof   (i) Since B is connected, i.e. $I(B)$ is a one element set, the connectedness of $\mathsf{Gal}\,[p]$ is equivalent to the fact that

$$ I(E \times_B E) \rightrightarrows I(E) \longrightarrow I(B) $$

is a coequalizer diagram. However, this follows from the fact that I is a left adjoint and

$$ E \times_B E \rightrightarrows E \longrightarrow B \qquad (**) $$

is a coequalizer diagram. The last assertion here will be explained for readers not familiar with descent theory.

First of all it uses the existence of coequalizers in C, otherwise we could only prove that $(**)$ is a coequalizer diagram in C/B (regarding B, E, $E \times_B E$ as objects in C/B in the obvious way) – but we could then use I^B instead of I. Now, as soon as we are in C/B, we could either use the description of the left adjoint of the comparison functor, or directly check that the image of $(**)$ under $p^*\colon C/B \longrightarrow C/E$ is a split coequalizer and apply the Beck criterion as formulated in section 4.4.

Part (ii) follows from (i) and theorem 6.6.7 since every connected groupoid is equivalent to a one object groupoid, i.e. to a group.

(iii) It suffices to show that there exists a non-initial object (A',α') in C/B from which there are a morphism to (A_1,α_1) and a morphism to (A_2,α_2). However, such an object does exist already in $\mathsf{Cov}(B)$ – this follows from (ii) and the fact that $B = (B,1_B)$ corresponds to a one element G-set under the equivalence $\mathsf{Cov}(B) \approx \mathsf{Set}^G$, since the product of two non-empty G-sets is always non-empty.

(iv) The existence of a morphism $(E,p) \longrightarrow (A,\alpha)$ is equivalent to the existence of a right inverse for the projection $\pi_1\colon E \times_B A \longrightarrow A$. On the other hand as we see from definition 6.5.9, π_1 has a right inverse if and only if the map

$$ I(\pi_1)\colon I(E \times_B A) \longrightarrow I(E) $$

is surjective. Since

$$E \times_B A \approx \coprod_{i \in I(E)} E_i \times_B A$$

(where E_i, $i \in I(E)$, are the connected components of E) the surjectivity of $I(\pi_1)$ follows from the fact that each $E_i \times_B A$ is non-initial, which itself follows from (iii) since each $E_i \longrightarrow B$ obviously is a covering morphism.

The fact that α is an effective descent morphism follows now from lemma 6.6.1. $\square$

Corollary 6.7.2 *Under the assumptions of lemma 6.7.1, the following conditions on a covering morphism* $\alpha \colon A \longrightarrow B$ *are equivalent:*

(i) α *is a universal covering morphism;*

(ii) A *is non-initial and there exists a morphism* $(A, \alpha) \longrightarrow (E, p)$ *in* $\mathcal{C}/B$;

(iii) A *is non trivial and for every covering morphism* $\alpha' \colon A' \longrightarrow B$ *with non-initial* A', *there exists a morphism* $(A, \alpha) \longrightarrow (A', \alpha')$;

(iv) (A, α) *corresponds to a non-empty free* G-*set under the equivalence* $\mathsf{Cov}(B) \approx \mathsf{Set}^G$ *of 6.7.1(ii);*

(v) *each connected component of* (A, α) *determines a universal covering morphism to* B.

Proof (i) $\Rightarrow$ (ii) follows from 6.7.1(iv): just replace (E, p) and (A, α) by each other.

(ii) $\Rightarrow$ (iii) also follows from 6.7.1(iv) and (iii) $\Rightarrow$ (ii) is trivial.

(iii) $\Rightarrow$ (iv) follows from the fact that only a free G-set can have a morphism to any other (non-empty) G-set.

(iv) $\Rightarrow$ (i) follows from the fact that (A, α) and (E, p) have isomorphic connected components (i.e. each connected component of (A, α) is isomorphic to each connected component of (E, p)), since (E, p) also corresponds to a free G-set by (i) $\Rightarrow$ (iv) applied to (E, p). However, we should also observe that α is an effective descent morphism by 6.7.1(iv).

Thus the conditions (i)–(iv) are equivalent to each other; this also implies that they are equivalent to (v). $\square$

Since there exists exactly one (up to an isomorphism) connected free G-set – namely G itself considered as a G-set – we also obtain

Corollary 6.7.3 *Every connected object* B *in* $\mathcal{C}$ *which admits a universal covering does admit a unique (up to isomorphism) connected universal covering, i.e. a universal covering* $p \colon E \longrightarrow B$ *with connected* E. $\square$

And finally we have

Theorem 6.7.4 *Let p: $E \longrightarrow B$ be a universal covering morphism with connected E (and therefore connected B). Then* Gal$[p]$ *is a group isomorphic to* Aut(p), *the group of all automorphisms u of E with $pu = p$.*

Proof First we recall that Gal$[p]$ is a group, i.e. one object groupoid, simply by the definition, since its set of objects is $I(E)$, which is a one element set when E is connected. We then take G as in 6.7.1(ii) and in 6.7.2(iv), and since

$$\mathsf{Set}^{\mathsf{Gal}\,[p]} \approx \mathsf{Cov}(B) \approx \mathsf{Set}^G,$$

we have Gal$[p] \approx G$. On the other hand since (E, p) corresponds to the G-set G via the equivalence $\mathsf{Cov}(B) \approx \mathsf{Set}^G$, we obtain

$$\mathsf{Aut}(p) \approx \mathsf{Aut}(G) \approx G$$

– here Aut(G) is the automorphism group of G as a G-set, which is isomorphic to the group G of course.

That is, Gal$[p] \approx$ Aut(p) as desired.   □

Remarks 6.7.5

(i) Since in the assumptions of theorem 6.7.4 there is a category equivalence

$$\mathsf{Cov}(B) \approx \mathsf{Set}^{\mathsf{Aut}(p)}$$

one easily establishes the standard Galois correspondence between the quotients of (E, p) which are coverings and the subgroups of Aut(p) – which we leave as a simple exercise for the reader.

(ii) According to the usual terminology in algebraic topology, only connected (universal) coverings should have been called universal coverings. However, this becomes less convenient in a general context, especially in the most general context of section 5.1.

6.8 Path and simply connected spaces

An obvious question which occurs when we apply the general results of previous sections to topological spaces (i.e. to $\mathcal{C} = \mathsf{LoCo}/B$ for some space B) is: given a covering map p: $E \longrightarrow B$, how can we find out if E is closed in the sense of definition 6.6.5? Since this is the case if and only if E has no non-trivial coverings in the classical sense, the classical theory provides a (well-known) partial answer, which is theorem 6.8.12 below. All the arguments we use in our detailed proof are very simple and can be found in any standard textbook which has a chapter on covering spaces, but they are generally mixed up in those books with a lot of other arguments needed for the "geometrical" proof of the appropriate special case of theorem 6.6.7. We will also recall here all the elementary notions of homotopy theory which we will need for that proof.

A path in a topological space B is a continuous map

$$f : [0,1] \longrightarrow B,$$

where

$$[0,1] = \{t \in \mathbb{R} \mid 0 \le t \le 1\}$$

is the (closed) interval of all real numbers between 0 and 1.

Two points b and b' in B are said to be connected by a path, if there exists a path f in B with $f(0) = b$ and $f(1) = b'$. This is clearly an equivalence relation, and the equivalence classes are called path connected components.

Two paths f and g are (homotopy) equivalent if there exists a continuous map

$$h : [0,1] \times [0,1] \longrightarrow B$$

with

$$h(t,0) = f(t), \quad h(t,1) = g(t),$$
$$f(0) = h(0,t) = g(0),$$
$$f(1) = h(1,t) = g(1),$$

and we then write $f \sim g$ or $[f] = [g]$, denoting the equivalence class of f by $[f]$; the relation $\sim$ is clearly an equivalence relation.

Definition 6.8.1 A topological space B is said to be

  (i) path connected, if it has exactly one path connected component,

  (ii) simply connected, if it is path connected and every two paths f and g in B with $f(0) = g(0)$ and $f(1) = g(1)$ are equivalent,

  (iii) locally path connected, if every open subset in B has open path connected components.

Consider the lifting problem

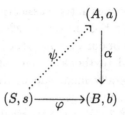

which is to find, for given $\alpha\colon A\longrightarrow B$ with $\alpha(a) = b$ and $\varphi\colon S\longrightarrow B$
with $\varphi(s) = b$, a map $\psi\colon S\longrightarrow A$ with $\alpha\psi = \varphi$ and $\psi(s) = a$. Here α,
φ, ψ are supposed to be continuous maps of topological spaces, and we
call ψ a solution of the lifting problem. If for a given $\alpha\colon A\longrightarrow B$ and
S such a lifting problem has a solution for all possible a, b, s, α, φ then
we say that α has the lifting property with respect to S. In particular
if this is the case for $S = [0,1]$, then we say that α has the path-lifting
property. We will also speak of the unique (path-) lifting property if the
ψ above is uniquely determined. Considering path-lifting we will always
use the following.

Remark 6.8.2 For the path-lifting and the unique path-lifting properties we can assume that $s = 0$ in $S = [0,1]$.

Proof This follows from the fact that for every s with $0 < s < 1$ there
are

(i) a canonical coequalizer diagram
$$\{s\} \rightrightarrows [0,s] \sqcup [s,1] \longrightarrow [0,1],$$

(ii) homeomorphisms $[0,1]\longrightarrow[0,s]$ and $[0,1]\longrightarrow[s,1]$, both with
$0\mapsto s$ (and there exists a homeomorphism $[0,1]\longrightarrow[0,1]$ with
$0\mapsto 1$). $\square$

Investigating the lifting property, we begin with

Proposition 6.8.3 *Every trivial covering map (as defined in 6.5.4(i))
has the unique lifting property with respect to every connected space.*

Proof In the notation above, let
$$A = \coprod_{\lambda\in\Lambda} A_\lambda$$

be a disjoint union of open subsets, each of which is mapped homeomorphically onto B by α, and let λ_0 be a fixed element in Λ for which

$a \in A_{\lambda_0}$. If S is connected, then since A is a coproduct of A_λ, every continuous map $\psi \colon S \longrightarrow A$ with $\psi(s) = a \in A_{\lambda_0}$ must have $\psi(S) \subseteq A_{\lambda_0}$. Therefore the problem of the unique lifting reduces to the trivial case of a homeomorphism (namely of $A_{\lambda_0} \longrightarrow B$). $\qquad\square$

The general covering maps do not have the same property; moreover, we have

Proposition 6.8.4 *Let* $\alpha \colon A \longrightarrow B$ *be a covering map of connected topological spaces. If B has a universal covering, then the following conditions are equivalent:*

  (i) α *has the unique lifting property with respect to every connected space;*
 (ii) α *has the unique lifting property with respect to B;*
(iii) α *has the lifting property with respect to B;*
 (iv) α *is a split epimorphism, that is, there exists a continuous map* $\beta \colon B \longrightarrow A$ *with* $\alpha \circ \beta = 1_B$;
  (v) α *is an isomorphism (i.e. a homeomorphism).*

Proof Since the implications (v) $\Rightarrow$ (i) $\Rightarrow$ (ii) $\Rightarrow$ (iii) $\Rightarrow$ (iv) are trivial (note that (iii) $\Rightarrow$ (iv) is trivial just because A is non-empty), we only need to prove (iv) $\Rightarrow$ (v). However, (iv) $\Rightarrow$ (v) follows from 6.7.1(ii) since

- the connectedness in the topological sense coincides with the categorical connectedness not only in Top but also in $\mathsf{Cov}(B)$;
- under the equivalence of categories $\mathsf{Cov}(B) \approx \mathsf{Set}^G$, the morphism $\alpha \colon (A, \alpha) \longrightarrow (B, 1_B)$ in $\mathsf{Cov}(B)$ corresponds to a map from a connected G-set to a one element G-set. $\qquad\square$

Yet, every covering map has the unique lifting property with respect to some special spaces S, in particular to $S = [0,1]^n$, which gives

- a trivial result for $n = 0$,
- the unique path-lifting property for $n = 1$,
- the unique path-equivalence-lifting property (described below) for $n = 2$.

We will prove this with all auxiliary results just for $n = 2$ since the path-equivalence-lifting is exactly what we need in order to show that certain spaces are Galois closed. However, the reader can easily reformulate all arguments for general n.

Lemma 6.8.5 *Let* $(B_\lambda)_{\lambda \in \Lambda}$ *be a family of open subsets in a topological space B with $B = \bigcup_{\lambda \in \Lambda} B_\lambda$ and $\varphi \colon [0,1]^2 \longrightarrow B$ is a continuous map. Then there exist finite sequences $x_1, \dots, x_k$ and $y_1, \dots, y_l$ of real numbers with*

(i) $0 = x_1 < \cdots < x_k = 1$ *and* $0 = y_1 < \cdots < y_l = 1$,

(ii) *for every* $(i,j) \in \{1, \dots, k-1\} \times \{1, \dots, l-1\}$ *there exists* $\lambda \in \Lambda$ *with*

$$f\big([x_i, x_{i+1}] \times [y_j, y_{j+1}]\big) \subseteq B_\lambda.$$

Proof We have to make a two step choice.

Step 1. For each $(x,y) \in [0,1]^2$ we choose four numbers x^-, x^+, y^-, y^+ in $[0,1]$ with the following properties:

- $x^- \leq x \leq x^+$ and $y^- \leq y \leq y^+$;
- $x^- = x$ only for $x = 0$, and $y^- = y$ only for $y = 0$;
- $x^+ = x$ only for $x = 1$, and $y^+ = y$ only for $y = 1$;
- there exists $\lambda \in \Lambda$ with

$$f\big([x^-, x^+] \times [y^-, y^+]\big) \subseteq B_\lambda.$$

This is possible because each (x,y) belongs to some $f^{-1}(B_\lambda)$, and each $f^{-1}(B_\lambda)$ is an open subset in $[0,1]^2$.

Step 2. Choose finite sequences $s_1, \dots, s_m$ and $t_1, \dots, t_n$ in $[0,1]$ with

$$[0,1]^2 = \bigcup_{i=1}^m \bigcup_{j=1}^n \mathrm{int}[s_i^-, s_i^+] \times [t_j^-, t_j^+],$$

using the following notation: for $u, v \in [0,1]$, $\mathrm{int}[u,v]$ is the interior of $[u,v]$ in $[0,1]$, i.e.

$$w \in \mathrm{int}[u,v] \Leftrightarrow \begin{cases} 0 \leq w < v & \text{when} \quad u = 0 \quad \text{and} \quad v \neq 1, \\ u < w < v & \text{when} \quad u \neq 0 \quad \text{and} \quad v \neq 1, \\ u < w \leq 1 & \text{when} \quad u \neq 0 \quad \text{and} \quad v = 1 \end{cases}$$

(note that the case $u = v$ never occurs above). This is possible since

$$[0,1]^2 = \bigcup_{(x,y) \in [0,1]^2} \mathrm{int}[x^-, x^+] \times \mathrm{int}[y^-, y^+]$$

and $[0,1]^2$ is a compact space.

After that we just take $x_1, \dots, x_k$ to be the set of all real numbers which occur either as s_i^- or as s_i^+, with the indices determined by the condition (i), and similarly define $y_1, \dots, y_l$ via t_j^- and t_j^+. Indeed, the condition (ii) holds since every rectangle of the form $[x_i, x_{i+1}] \times [y_j, y_{j+1}]$ is a subset of some rectangle of the form $[s_{i'}^-, s_{i'}^+] \times [t_{i'}^-, t_{i'}^+]$. $\qquad \square$

Corollary 6.8.6 *Let* $\alpha\colon A \longrightarrow B$ *be a covering map of topological spaces and* $\varphi\colon [0,1]^2 \longrightarrow B$ *a continuous map. Then there exist finite sequences* $x_1,\ldots,x_k$ *and* $y_1,\ldots,y_l$ *of real numbers with*

(a) $0 = x_1 < \cdots < x_k = 1$ *and* $0 = y_1 < \cdots < y_l = 1$,

(b) *for every* $(i,j) \in \{1,\ldots,k-1\} \times \{1,\ldots,l-1\}$ *there exists an open subset* B' *in* B *such that*

$$f\big([x_i, x_{i+1}] \times [y_j, y_{j+1}]\big) \subseteq B'$$

and the map $\alpha^{-1}(B') \longrightarrow B'$ *induced by* α *is a trivial covering map.* □

Lemma 6.8.7 *Let*

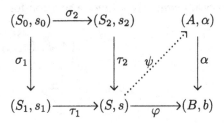

be a diagram in the category of pointed topological spaces in which the square part is a pushout and the triangle represents a lifting problem. Then this lifting problem has a unique solution provided the lifting problems

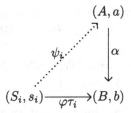

$(i = 0, 1, 2,$ *where* τ_0 *is defined as* $\tau_1\sigma_1 = \tau_2\sigma_2$) *have a unique solution.*

Proof is obvious – just note ψ_1 and ψ_2 must agree on S_0 (i.e. $\psi_1\sigma_1 = \psi_2\sigma_2$) by the uniqueness of ψ_0. □

Now consider the lifting problem

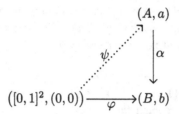

where $\alpha \colon A \longrightarrow B$ is a covering map.

We can choose $x_1, \dots, x_k, y_1, \dots, y_l$ as in corollary 6.8.6 and find a unique solution ψ using proposition 6.8.3 and lemma 6.8.7 as follows:

- we first restrict the original problem from $[0,1]^2$ to the "first small rectangle", i.e. to $[x_1, x_2] \times [y_1, y_2]$ (recall that $x_1 = 0 = y_1$), and then proposition 6.8.3 gives a unique solution ψ_{11};

- if $k \neq 2$, we obtain similarly a unique solution for

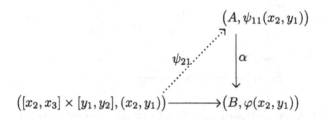

- then we apply lemma 6.8.7 to obtain a unique solution for the lifting problem above extended from $[x_2, x_3] \times [y_1, y_2]$ to $[x_1, x_3] \times [y_1, y_2]$ – and since that solution also extends ψ_{11}, it is also a unique solution for the original problem restricted to $[x_1, x_3] \times [y_1, y_2]$;

- then repeating similar arguments "horizontally", i.e. moving from $[x_1, x_i]$ to $[x_1, x_{i+1}]$, $i = 2, \dots, k_1$, we extend the (unique) lifting to $[0, 1] \times [y_1, y_2]$;

- finally we repeat the same "vertically", i.e. using the rectangles $[0, 1] \times [y_i, y_{i+1}]$, $j = 1, \dots, l-1$.

Moreover, using now a "two dimensional version" of remark 6.8.2 (or similar arguments directly), we can do the same with $\big([0,1]^2, (u,v)\big)$ instead of $\big([0,1]^2, (0,0)\big)$. That is, we obtain

Lemma 6.8.8 *Every covering map has the unique lifting property with respect to* $[0,1]^2$. $\qquad\qquad\qquad\qquad\qquad\qquad\qquad\square$

Of course it is even simpler to carry out the same arguments for $[0,1]$ instead of $[0,1]^2$ and therefore obtain the unique path-lifting property for the covering maps. However, it also formally follows from lemma 6.8.8 (see the assertion (i) in corollary 6.8.10 below).
We need one more obvious lemma:

Lemma 6.8.9 *If a continuous map* $\alpha\colon A \longrightarrow B$ *has the unique lifting property with respect to a space* S, *then it has the same property with respect to any retract of* S. □

Corollary 6.8.10 *Let* $\alpha\colon A \longrightarrow B$ *be a continuous map which has the unique lifting property with respect to* $[0,1]^2$. *Then* α *also has*

(i) *the unique path-lifting property;*
(ii) *the path-equivalence-lifting property, i.e. if* f *and* g *are paths in* A *with* $f(0) = g(0)$ *and* $\alpha f \sim \alpha g$, *then* $f \sim g$ *(and in particular* $f(1) = g(1)$*).*

Proof Part (i) immediately follows from lemma 6.8.9 since $[0,1]$ is a retract of $[0,1]^2$.

(ii) Let S be the quotient space $[0,1]^2/R$, where $(x,y)R(x',y')$ if either $x = 0 = x'$, or $x = 1 = x'$ (or $(x,y) = (x',y')$), and let $\kappa_i\colon [0,1] \longrightarrow S$, $i = 0, 1$, be the composites

$$[0,1] \xrightarrow{\;\;x \mapsto (x,i)\;\;} [0,1]^2 \xrightarrow{\;\;\text{canonical map}\;\;} S.$$

Then $\alpha f \sim \alpha g$ produces a continous map $\varphi\colon S \longrightarrow B$ with $\varphi\kappa_0 = \alpha f$ and $\varphi\kappa_1 = \alpha g$, and since S is obviously a retract of $[0,1]^2$ lemma 6.8.9 gives a (unique) solution of the lifting problem

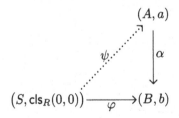

where $f(0) = a = g(0)$ and $b = \alpha(a)$; $\mathsf{cls}_R(0,0)$ denotes the equivalence class of $(0,0)$ for the relation R. We also have $\psi\kappa_0 = f$ by (a) (since both $\psi\kappa_0$ and f are liftings of $\varphi\kappa_0 = \alpha f$), and similarly $\varphi\kappa_1 = g$. Therefore ψ makes $f \sim g$. □

Corollary 6.8.11 *If* $\alpha\colon A \longrightarrow B$ *is as in corollary 6.8.10, A is path connected, and B is simply connected, then* α *is a bijection.*

Proof Injectivity: suppose $\alpha(a) = \alpha(a')$. Since A is path connected there is a path f in A with $f(0) = a$ and $f(1) = a'$. Compare f with the constant path $g\colon t \mapsto a$, $(t \in [0,1])$: since their composites with α coincide on 0 and 1 and B is simply connected, we conclude that those composites are equivalent and so $a' = f(1) = g(1) = a$ by 6.8.10(ii).

Surjectivity is also obvious since $\alpha(A) \neq \emptyset$, B is path connected, and α has the path-lifting property. $\qquad\square$

As we see from this result, in order to prove Galois closedness, we need a property of a space B which will make any connected A path connected when there is a covering map $A \longrightarrow B$. For, we observe,

- if $\alpha\colon A \longrightarrow B$ is a covering map (or just étale), and B is locally path connected, then so is A,
- if A is locally path connected, then it is a disjoint union of path connected open subsets,
- therefore if A is connected and locally path connected, then it is path connected.

Theorem 6.8.12 *Every simply connected locally path connected topological space is Galois closed.* $\qquad\square$

7

Non-galoisian Galois theory

This chapter presents a further generalization of the Galois theorem. Our new theorem holds for every effective descent morphism, not necessarily of Galois descent.

This generalized Galois theorem applies in particular to the case of an arbitrary field extension, not necessarily galoisian. But as the reader can guess, there is a price to pay for this generalization: the classical Galois group is now replaced by a weaker structure, namely, by what we call a "Galois pregroup".

The famous Joyal–Tierney theorem extending the Galois theory of Grothendieck to the context of toposes enters the scope of the present chapter.

7.1 Internal presheaves on an internal groupoid

In view of further generalizations, it is now useful to make more precise the notions of internal categories, groupoids and presheaves already introduced in section 4.6.

An internal category $\mathbb{C}$ in a category $\mathcal{C}$ with pullbacks consists in giving

$$
C_2 \xrightarrow[\substack{\xrightarrow{f_0} \\ \xrightarrow{m} \\ \xrightarrow{f_1}}]{} C_1 \underset{\substack{\xleftarrow{d_0} \\ \xleftarrow{n} \\ \xleftarrow{d_1}}}{} C_0
$$

where

(G1) the square

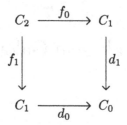

is a pullback,

(G2) the triangles

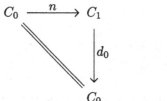   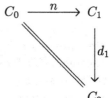

are commutative,

(G3) the squares

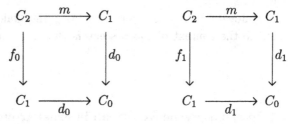

are commutative,

(G4) the triangles

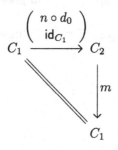   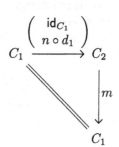

are commutative,

(G5) considering further the pullback

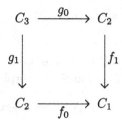

the diagram

$$C_3 \xrightarrow{\begin{pmatrix} f_0 \circ g_0 \\ m \circ g_1 \end{pmatrix}} C_2$$

$$\begin{pmatrix} m \circ g_0 \\ f_1 \circ g_1 \end{pmatrix} \Big\downarrow \qquad\qquad \Big\downarrow m$$

$$C_2 \xrightarrow{\quad m \quad} C_1$$

is commutative.

In the case $\mathcal{C} = $ **Set**, this reduces to the usual definition of a small category $\mathbb{C}$, with

- C_0 the set of objects,
- C_1 the set of arrows,
- C_2 the set of composable pairs

$$A \xrightarrow{\alpha} B \xrightarrow{\beta} C,$$

- C_3 the set of composable triples

$$A \xrightarrow{\alpha} B \xrightarrow{\beta} C \xrightarrow{\gamma} D,$$

Using the notation

- $d_0(\alpha) = A$, $d_1(\alpha) = B$,
- $n(A) = \mathrm{id}_A$,
- $f_0(\alpha, \beta) = \alpha$, $f_1(\alpha, \beta) = \beta$,
- $m(\alpha, \beta) = \beta \circ \alpha$.

the axioms reduce then to the following assertions:

(G1) is just the definition of C_2;
(G2) id_A is an arrow from A to A;
(G3) $\beta \circ \alpha$ is an arrow from A to C;

(G4) $\mathrm{id}_B \circ \alpha = \alpha$, $\alpha \circ \mathrm{id}_A = \alpha$;
(G5) $\gamma \circ (\beta \circ \alpha) = (\gamma \circ \beta) \circ \alpha$.

An internal groupoid $\mathbb{G}$ is an internal category $\mathbb{C}$ as above, together with an additional datum, namely, a morphism

$$\tau : C_1 \longrightarrow C_1$$

such that

(G6) the triangles

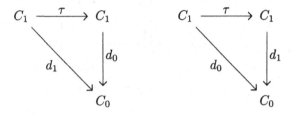

are commutative,

(G7) the squares

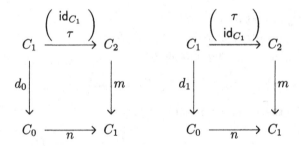

are commutative.

These conditions imply at once $\tau \circ \tau = \mathrm{id}_{C_1}$.

In the case $\mathcal{C} = \mathsf{Set}$ and using again, without further notice, the previous notation, the additional requirements for a groupoid become

- $\tau(\alpha) = \alpha^{-1}$,

with the axioms

(G6) α^{-1} is an arrow from B to A,
(G7) $\alpha^{-1} \circ \alpha = \mathrm{id}_A$, $\alpha \circ \alpha^{-1} = \mathrm{id}_B$.

It is also well known, and left to the reader as an exercise, that an internal category is an internal groupoid precisely when the two commutative squares in (G3) turn out to be pullbacks. Thus for an internal category, being an internal groupoid is a property, not an additional structure.

A covariant internal presheaf on the internal category $\mathbb{C}$ consists in giving the data

$$P_1 \xrightarrow{\ \delta_1\ } P_0 \xrightarrow{\ p_0\ } C_0$$

where

(P1) the object P_1 is defined by the pullback

$$
\begin{array}{ccc}
P_1 & \xrightarrow{\ \delta_0\ } & P_0 \\
{\scriptstyle p_1}\big\downarrow & & \big\downarrow{\scriptstyle p_0} \\
C_1 & \xrightarrow[\ d_0\ ]{} & C_0
\end{array}
$$

(P2) the square

$$
\begin{array}{ccc}
P_1 & \xrightarrow{\ \delta_1\ } & P_0 \\
{\scriptstyle p_1}\big\downarrow & & \big\downarrow{\scriptstyle p_0} \\
C_1 & \xrightarrow[\ d_1\ ]{} & C_0
\end{array}
$$

is commutative,

(P3) the triangle

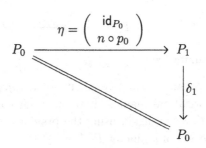

is commutative,

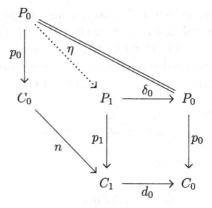

Diagram 7.1

(P4) given the pullback

$$
\begin{array}{ccc}
P_2 & \xrightarrow{\;\varphi_0\;} & P_1 \\
\downarrow{p_2} & & \downarrow{p_1} \\
C_2 & \xrightarrow[\;f_0\;]{} & C_1
\end{array}
$$

the diagram

$$
\begin{array}{ccc}
& \mu = \begin{pmatrix} \delta_0 \circ \varphi_0 \\ m \circ p_2 \end{pmatrix} & \\
P_2 & \xrightarrow{\hspace{6cm}} & P_1 \\
\varphi_1 = \begin{pmatrix} \delta_1 \circ \varphi_0 \\ f_1 \circ p_2 \end{pmatrix} \Big\downarrow & & \Big\downarrow \delta_1 \\
P_1 & \xrightarrow[\;\delta_1\;]{} & P_0
\end{array}
$$

is commutative.

Diagrams 7.1, 7.2 and 7.3 make explicit the way η, μ and φ_1 are defined, as unique factorizations making those diagrams commutative.

In the case $\mathcal{C} = \mathsf{Set}$, again using the previous notation, this reduces to the usual notion of a functor $P \colon \mathbb{C} \longrightarrow \mathsf{Set}$:

- $P_0 = \big\{ (a, A) \big| a \in P(A) \big\}$;

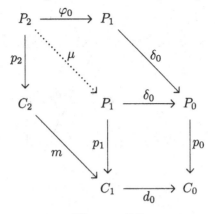

Diagram 7.2

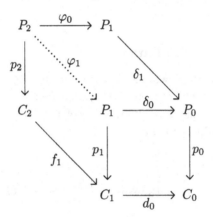

Diagram 7.3

- $p_0(a, A) = A$;
- $P_1 = \{(a, \alpha) \mid \alpha \colon A \longrightarrow B, \ a \in P(A)\}$;
- $p_1(a, \alpha) = \alpha$; $\delta_0(a, \alpha) = (a, A)$;
- $\delta_1(a, \alpha) = P(\alpha)(a)$;
- $P_2 = \{(a, \alpha, \beta) \mid A \xrightarrow{\alpha} B \xrightarrow{\beta} C, \ a \in P(A)\}$;
- $p_2(a, \alpha, \beta) = (\alpha, \beta)$, $\varphi_0(a, \alpha, \beta) = (a, \alpha)$;
- $\eta(a, A) = (a, \mathrm{id}_A)$, $\mu(a, \alpha, \beta) = (a, \beta \circ \alpha)$;
- $\varphi_1(a, \alpha, \beta) = \big(P(\alpha)(a), \beta\big)$.

The axioms reduce to the following assertions:

(P1) this is just the definition of P_1;
(P2) $P(\alpha)(a)$ is an element of $P(B)$;

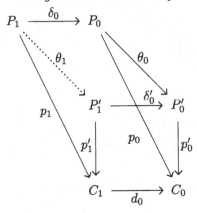

Diagram 7.4

(P3) $P(\mathrm{id}_A)(a) = a$;
(P4) $P(\beta \circ \alpha)(a) = P(\beta)P(\alpha)(a)$.

It remains to write down the definition of an internal natural transformation $\theta \colon P \Rightarrow P'$ between two internal presheaves on the internal category $\mathbb{C}$. This consists in giving a morphism $\theta_0 \colon P_0 \longrightarrow P'_0$ such that the following conditions hold:

(N1) the triangle

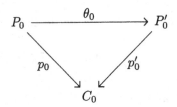

is commutative;
(N2) considering the factorization θ_1 in the commutative diagram 7.4, the following diagram is commutative:

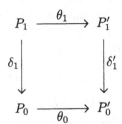

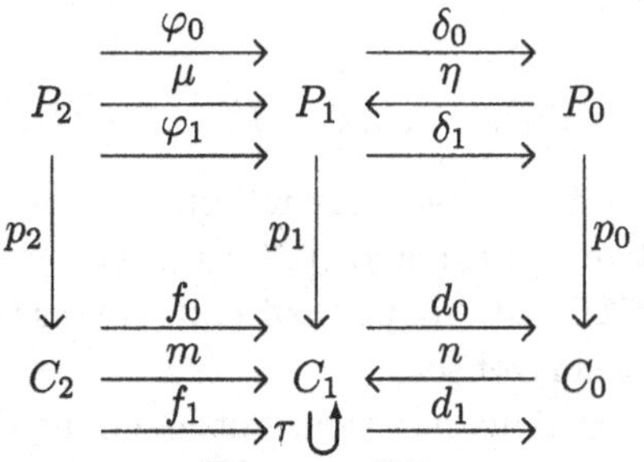

$$P_2 \underset{\underset{\varphi_1}{\longrightarrow}}{\overset{\overset{\varphi_0}{\longrightarrow}}{\xrightarrow{\mu}}} P_1 \underset{\underset{\delta_1}{\longrightarrow}}{\overset{\overset{\delta_0}{\longrightarrow}}{\xleftarrow{\eta}}} P_0$$

$$\left\downarrow p_2 \qquad \left\downarrow p_1 \qquad \left\downarrow p_0\right.$$

$$C_2 \underset{\underset{f_1}{\longrightarrow}{}_{\tau}}{\overset{\overset{f_0}{\longrightarrow}}{\xrightarrow{m}}} C_1 \underset{\underset{d_1}{\longrightarrow}}{\overset{\overset{d_0}{\longrightarrow}}{\xleftarrow{n}}} C_0$$

Diagram 7.5

Again in the case $\mathcal{C} = \mathsf{Set}$ and two presheaves $P, P' \colon \mathbb{C} \longrightarrow \mathsf{Set}$, this reduces to the definition of an ordinary natural transformation $\theta \colon P \Rightarrow P'$:

- $\theta_0(a, A) = \big(\theta_A(a), A\big),$
- $\theta_1(a, \alpha) = \big(\theta_A(a), \alpha\big),$

with axioms becoming

(N1) $\theta_A(a)$ is an element of $P'(A)$,
(N2) $P'(\alpha)\big(\theta_A(a)\big) = \theta_B\big(P(\alpha)(a)\big).$

The notion of internal presheaf on an internal groupoid $\mathbb{G}$ is simply the notion of internal presheaf on the underlying internal category $\mathbb{C}$. But in the case of an internal groupoid, internal presheaves admit a much more elegant characterization, given by the following lemma.

Lemma 7.1.1 *With the previous notation, a covariant presheaf on the internal groupoid $\mathbb{G}$ given by the bottom line of diagram 7.5 consists in the other data constituting the diagram, where all the squares of arrows "corresponding to each other by the notation" are pullbacks. More precisely, the squares*

$$p_1 \circ \varphi_i = f_i \circ p_2, \quad p_0 \circ \delta_i = d_i \circ p_1, \quad p_1 \circ \mu = m \circ p_2, \quad p_1 \circ \eta = n \circ p_0$$

are pullbacks.

Proof Since the problem is entirely expressed in terms of commutative diagrams and pullbacks, it suffices to give a proof of the result in the category of sets. Indeed, write $\mathcal{D} \subseteq \mathcal{C}$ for the full subcategory of $\mathcal{C}$ obtained as closure under pullbacks of the finitely many objects which enter the problem. We can equivalently develop the proof in this category $\mathcal{D}$, which is now small and still has pullbacks. But the Yoneda embedding

$$\mathcal{D} \xrightarrow{\;Y\;} [\mathcal{D}^*, \mathsf{Set}], \quad D \mapsto \mathcal{D}(-, D)$$

is full and faithful and preserves and reflects pullbacks. It thus suffices to prove the result in the category $[\mathcal{D}^*, \mathsf{Set}]$. But since pullbacks and commutativities in $[\mathcal{D}^*, \mathsf{Set}]$ are pointwise, it suffices to prove the result pointwise in the category of sets.

Thus we use the previous notation, but assuming $\mathcal{C} = \mathsf{Set}$. For simplicity, we use the standard notation $m(\alpha, \beta) = \beta \circ \alpha$, $n(A) = \mathrm{id}_A$ and $\tau(\alpha) = \alpha^{-1}$.

Let us start with an internal presheaf on the groupoid $\mathbb{G}$ and let us prove that the squares of the statement are pullbacks. Two of them are pullbacks by definition of P_1 and P_2. We have already observed that when $\mathcal{C} = \mathsf{Set}$, the internal presheaf reduces to an ordinary covariant functor $P\colon \mathbb{G} \longrightarrow \mathsf{Set}$ and we use the corresponding set theoretical notation, made explicit above in this section.

Let us verify first that the square

$$
\begin{array}{ccc}
P_1 & \xrightarrow{\;\delta_1\;} & P_0 \\
\Big\downarrow{\scriptstyle p_1} & & \Big\downarrow{\scriptstyle p_0} \\
C_1 & \xrightarrow[\;d_1\;]{} & C_0
\end{array}
$$

is a pullback. If

$$(\alpha\colon A \longrightarrow B) \in C_1, \quad b \in P(B), \quad d_1(\alpha) = p_0(b),$$

then $P(\alpha^{-1})(b) \in A$ and thus

$$\left(P(\alpha^{-1})(b), \alpha \right) \in P_1,$$

with the expected properties

$$p_1\left(P(\alpha^{-1})(b), \alpha \right) = \alpha, \quad \delta_1\left(P(\alpha^{-1})(b), \alpha \right) = P(\alpha)P(\alpha^{-1})(b) = b.$$

The pair $\left(P(\alpha^{-1})(b), \alpha \right)$ is unique with these properties. Indeed given

$$(a, \beta) \in P_1, \quad p_1(a, \beta) = \alpha, \quad \delta_1(a, \beta) = b,$$

we get at once $\alpha = \beta$ and $b = P(\beta)(a)$, that is, $a = P(\beta^{-1})(a) = P(\alpha^{-1})(a)$.

To verify that the square

$$P_1 \xleftarrow{\quad \eta \quad} P_0$$

$$p_1 \Big\downarrow \qquad\qquad \Big\downarrow p_0$$

$$C_1 \xleftarrow{\quad n \quad} C_0$$

is a pullback, choose

$$(a, \alpha) \in P_1, \quad A \in C_0, \quad p_1(a, \alpha) = n(A).$$

This implies $\alpha = \mathrm{id}_A$ and one has indeed

$$(a, A) \in P_0, \quad \eta(a, A) = (a, \mathrm{id}_A) = (a, \alpha), \quad p_0(a, A) = A.$$

Such a pair (a, A) is unique, since given

$$(b, B) \in P_0, \quad \eta(b, B) = (a, \alpha), \quad p_0(b, B) = A,$$

we get at once $A = B$ and $(b, \mathrm{id}_A) = \eta(b) = (a, \alpha)$ and thus $(b, B) = (a, A)$.

We prove now that the square

$$P_2 \xrightarrow{\quad \mu \quad} P_1$$

$$p_2 \Big\downarrow \qquad\qquad \Big\downarrow p_1$$

$$C_2 \xrightarrow{\quad m \quad} C_1$$

is a pullback. Given

$$(\alpha, \beta) \in C_2, \quad (a, \gamma) \in P_1, \quad m(\alpha, \beta) = p_1(a, \gamma),$$

one has $\gamma = \beta \circ \alpha$ and thus

$$(a, \alpha, \beta) \in P_2, \quad \mu(a, \alpha, \beta) = (a, \beta \circ \alpha) = (a, \gamma), \quad p_2(a, \alpha, \beta) = (\alpha, \beta).$$

The uniqueness of a triple with these properties holds, since given

$$(b, \delta, \varepsilon) \in P_2, \quad \mu(b, \delta, \varepsilon) = (a, \gamma), \quad p_2(b, \delta, \varepsilon) = (\alpha, \beta),$$

we deduce

$$(a, \gamma) = \mu(b, \delta, \varepsilon) = (b, \varepsilon \circ \delta), \quad (\alpha, \beta) = p_2(b, \delta, \varepsilon) = (\delta, \varepsilon),$$

from which $a = b$, $\alpha = \delta$ and $\beta = \epsilon$.

There remains the case of the square

$$P_2 \xrightarrow{\varphi_1} P_1$$

$$p_2 \downarrow \qquad \qquad \downarrow p_1$$

$$C_2 \xrightarrow[f_1]{} C_1$$

Consider

$$(\alpha, \beta) \in C_2, \quad (b, \gamma) \in P_1, \quad f_1(\alpha, \beta) = p_1(b, \gamma).$$

This forces $\beta = \gamma$ and thus

$$\left(P(\alpha^{-1})(b), \alpha, \beta \right) \in P_2$$

with the expected properties

$$\varphi_1\left(P(\alpha^{-1})(b), \alpha, \beta \right) = \left(P(\alpha)P(\alpha^{-1})(b), \beta \right) = (b, \beta) = (b, \gamma),$$

$$p_2\left(P(\alpha^{-1})(b), \alpha, \beta \right) = (\alpha, \beta).$$

To prove the uniqueness of such a triple, consider

$$(a, \delta, \varepsilon) \in P_2, \quad \varphi_1(a, \delta, \varepsilon) = (b, \gamma), \quad p_2(a, \delta, \varepsilon) = (\alpha, \beta).$$

This yields

$$(b, \gamma) = \varphi_1(a, \delta, \varepsilon) = \left(P(\delta)(a), \varepsilon \right), \quad (\alpha, \beta) = p_2(a, \delta, \varepsilon) = (\delta, \varepsilon)$$

and therefrom the equalities $\alpha = \delta$, $\beta = \varepsilon$, $b = P(\delta)(a)$. This also implies $a = P(\delta^{-1})(b) = P(\alpha^{-1})(b)$.

We must now prove the converse implication. Thus we start from the situation of the statement and prove that (P_0, p_0, δ_1) is an internal presheaf. By assumption, we have at once conditions (P1) and (P2). Considering the pullbacks

$$P_0 \xrightarrow{\eta} P_1 \xrightarrow{\delta_1} P_0$$

$$p_0 \downarrow \qquad \quad p_1 \downarrow \qquad \quad \downarrow p_0$$

$$C_0 \xrightarrow[n]{} C_1 \xrightarrow[d_1]{} C_0$$

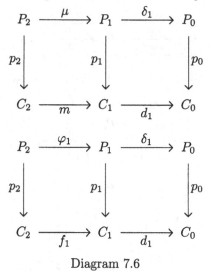

Diagram 7.6

with $d_1 \circ n = \mathrm{id}_{C_0}$, we get $\delta_1 \circ \eta = \mathrm{id}_{P_0}$, which is condition (P3). Finally consider the pullbacks of diagram 7.6. The equality $d_1 \circ m = d_1 \circ f_1$ indicates that those pullbacks have the same bottom composite, from which $\delta_1 \circ \mu = \delta_1 \circ \varphi_1$, which is condition (P4). $\qquad \square$

Lemma 7.1.2 *Let C be a category with pullbacks and $\mathbb{G}$ an internal groupoid in C. Given two presheaves P, P' on $\mathbb{G}$, an internal natural transformation $\theta \colon P \Rightarrow P'$ is equivalently given by three arrows $\theta_i \colon P_i \longrightarrow P_i'$ such that the following triangles commute –*

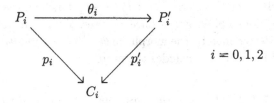

– and, in diagram 7.7 where the notation is borrowed from axioms (P3) *and* (P4), *all the squares of "corresponding arrows" are commutative.*

Proof   An internal natural transformation $\theta \colon P \Rightarrow P'$ yields the equality

$$p_1' \circ (\theta_1 \circ \varphi_0) = p_1 \circ \varphi_0 = f_0 \circ p_2.$$

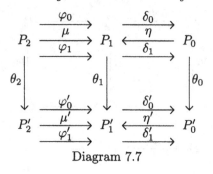

Diagram 7.7

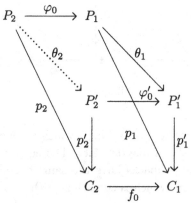

Diagram 7.8

This implies the existence of a unique factorization θ_2 through the pullback defining P_2, as in diagram 7.8. This yields at once the commutativity of diagram 7.7.

Conversely given three arrows $\theta_i \colon P_i \longrightarrow P_i'$ which make commutative diagrams 7.7 and 7.8, we get at once an internal natural transformation $\theta \colon P \Rightarrow P'$ determined by the morphism θ_0. Since θ_1 and θ_2 are uniquely determined by θ_0, this concludes the proof.   □

Corollary 7.1.3 *Let $\mathcal{C}$ be a category with pullbacks and let*

$$C_2 \xrightarrow[\;\;\;\;\underset{f_1}{\overset{m}{\longrightarrow}}\;\;\;\;]{f_0} C_1 \xleftarrow[\;\;\;\underset{d_1}{\overset{n}{\longleftarrow}}\;\;\;]{d_0} C_0$$

be an internal groupoid in $\mathcal{C}$. The category of internal presheaves on this internal groupoid is, in the category CAT of all categories and functors, the two dimensional limit of the diagram

$$\mathcal{C}/C_2 \xleftarrow[\xleftarrow{\quad f_1^* \quad}]{\xleftarrow{\quad f_0^* \quad}{\quad m^* \quad}} \mathcal{C}/C_1 \xrightarrow[\xleftarrow{\quad d_1^* \quad}]{\xleftarrow{\quad d_0^* \quad}{\quad n^* \quad}} \mathcal{C}/C_0$$

where, as usual, f^* indicates the functor "pullback along f".

Before proving this corollary, we comment on the notion of "two dimensional limit". Roughly speaking, it is the usual notion of limit "where equalities are replaced by isomorphisms". A formal theory is developed in section 7.9, but the following more intuitive approach will be sufficient to handle the generalized Galois theory of section 7.5.

Most constructions in category theory allow us just to define a category up to an equivalence and a functor up to an isomorphism, as in the statement of corollary 7.1.3, where the pullback functors are only defined up to isomorphism. It is sensible to expect that in a diagram of categories and functors of which one computes the limit, replacing the categories by equivalent ones and the functors by isomorphic ones will finally yield a limit category which is equivalent to the original one. Unfortunately, this sensible expectation is just wrong!

Write **1** for the unit category, with one object and the identity on it. The equalizer of twice the identity on **1** is again the category **1**. Now replace **1** by the equivalent category **1**$_2$, having two isomorphic objects. Write $F_1, F_2 \colon \mathbf{1}_2 \xrightarrow{\quad\quad} \mathbf{1}_2$ for the two possible constant functors, both isomorphic to the identity on **1**$_2$. This time, $\mathsf{Ker}\,(F_1, F_2)$ is the empty category, which is not at all equivalent to the original equalizer **1**.

An object in a limit category $\lim_{i \in I} \mathcal{X}_i$ is a compatible family $(X_i)_{i \in I}$ of objects $X_i \in \mathcal{X}_i$, where the compatibility means that given a functor $F \colon \mathcal{X}_i \longrightarrow \mathcal{X}_j$ in the diagram, $F(X_i) = X_j$. Such an equality is to be considered somehow unnatural in category theory, since most often F has only been defined up to isomorphism, like the pullback functors f^* in the statement of 7.1.3. Thus requiring an isomorphism $F(X_i) \cong X_j$ in the category $\mathcal{X}_j$ would be more sensible. This is the spirit of what a two dimensional limit is.

Now observe that given a commutative diagram

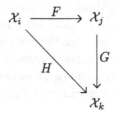

in a diagram of categories and functors, we have of course

$$F(X_i) = X_j \text{ and } G(X_j) = X_k \Rightarrow H(X_i) = X_k.$$

But when F, G, H have been defined "up to an isomorphism", as in corollary 7.1.3, it no longer makes sense to expect an actual commutative diagram, but only a commutativity "up to an isomorphism", like the equality $d_0 \circ m = d_0 \circ f_0$ in 7.1.3 which yields the isomorphism (not the equality) $m^* \circ d_0^* \cong f_0^* \circ d_0^*$. So imagine now that an isomorphism

$$\theta: G \circ F \Rightarrow H$$

is given in the previous triangle, instead of plain commutativity. In our two dimensional limit we require isomorphisms

$$\alpha: F(X_i) \longrightarrow X_j, \quad \beta: G(X_j) \longrightarrow X_k, \quad \gamma: H(X_i) \longrightarrow X_k.$$

We must of course require a coherence condition between these various isomorphisms, namely, the commutativity of the diagram

$$
\begin{array}{ccc}
GF(X_i) & \xrightarrow{\ G(\alpha)\ } & G(X_j) \\
\theta_{X_i} \downarrow & & \downarrow \beta \\
H(X_i) & \xrightarrow[\ \gamma\ ]{} & X_k
\end{array}
$$

Such a coherence condition is obviously satisfied in the special case where θ, α, β, γ are identities, that is, in the case of an ordinary limit.

In corollary 7.1.3, the diagram of pullback functors must thus be understood as a diagram of categories and functors, together with commutativities "up to isomorphisms" between some composite functors. That is, every equality like $d_0 \circ m = d_0 \circ f_0$ in the original diagram yields a corresponding "commutativity up to an isomorphism" in the diagram of pullback functors. The two dimensional limit of this diagram is then constituted of the families $(P_i, p_i) \in \mathcal{C}/C_i$, together with an isomorphism $\alpha: h^*(P_i, p_i) \longrightarrow (P_j, p_j)$ for every arrow $h: C_j \longrightarrow C_i$ in the diagram we consider; those isomorphisms are required to satisfy coherence conditions with respect to all "commutativities up to isomorphisms".

Such a two dimensional limit has the advantage of being independent (up to an equivalence) of isomorphic choices of the pullback functors h^*.

Of course there is a corresponding notion of two dimensional colimit of a diagram $(\mathcal{X}_i)_{i \in I}$ of categories and functors. One first considers the coproduct $\coprod_{i \in I} \mathcal{X}_i$ of the categories involved and, for every functor $F \colon \mathcal{X}_i \longrightarrow \mathcal{X}_j$ in the diagram, one forces the existence of coherent isomorphisms (instead of equalities) between each $X \in \mathcal{X}_i$ and the corresponding object $F(X) \in \mathcal{X}_j$. We refer again to 7.9 for a formal approach.

With these comments in mind, we switch now to the proof of corollary 7.1.3.

Proof of corollary An object of the two dimensional limit thus yields three arrows

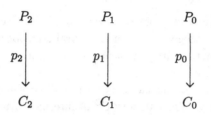

together with compatibility isomorphisms
$$p_2 \cong f_0^*(p_1) \cong m^*(p_1) \cong f_1^*(p_1),$$
$$p_1 \cong d_0^*(p_0) \cong d_1^*(p_0),$$
$$p_0 \cong n^*(p_1).$$
This means precisely giving arrows φ_0, μ, φ_1, δ_0, δ_1, η yielding pullbacks as in lemma 7.1.1. Thus the objects of the two dimensional limit category correspond precisely to the internal presheaves on the internal groupoid.

The case of morphisms follows at once from lemma 7.1.2. $\qquad\square$

7.2 Internal precategories and their presheaves

Very roughly speaking, a precategory is what remains from the definition of an internal category when you cancel all references to pullbacks.

Definition 7.2.1 An internal precategory $\mathbb{C}$ in a category $\mathcal{C}$ consists in giving the data
$$C_2 \xrightarrow[\substack{\ f_1\ }]{\substack{\ f_0\ \\ m}} C_1 \underset{\substack{\ d_1\ }}{\overset{\substack{\ d_0\ \\ n}}{\longleftarrow}} C_0$$
where the following relations hold:
$$d_0 \circ f_1 = d_1 \circ f_0, \quad d_1 \circ m = d_1 \circ f_1, \quad d_0 \circ m = d_0 \circ f_0,$$
$$d_0 \circ n = \mathrm{id}_{C_0}, \quad d_1 \circ n = \mathrm{id}_{C_0}.$$

For simplicity of notation, let us write $\mathbb{P}$ for the (ordinary) category with three objects P_0, P_1, P_2 and generated by six arrows

$$P_2 \xrightarrow[\substack{f_0 \\ m \\ f_1}]{} P_1 \xleftarrow[\substack{d_0 \\ n \\ d_1}]{} P_0$$

on which the following conditions are imposed:

$$d_0 \circ f_1 = d_1 \circ f_0, \quad d_1 \circ m = d_1 \circ f_1, \quad d_0 \circ m = d_0 \circ f_0,$$
$$d_0 \circ n = \mathsf{id}_{C_0}, \quad d_1 \circ n = \mathsf{id}_{C_0}.$$

A precategory $\mathbb{C}$ in a category $\mathcal{C}$ is thus simply a functor $\mathbb{C} \colon \mathbb{P} \longrightarrow \mathcal{C}$. A morphism of precategories is then a natural transformation. We shall write $\mathsf{PreCat}(\mathcal{C})$ for the category of precategories in $\mathcal{C}$.

Definition 7.2.2 Let $\mathcal{C}$ be a category with pullbacks and $\mathbb{C} \colon \mathbb{P} \longrightarrow \mathcal{C}$ an internal precategory. The category $\mathcal{C}^{\mathbb{C}}$ of internal covariant presheaves on $\mathbb{C}$ is, by definition, the two dimensional limit of the diagram

$$\mathcal{C}/\mathbb{C}(P_2) \xleftarrow[\substack{f_0^* \\ m^* \\ f_1^*}]{} \mathcal{C}/\mathbb{C}(P_1) \xrightarrow[\substack{d_0^* \\ n^* \\ d_1^*}]{} \mathcal{C}/\mathbb{C}(P_0).$$

Let us make a strong point that definition 7.2.2 is not at all a generalization of the definition of internal presheaf on an internal category. But when the internal category turns out to be an internal groupoid, corollary 7.1.3 indicates that definition 7.2.2 is equivalent to the usual definition.

Now given a category $\mathcal{A}$ and a precategory $\mathbb{C}$ in $[\mathcal{A}^{\mathsf{op}}, \mathsf{CAT}]$, for every object $A \in \mathcal{A}$ we can consider the composite

$$\mathbb{P} \xrightarrow{\;\mathbb{C}\;} [\mathcal{A}^{\mathsf{op}}, \mathsf{CAT}] \xrightarrow{\;\mathsf{ev}_A\;} \mathsf{CAT}$$

where ev_A is the evaluation functor at the object A. This yields a diagram in CAT of which we can take the two dimensional limit, written $(\mathsf{Lim}\,\mathbb{C})(A)$, or the two dimensional colimit, written $(\mathsf{Colim}\,\mathbb{C})(A)$. This yields two functors

$$\mathsf{Colim}\,\mathbb{C} \colon \mathcal{A}^{\mathsf{op}} \longrightarrow \mathsf{CAT}, \quad A \mapsto (\mathsf{Colim}\,\mathbb{C})(A),$$
$$\mathsf{Lim}\,\mathbb{C} \colon \mathcal{A}^{\mathsf{op}} \longrightarrow \mathsf{CAT}, \quad A \mapsto (\mathsf{Lim}\,\mathbb{C})(A),$$

and thus finally two other functors

$$\mathsf{Colim} \colon \mathsf{PreCat}[\mathcal{A}^{\mathsf{op}}, \mathsf{CAT}] \longrightarrow [\mathcal{A}^{\mathsf{op}}, \mathsf{CAT}],$$

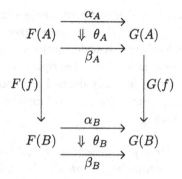

Diagram 7.9

$$\mathsf{Lim}\colon \mathsf{PreCat}[\mathcal{A}^{\mathrm{op}}, \mathsf{CAT}] \longrightarrow [\mathcal{A}^{\mathrm{op}}, \mathsf{CAT}]$$

sending a precategory $\mathbb{C}\colon \mathbb{P} \longrightarrow [\mathcal{A}^{\mathrm{op}}, \mathsf{CAT}]$ to its "pointwise 2-dimensional colimit" or its "pointwise two dimensional limit".

Let us now recall that given two functors $F, G\colon \mathcal{A} \Longrightarrow \mathsf{CAT}$ and two natural transformations $\alpha, \beta\colon F \to G$, there is a notion of modification $\theta\colon \alpha \rightsquigarrow \beta$. For each object $A \in \mathcal{A}$, α and β induce two functors $\alpha_A, \beta_A\colon F(A) \Longrightarrow G(A)$. A modification θ consists in a natural transformation $\theta_A\colon \alpha_A \Rightarrow \beta_A$ for each $A \in \mathcal{A}$, with the naturality requirement $\theta_B \star 1_{F(f)} = 1_{G(f)} \star \theta_A$ for each arrow $f\colon A \longrightarrow B$ in $\mathcal{A}$, where $\star$ indicates the horizontal composition of natural transformations, also called "Godement product" (see diagram 7.9). The natural transformations between F and G and the modifications between them constitute a category which we still write $\mathsf{Nat}(F, G)$.

That situation can easily be generalized to the context of "pseudo-functors". The prototype of a contravariant pseudo-functor on a category $\mathcal{C}$ with pullbacks is given by

$$Z_{\mathcal{C}}\colon \mathcal{C} \longrightarrow \mathsf{CAT}, \quad Z_{\mathcal{C}}(C) = \mathcal{C}/C, \quad Z_{\mathcal{C}}(f) = f^*.$$

Each $Z_{\mathcal{C}}(f)$ is only defined up to an isomorphism, thus the requirements $Z_{\mathcal{C}}(f \circ g) = Z_{\mathcal{C}}(g) \circ Z_{\mathcal{C}}(f)$ for a functor do not hold, but must be replaced by corresponding coherent isomorphisms $Z_{\mathcal{C}}(f \circ g) \cong Z_{\mathcal{C}}(g) \circ Z_{\mathcal{C}}(f)$. We have considered such a situation already many times. Observe that in this specific case, we can make the canonical choice that pulling back along an identity is the identity functor. A pseudo-functor $F\colon \mathcal{C} \longrightarrow \mathsf{CAT}$ on a category $\mathcal{C}$ generalizes the previous situation: in the definition of a functor, one replaces the preservation of the composition law by a preservation up to coherent isomorphisms. The notions of natural transformation and modification extend at once to this context. We write $\mathsf{Ps}(\mathcal{C}, \mathsf{CAT})$ for the 2-category (see definition 7.9.1) of pseudo-functors, pseudo-natural-transformations and modifications.

The reader who needs a more precise description should refer to section 7.9, where these notions are formally defined in a more general context.

The following Yoneda lemma for pseudo-functors is classical (see for example [8], volume 2, chapter 8).

Lemma 7.2.3 *Consider a pseudo-functor $F\colon \mathcal{A}^{\mathrm{op}} \longrightarrow \mathsf{CAT}$, an object $A \in \mathcal{A}$ and the corresponding "representable" functor*

$$\mathcal{A}(-, A)\colon \mathcal{A}^{\mathrm{op}} \longrightarrow \mathsf{CAT}, \quad B \mapsto \mathcal{A}(B, A)$$

where $\mathcal{A}(B, A)$ is viewed as a discrete category. The following equivalence of categories holds –

$$F(A) \approx \mathsf{PsNat}\big(\mathcal{A}(-, A), F\big)$$

– between the category $F(A)$ and the category of pseudo-natural-transformations and modifications.

Proof In one direction, one considers the functor

$$M\colon F(A) \longrightarrow \mathsf{PsNat}\big(\mathcal{A}(-, A), F\big)$$

which sends an object $X \in F(A)$ to the pseudo-natural-transformation $M(X)$ determined by the functors

$$M(X)_B\colon \mathcal{A}(B, A) \longrightarrow F(B), \quad f \mapsto F(f)(X).$$

Conversely, one considers the functor

$$N\colon \mathsf{PsNat}\big(\mathcal{A}(-, A), F\big) \longrightarrow F(A)$$

which sends $\alpha\colon \mathcal{A}(-, A) \Rightarrow F$ to $\alpha_A(\mathrm{id}_A)$ and a modification $\theta\colon \alpha \rightsquigarrow \beta$ to $(\theta_A)_{\mathrm{id}_A}$. It is routine to check that this determines an equivalence of categories. $\qquad\square$

Proposition 7.2.4 *Consider*

- *a category $\mathcal{A}$,*
- *a precategory $\mathbb{C}\colon \mathbb{P} \longrightarrow [\mathcal{A}^{\mathrm{op}}, \mathsf{CAT}]$,*
- *a pseudo-functor $F\colon \mathcal{A}^{\mathrm{op}} \longrightarrow \mathsf{CAT}$.*

There is an equivalence of categories

$$\mathsf{PsNat}(\mathsf{Colim}\,\mathbb{C}, F) \approx \underset{i}{\mathsf{Lim}}\,\mathsf{PsNat}\big(\mathbb{C}(P_i), F\big).$$

Proof Since $\mathrm{Colim}\,\mathbb{C}$ is the two dimensional colimit in $\mathrm{Ps}(\mathcal{A}^{\mathrm{op}}, \mathrm{CAT})$ of the diagram

$$\mathbb{C}(P_2) \underset{\underset{\mathbb{C}(f_1)}{\xrightarrow{\hspace{1cm}}}}{\overset{\overset{\mathbb{C}(f_0)}{\xrightarrow{\hspace{1cm}}}}{\xrightarrow[\mathbb{C}(m)]{\hspace{1cm}}}} \mathbb{C}(P_1) \underset{\underset{\mathbb{C}(d_1)}{\xleftarrow{\hspace{1cm}}}}{\overset{\overset{\mathbb{C}(d_0)}{\xrightarrow{\hspace{1cm}}}}{\xleftarrow[\mathbb{C}(n)]{\hspace{1cm}}}} \mathbb{C}(P_0)$$

the result reduces to saying that the "representable" functor

$$\mathrm{PsNat}(-, F) \colon \mathrm{Ps}(\mathcal{A}^{\mathrm{op}}, \mathrm{CAT}) \longrightarrow \mathrm{CAT}$$

transforms this two dimensional colimit into a two dimensional limit, which is a classical result (see [8], volume 2). □

Corollary 7.2.5 *Let us consider*

- *a category $\mathcal{A}$,*
- *an internal precategory $\mathbb{C} \colon \mathbb{P} \longrightarrow \mathcal{A}$,*
- *the "Yoneda embedding" $Y_{\mathcal{A}} \colon \mathcal{A} \longrightarrow \mathrm{Ps}(\mathcal{A}^{\mathrm{op}}, \mathrm{CAT})$ mapping $A \in \mathcal{A}$ onto the discrete representable functor $\mathcal{A}(-, A)$ of lemma 7.2.3,*
- *a pseudo-functor $F \colon \mathcal{A}^{\mathrm{op}} \longrightarrow \mathrm{CAT}$.*

The following equivalence of categories holds:

$$\mathrm{PsNat}(\mathrm{Colim}\, Y_{\mathcal{A}} \circ \mathbb{C}, F) \approx \mathrm{Lim}(F \circ \mathbb{C}).$$

Proof In proposition 7.2.4, it suffices to observe that the equivalence

$$\mathrm{PsNat}\Big(\mathcal{A}\big(-, \mathbb{C}(P_i)\big), P\Big) \approx F\big(\mathbb{C}(P_i)\big)$$

holds by the Yoneda lemma 7.2.3. □

Definition 7.2.6 Let us consider

- a category $\mathcal{A}$,
- an internal precategory $\mathbb{C} \colon \mathbb{P} \longrightarrow \mathcal{A}$,
- a pseudo-functor $F \colon \mathcal{A}^{\mathrm{op}} \longrightarrow \mathrm{CAT}$.

The category $\mathrm{Lim}(F \circ \mathbb{C})$ will be written $F^{\mathbb{C}}$ and called "category of covariant internal F-presheaves on $\mathbb{C}$".

Lemma 7.2.7 *Let us consider*

- *a category $\mathcal{A}$ with pullbacks,*
- *an internal groupoid $\mathbb{C} \colon \mathbb{P} \longrightarrow \mathcal{A}$.*

The category $Z_{\mathcal{A}}^{\mathbb{C}}$ is equivalent to the usual category of covariant internal presheaves on the internal groupoid $\mathbb{C}$, where

$$Z_{\mathcal{A}}\colon \mathcal{A} \longrightarrow \mathsf{CAT}, \quad Z_{\mathcal{A}}(A) = \mathcal{A}/A, \quad Z_{\mathcal{A}}(f) = f^*$$

is the "pulling back" pseudo-functor.

Proof   This is a reformulation of corollary 7.1.3, with the notation of definition 7.2.2.   □

Definition 7.2.8 Let $\mathcal{A}$ be a category. The discrete internal (pre)category on an object $A \in \mathcal{A}$, written $\mathbb{C}_A$, is the precategory

$$A \;\substack{\longrightarrow \\ \longrightarrow \\ \longrightarrow}\; A \;\substack{\longrightarrow \\ \longleftarrow \\ \longrightarrow}\; A$$

where all six arrows are identity on A.

Lemma 7.2.9 *Let us consider*

- *a category $\mathcal{A}$,*
- *the discrete precategory $\mathbb{C}_A\colon \mathbb{P} \longrightarrow \mathcal{A}$ on an object $A \in \mathcal{A}$,*
- *a pseudo-functor $F\colon \mathcal{A}^{\mathrm{op}} \longrightarrow \mathsf{CAT}$.*

In these conditions, one has the equivalence

$$F^{\mathbb{C}_A} \approx F(A).$$

Proof   The pseudo-functor $F \circ \mathbb{C}_A$ is the constant pseudo-functor on $F(A)$ and the diagram $\mathbb{P}$ is connected. This argument extends easily to the two dimensional aspects.   □

7.3 A factorization system for functors

This section introduces a factorization system on CAT, the category of categories and functors, in the spirit of section 5.3.

Proposition 7.3.1 *Every functor $F\colon \mathcal{A} \longrightarrow \mathcal{B}$ between categories factors, uniquely up to an equivalence, as $F = H \circ G$:*

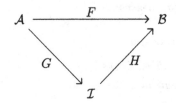

where

(i) *G is essentially surjective on objects, that is, every object of $\mathcal{I}$ is isomorphic to an object of the form $G(A)$,*

(ii) *H is full and faithful.*

Proof It suffices to take for $\mathcal{I}$ the full subcategory of $\mathcal{B}$ generated by the objects of the form $G(A)$, for $A \in \mathcal{A}$. Notice that if

$$\mathcal{A} \xrightarrow{\;G'\;} \mathcal{I}' \xrightarrow{\;H'\;} \mathcal{B}$$

is another such factorization of F, the category $H'(\mathcal{I}')$ is equivalent to $\mathcal{I}'$ because H' is full and faithful. But the objects of $H'(\mathcal{I}')$ are exactly all the objects of the form $F(A)$ and, by assumption on G', possibly some objects isomorphic to an object of the form $F(A)$. Thus $H'(\mathcal{I}')$ is equivalent to $\mathcal{I}$ and therefore $\mathcal{I}'$ is equivalent to $\mathcal{I}$. $\qquad\square$

Proposition 7.3.1 yields in fact a factorization system "up to an equivalence" in the spirit of definition 5.3.1.

Proposition 7.3.2 *In the category* CAT *of all categories and functors, let us write*

- $\mathcal{E}$ *for the class of those functors which are essentially surjective on objects,*
- $\mathcal{M}$ *for the class of fully faithful functors.*

The following properties hold:

(i) *$\mathcal{E}$ and $\mathcal{M}$ contain all equivalences;*

(ii) *$\mathcal{E}$ and $\mathcal{M}$ are closed under composition;*

(iii) *every functor F factors as $F = H \circ G$, with $H \in \mathcal{M}$ and $G \in \mathcal{E}$;*

(iv) *consider a square*

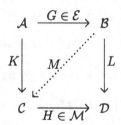

where $G \in \mathcal{E}$, $H \in \mathcal{M}$ and the square commutes up to an isomorphism; in these conditions, there exists a functor M, unique up to an isomorphism, which makes the whole diagram commutative up to isomorphisms.

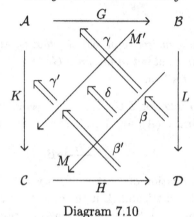

Diagram 7.10

Let us make more explicit condition (iv) of this statement; we refer to diagram 7.10. Since only isomorphisms are involved, we find it
convenient to write them down in the direction which makes the formulæ as simple as possible. First, an isomorphic natural transformation
$\alpha\colon L\circ G\overset{\cong}{\Longrightarrow}H\circ K$ is given. Together with M, we require the existence of two isomorphic natural transformations $\beta\colon L\overset{\cong}{\Longrightarrow}H\circ M$ and
$\gamma\colon M\circ G\overset{\cong}{\Longrightarrow}K$ with the property $(H\star\gamma)\circ(\beta\star G)=\alpha$. If M', β',
γ' have analogous properties, then there exists an isomorphic natural
transformation $\delta\colon M\overset{\cong}{\Longrightarrow}M'$ such that $\beta'=(H\star\delta)\circ\beta$, $\gamma=\gamma'\circ(\delta\star G)$.

Proof of proposition   Conditions (i) and (ii) are obviously satisfied,
while condition (iii) follows from proposition 7.3.1. For condition (iv),
observe that when $B\in\mathcal{B}$, then $B\cong G(A)$ for some $A\in\mathcal{A}$, because
$G\in\mathcal{E}$. We put $M(B)=K(A)$, from which we get an isomorphism
$K(A)=M(B)\cong MG(A)$. Also observe that

$$L(B)\cong LG(A)\cong HK(A)\cong HMG(A)\cong HM(B).$$

Defining M on the arrows reduces to giving a map

$$\mathcal{B}(B,B')\longrightarrow\mathcal{C}\big(M(B),M(B')\big)\cong\mathcal{D}\big(HM(B),HM(B')\big)$$
$$\cong\mathcal{D}\big(L(B),L(B')\big)$$

since H is full and faithful. This composite is defined as being the action
of the functor L. One concludes the proof with routine verifications. $\square$

Corollary 7.3.3 *With the notation of proposition 7.3.2, if*

$$(G\colon \mathcal{A}\longrightarrow \mathcal{B}) \in \mathcal{E}, \quad (H\colon \mathcal{C}\longrightarrow \mathcal{D}) \in \mathcal{M},$$

the following square is a two dimensional pullback, that is, commutes up to an isomorphism and is universal for that property.

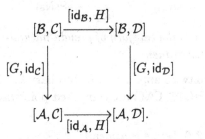

Here, $[\mathcal{A}, \mathcal{B}]$ denotes the category of functors from $\mathcal{A}$ to $\mathcal{B}$ and natural transformations between them.

Proof This is just a reformulation of condition 7.3.2(iv).   □

The previous results translate to the categories $\mathsf{Ps}(\mathcal{A}^{\mathrm{op}}, \mathsf{CAT})$, that is, the categories of pseudo-functors on a given category $\mathcal{A}$.

Proposition 7.3.4 *For a category $\mathcal{A}$, we consider the corresponding 2-category $\mathsf{Ps}(\mathcal{A}^{\mathrm{op}}, \mathsf{CAT})$ (see section 7.9) of pseudo-functors, pseudo-natural-transformations and modifications between them. We denote by*

- *$\mathcal{E}$ the class of pseudo-natural-transformations in $\mathsf{Ps}(\mathcal{A}^{\mathrm{op}}, \mathsf{CAT})$ all of whose components are functors essentially surjective on objects,*

- *$\mathcal{M}$ the class of pseudo-natural-transformations in $\mathsf{Ps}(\mathcal{A}^{\mathrm{op}}, \mathsf{CAT})$ all of whose components are full and faithful functors.*

The classes $\mathcal{E}$, $\mathcal{M}$ of $[\mathcal{A}^{\mathrm{op}}, \mathsf{CAT}]$ satisfy all axioms (i), (ii), (iii), (iv) of proposition 7.3.2.

Proof The proof is just a routine pointwise application of proposition 7.3.2.   □

Corollary 7.3.5 *With the notation of proposition 7.3.4, if*

$$(\varepsilon\colon F \Rightarrow G) \in \mathcal{E}, \quad (\mu\colon H \Rightarrow K) \in \mathcal{M}$$

in $\mathsf{Ps}(\mathcal{A}^{\mathrm{op}}, \mathsf{CAT})$, the following square is a two dimensional pullback:

$$\mathsf{PsNat}(G,H) \xrightarrow{(\mathrm{id}_G,\mu)} \mathsf{PsNat}(G,K)$$

$$(\varepsilon,\mathrm{id}_H) \Big\downarrow \qquad\qquad \Big\downarrow (\varepsilon,\mathrm{id}_K)$$

$$\mathsf{PsNat}(F,H) \xrightarrow[(\mathrm{id}_F,\mu)]{} \mathsf{PsNat}(F,K)$$

where PsNat *denotes the category of pseudo-natural-transformations and modifications between them.*

Proof By a pointwise application of corollary 7.3.3, since two dimensional limits in $\mathsf{Ps}(\mathcal{A}^{\mathrm{op}}, \mathsf{CAT})$ are computed pointwise. $\qquad\square$

Proposition 7.3.6 *Let us consider*

- *a category $\mathcal{A}$,*
- *a precategory $\mathbb{C}\colon \mathbb{P}\longrightarrow \mathsf{Ps}(\mathcal{A}^{\mathrm{op}}, \mathsf{CAT})$.*

The canonical morphism

$$s_0 \colon \mathbb{C}(P_0) \Longrightarrow \mathsf{Colim}\,\mathbb{C}$$

is, in each component, essentially surjective on the objects.

Proof The composite

$$\mathbb{P} \xrightarrow{\ \mathbb{C}\ } \mathsf{Ps}(\mathcal{A}^{\mathrm{op}}, \mathsf{CAT}) \xrightarrow{\ \mathsf{ev}_A\ } \mathsf{CAT}$$

yields a diagram in CAT which we write simply as

$$C_2 \ \begin{array}{c} \xrightarrow{\ f_0\ } \\ \xrightarrow{\ m\ } \\ \xrightarrow{\ f_1\ } \end{array}\ C_1 \ \begin{array}{c} \xrightarrow{\ d_0\ } \\ \xleftarrow{\ n\ } \\ \xrightarrow{\ d_1\ } \end{array}\ C_0.$$

We consider the two dimensional colimit of this diagram in CAT. For every object $A \in \mathcal{A}$ we must prove that the canonical functor

$$s_{0,A} \colon \mathbb{C}(P_0)(A) \longrightarrow (\mathsf{Colim}\,\mathbb{C})(A)$$

is essentially surjective on the objects.

This is an immediate consequence of the construction of a 2-dimensional colimit in CAT. Since there are morphisms

$$C_2 \longrightarrow C_1 \longrightarrow C_0,$$

every object of C_2 or C_1 is, up to isomorphism, isomorphic in the two dimensional colimit to an object of C_0. Thus every object of the two dimensional colimit is isomorphic to an object arising from C_0. $\qquad\square$

7.4 Generalized descent theory

In the category of sets or, more generally, in a regular category (see [4]), the image factorization of a morphism $f : A \longrightarrow B$ is obtained by taking the quotient of A by the kernel pair of f, that is, in the case of sets, by the equivalence relation

$$R_f = \{(a, a') | a, a' \in A, \ f(a) = f(a')\}.$$

Thus this yields the situation

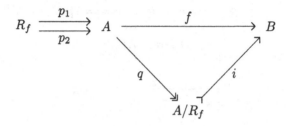

Writing Δ_A, Δ_B for the diagonals of A and B – which are themselves the kernel pairs of id_A and id_B – the image factorization of f can be rewritten

$$A/\Delta_A \xrightarrow{\quad q \quad} A/R_f \xrightarrow{\quad i \quad} B/\Delta_B$$

and these morphisms are induced by the corresponding arrows between the kernel pairs

$$\Delta_A \rightarrowtail R_f \xrightarrow{\ f \times f\ } \Delta_B.$$

In example 4.6.2 we have seen how to view a kernel pair as an internal groupoid. Thus finally, the image factorization of f can be translated in terms of a factorization property involving the groupoids Δ_A, R_f and Δ_B.

Definition 7.4.1 generalizes this situation, allowing internal precategories instead of internal groupoids.

Definition 7.4.1 Let $\sigma : S \longrightarrow R$ be an arrow in a category $\mathcal{A}$. A precategorical decomposition of σ is a factorization in $\mathsf{PreCat}(\mathcal{A})$

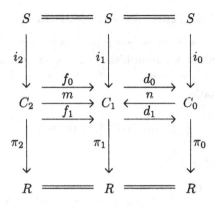

Diagram 7.11

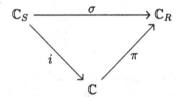

of the morphism σ, seen as an internal functor between the discrete internal categories $\mathbb{C}_S$ and $\mathbb{C}_R$ (see diagram 7.11), with, moreover, the two additional requirements

$$C_0 = S, \quad i_0 = \mathrm{id}_S.$$

The factorization of σ takes place in the category of internal precategories; thus in diagram 7.11, all vertical composites $\pi_k \circ i_k$ are equal to σ and all squares are commutative. The two additional requirements $C_0 = S$, $i_0 = \mathrm{id}_S$ extend the idea that, in the example of the relation R_f at the beginning of this section, R_f is a relation on A, the domain of f.

Lemma 7.4.2 *Let* $\sigma \colon S \longrightarrow R$ *be a morphism in a category* $\mathcal{A}$ *with pullbacks. Consider a precategorical decomposition of* σ

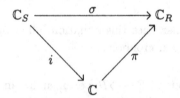

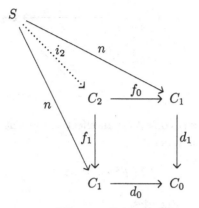

Diagram 7.12

in which $\mathbb{C}$ turns out to be an actual internal category. In these conditions, i and π are uniquely determined by the knowledge of σ and $\mathbb{C}$.

Proof With the notation of definition 7.4.1, we get at once, by commutativity of diagram 7.11,

$$i_0 = \mathsf{id}_S \quad \text{(by assumption)},$$
$$i_1 = n \circ i_0 = n,$$
$$\pi_0 = \pi_0 \circ \mathsf{id}_S = \pi_0 \circ i_0 = \sigma,$$
$$\pi_1 = \pi_0 \circ d_0 = \sigma \circ d_0,$$
$$\pi_2 = \pi_1 \circ f_0 = \sigma \circ d_0 \circ f_0.$$

There remains the case of i_2, which will use the fact that $\mathbb{C}$ is an internal category. We have at once

$$f_0 \circ i_2 = i_1 = n, \quad f_1 \circ i_2 = i_1 = n.$$

This proves that i_2 is the unique factorization making diagram 7.12 commutative, where the square is now a pullback by assumption. Thus i_2 is entirely determined by $\mathbb{C}$. $\qquad\square$

Definition 7.4.3 Let us consider

- a category $\mathcal{A}$,
- a morphism $\sigma \colon S \longrightarrow R$ in $\mathcal{A}$,
- a pseudo-functor $F \colon \mathcal{A}^{\mathrm{op}} \longrightarrow \mathsf{CAT}$,
- a precategorical decomposition of σ

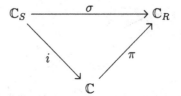

The data $\big(\sigma, (i, \mathbb{C}, \pi)\big)$ constitute an effective descent structure with respect to F when the functor

$$F^{\pi} \colon F^{\mathbb{C}_R} \longrightarrow F^{\mathbb{C}}$$

is an equivalence of categories.

Let us recall that we have the situation

$$\mathbb{P}^{\mathrm{op}} \underset{\underset{\mathbb{C}^{\mathrm{op}} \| \pi^{\mathrm{op}}}{\xrightarrow{\hspace{1cm}}}}{\overset{\mathbb{C}_R^{\mathrm{op}}}{\underset{\xrightarrow{\hspace{1cm}}}{\xrightarrow{\hspace{1cm}}}}} \mathcal{A}^{\mathrm{op}} \xrightarrow{\ F\ } \mathsf{CAT}$$

while, by definition 7.2.6,

$$F^{\mathbb{C}_R} = \mathrm{Lim}\, F \circ \mathbb{C}_R^{\mathrm{op}}, \quad F^{\mathbb{C}} = \mathrm{Lim}\, F \circ \mathbb{C}^{\mathrm{op}}.$$

The existence of the natural transformation

$$F \star \pi^{\mathrm{op}} \colon F \circ \mathbb{C}_R^{\mathrm{op}} \longrightarrow F \circ \mathbb{C}^{\mathrm{op}}$$

induces the existence of the factorization

$$F^{\pi} \colon \mathrm{Lim}\, F \circ \mathbb{C}_R^{\mathrm{op}} \longrightarrow \mathrm{Lim}\, F \circ \mathbb{C}^{\mathrm{op}}$$

involved in the statement. Let us recall further that, by lemma 7.2.9, one has an equivalence $F^{\mathbb{C}_R} \approx F(R)$.

Now we make more explicit, and develop further, the example sketched at the beginning of this section. At the same time, we justify our extension of the terminology "effective descent" in definition 7.4.3.

Proposition 7.4.4 *Consider an arrow $\sigma \colon S \longrightarrow R$ in a category $\mathcal{A}$ with pullbacks and the pseudo-functor*

$$Z_{\mathcal{A}} \colon \mathcal{A}^{\mathrm{op}} \longrightarrow \mathsf{CAT}, \quad A \mapsto \mathcal{A}/A, \quad f \mapsto f^{*}$$

of lemma 7.2.7. The kernel pair of σ, seen as an internal groupoid $\mathbb{G}_{\sigma}$ as in example 4.6.2, determines a precategorical decomposition of σ

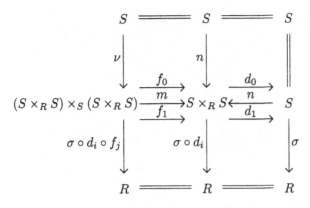

Diagram 7.13

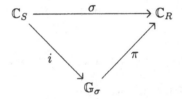

and the following conditions are equivalent:

(i) $\big(\sigma, (i, \mathbb{G}_\sigma, \pi)\big)$ is an effective descent structure with respect to the pseudo-functor $Z_{\mathcal{A}}$;

(ii) $\sigma \colon S \longrightarrow R$ is an effective descent morphism in the sense of definition 4.4.1.

Proof The precategorical decomposition of σ is simply given by diagram 7.13. Notice that this notation with some unspecified indices makes sense. Indeed $\sigma \circ d_0 = \sigma \circ d_1$, because (d_0, d_1) is by definition the kernel pair of σ. Moreover $d_0 \circ f_1 = d_1 \circ f_0$ since the corresponding square is a pullback by definition. So all $\sigma \circ d_i \circ f_j$, for all combinations of $i, j \in \{0, 1\}$, are the same. On the other hand n is the diagonal and ν is the unique factorization through the pullback, resulting from the relation $d_0 \circ n = \mathrm{id}_S = d_1 \circ n$. The commutativity of the diagram follows at once from the fact that m is given by the first and the fourth projections.

Lemma 7.2.7 implies that the category $Z_{\mathcal{A}}^{\mathbb{G}_\sigma}$ is equivalent to the category $\mathcal{A}^{\mathbb{G}_\sigma}$ of the usual internal presheaves on the internal groupoid $\mathbb{G}_\sigma$. On the other hand proposition 4.6.1 asserts that this category $\mathcal{A}^{\mathbb{G}_\sigma}$ is

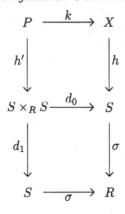

Diagram 7.14

monadic over $\mathcal{A}/S$, the functorial part of the corresponding monad $\mathbb{T}$ being the composite

$$T\colon \mathcal{A}/S \xrightarrow{\;d_0^*\;} \mathcal{A}/S \times_R S \xrightarrow{\;\Sigma_{d_1}\;} \mathcal{A}/S.$$

Thus this yields the following situation:

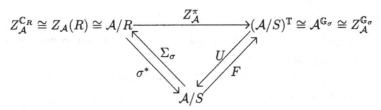

where $F \dashv U$ is the Eilenberg–Moore adjunction of the monad $\mathbb{T}$.

Let us first observe that

$$\sigma^* \circ \Sigma_\sigma \cong T \cong \Sigma_{d_1} \circ d_0^*.$$

Indeed, consider $h\colon X \longrightarrow S$. In diagram 7.14, where the squares are pullbacks, we have

$$
\begin{aligned}
(P, h') &\cong d_0^*(X, h), & (P, d_1 \circ h') &\cong (\Sigma_{d_1} \circ d_0^*)(X, h), \\
(X, \sigma \circ h) &\cong \Sigma_\sigma(X, h), & (P, d_1 \circ h') &\cong (\sigma^* \circ \Sigma_\sigma)(X, h).
\end{aligned}
$$

This proves already that the functors T and $\sigma^* \circ \Sigma_\sigma$ are isomorphic on the objects, from which routine verifications show that the monad $\mathbb{T}$ is isomorphic to the monad induced by the adjunction $\Sigma_\sigma \dashv \sigma^*$.

Next we observe that $U \circ Z_{\mathcal{A}}^{\pi} \cong \sigma^*$. Writing for simplicity

$$G_2 = (S \times_R S) \times_S (S \times_R S), \quad G_1 = S \times_R S, \quad G_0 = S,$$

the category $Z_{\mathcal{A}}^{G_\sigma}$ of lemma 7.2.7 is the two dimensional limit $\mathcal{L}$ of the bottom line in the following diagram:

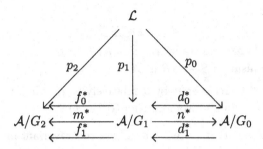

The forgetful functor U maps an object $L \in \mathcal{L}$ onto $p_0(L)$. On the other hand the category $Z_{\mathcal{A}}^{C_R}$ yields the trivial two dimensional limit

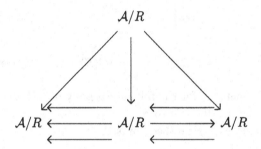

where all arrows are identities. The factorization $Z_{\mathcal{A}}^{\pi} : \mathcal{A}/R \longrightarrow \mathcal{L}$ induced by this situation is such that

$$p_0 \circ Z_{\mathcal{A}}^{\pi} \cong \sigma^*,$$
$$p_1 \circ Z_{\mathcal{A}}^{\pi} \cong (\sigma \circ d_i)^*,$$
$$p_2 \circ Z_{\mathcal{A}}^{\pi} \cong (\sigma \circ d_i \circ f_j)^*.$$

The first of these relations is precisely the expected relation $U \circ Z_{\mathcal{A}}^{\pi} = \sigma^*$.

Since the adjunction $\Sigma_\sigma \dashv \sigma^*$ induces the monad $\mathbb{T}$ and $U \circ Z_{\mathcal{A}}^{\pi} = \sigma^*$, it follows at once that σ^* is monadic precisely when the comparison functor $Z_{\mathcal{A}}^{\pi}$ is an equivalence of categories. This yields the result, via definitions 4.4.1 and 7.4.3. $\qquad \square$

7.5 Generalized Galois theory

This section establishes a Galois theorem for effective descent structures in the sense of definition 7.4.3, without any further assumption of Galois descent.

Definition 7.5.1 Let us consider

- a category $\mathcal{A}$,
- two pseudo-functors $F, G\colon \mathcal{A}^{\mathrm{op}} \rightrightarrows \mathsf{CAT}$,
- a pseudo-natural-transformation $\alpha\colon F \Rightarrow G$,
- a morphism $\sigma\colon S \longrightarrow R$ in $\mathcal{A}$.

An object $M \in G(R)$ is split by σ relatively to α, when there exist an object $X \in F(S)$ and an isomorphism $G(\sigma)(M) \cong \alpha_S(X)$ in $G(S)$.

Let us point out that since F and G are contravariant on $\mathcal{A}$, we have the following situation:

$$
\begin{array}{ccc}
F(R) & \xrightarrow{\ F(\sigma)\ } & F(S) \ni X \\
{\scriptstyle \alpha_R}\big\downarrow & & \big\downarrow{\scriptstyle \alpha_S} \\
M \in G(R) & \xrightarrow[\ G(\sigma)\ ]{} & G(S) \ni G(\sigma)(M) \cong \alpha_S(X)
\end{array}
$$

We shall write $\mathsf{Split}_\alpha(\sigma)$ for the full subcategory of $G(R)$ whose objects are those split by σ relatively to α.

The corresponding Galois theorem is then

Theorem 7.5.2 (Galois theorem) *Let us consider*

- *a category $\mathcal{A}$,*
- *two pseudo-functors $F, G\colon \mathcal{A}^{\mathrm{op}} \rightrightarrows \mathsf{CAT}$,*
- *a pseudo-natural-transformation $\alpha\colon F \Rightarrow G$, all of whose components α_A are full and faithful,*
- *a morphism $\sigma\colon S \longrightarrow R$ in $\mathcal{A}$,*
- *a precategorical decomposition of σ*

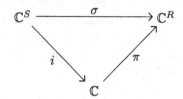

When $\big(\sigma, (i, \mathbb{C}, \pi)\big)$ is an effective descent structure with respect to the pseudo-functor G, one has an equivalence of categories

$$\mathsf{Split}_\alpha(\sigma) \approx F^{\mathbb{C}}.$$

Proof First of all, by definition 7.5.1, an object M is in $\mathsf{Split}_\alpha(\sigma)$ when an isomorphism $G(\sigma)(M) \cong \alpha_S(X)$ exists in $G(S)$ for some object $X \in F(S)$. In other words, the left hand square below is a two dimensional pullback (a "bipullback").

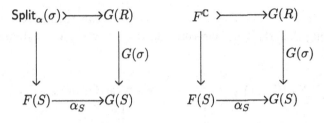

Notice that up to an equivalence, this is a special case of the construction given in the proof of proposition 7.9.6. It remains to prove that $F^{\mathbb{C}}$ can make the right hand square a two dimensional pullback as well.

By assumption, we have a structure of effective descent, thus in particular, by definition 7.4.3 and lemma 7.2.9, a commutative diagram, up to isomorphisms,

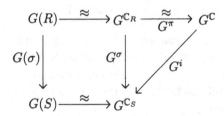

where all horizontal arrows are equivalences of categories. The equivalence $\mathsf{Split}_\alpha(\sigma) \approx F^{\mathbb{C}}$ thus reduces to proving the existence of a two dimensional pullback

$$
\begin{array}{ccc}
F^{\mathbb{C}} & \longrightarrow & G^{\mathbb{C}} \\
\downarrow & & \downarrow{\scriptstyle G^i} \\
F(S) & \xrightarrow[\alpha_S]{} & G^{\mathbb{C}_S}
\end{array}
$$

For this we apply successively the Yoneda lemma of 7.2.3, definition 7.2.6 and corollary 7.2.5, writing

$$Y_{\mathcal{A}} \colon \mathcal{A} \longrightarrow \mathsf{Ps}(\mathcal{A}^{\mathrm{op}}, \mathsf{CAT}), \quad A \mapsto \mathcal{A}(-, A)$$

for the "discrete Yoneda embedding" of $\mathcal{A}$, as in 7.2.5. We get

$$F(S) \approx \mathsf{PsNat}\big(Y_{\mathcal{A}}(S), F\big),$$
$$F^{\mathbb{C}} \approx \mathsf{Lim}\, F \circ \mathbb{C} \approx \mathsf{PsNat}(\mathsf{Colim}\, Y_{\mathcal{A}} \circ \mathbb{C}, F),$$

and similarly for G. The desired equivalence can then reduce further to proving that the following commutative square is a two dimensional pullback.

$$
\begin{array}{ccc}
\mathsf{PsNat}(\mathsf{Colim}\, Y_{\mathcal{A}} \circ \mathbb{C}, F) & \xrightarrow{(\mathrm{id},\,\alpha)} & \mathsf{PsNat}(\mathsf{Colim}\, Y_{\mathcal{A}} \circ \mathbb{C}, G) \\
\Big\downarrow {\scriptstyle (\varepsilon,\,\mathrm{id})} & & \Big\downarrow {\scriptstyle (\varepsilon,\,\mathrm{id})} \\
\mathsf{Nat}\big(Y_{\mathcal{A}}(S), F\big) & \xrightarrow[(\mathrm{id},\,\alpha)]{} & \mathsf{Nat}\big(Y_{\mathcal{A}}(S), G\big)
\end{array}
$$

In this square, $\varepsilon \colon Y_{\mathcal{A}}(S) \longrightarrow \mathsf{Colim}\, Y_{\mathcal{A}} \circ \mathbb{C}$ is the canonical morphism

$$(Y_{\mathcal{A}} \circ \mathbb{C})(P_0) \longrightarrow \mathsf{Colim}\, Y_{\mathcal{A}} \circ \mathbb{C}$$

of proposition 7.3.6, in the case of the precategory $Y_{\mathcal{A}} \circ \mathbb{C}$ in the category $\mathsf{Ps}(\mathcal{A}^{\mathrm{op}}, \mathsf{CAT})$. By proposition 7.3.6, each component of this morphism is surjective on the objects. On the other hand, by assumption, each component of α is full and faithful. Our square is thus a two dimensional pullback, by corollary 7.3.5. $\qquad\square$

Corollary 7.5.3 *Let us consider*

- *two categories $\mathcal{A}$, $\mathcal{X}$,*
- *a functor $H \colon \mathcal{A} \longrightarrow \mathcal{X}$,*
- *two pseudo-functors $K \colon \mathcal{X}^{\mathrm{op}} \longrightarrow \mathsf{CAT}$ and $G \colon \mathcal{A}^{\mathrm{op}} \longrightarrow \mathsf{CAT}$,*
- *a pseudo-natural-transformation $\alpha \colon K \circ H \Rightarrow G$, all of whose components α_A are full and faithful,*
- *a morphism $\sigma \colon S \longrightarrow R$ in $\mathcal{A}$,*
- *a precategorical decomposition of σ*

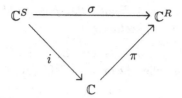

When $\big(\sigma, (i, \mathbb{C}, \pi)\big)$ is an effective descent structure with respect to the pseudo-functor G, one has an equivalence of categories

$$\mathsf{Split}_\alpha(\sigma) \approx K^{H \circ \mathbb{C}}.$$

Proof By definition 7.2.6 one has

$$K^{H \circ \mathbb{C}} = \mathsf{Lim}\, K \circ H \circ \mathbb{C} = (K \circ H)^{\mathbb{C}}$$

and it remains to apply theorem 7.5.2 with $F = K \circ H$. $\qquad\square$

7.6 Classical Galois theories

First we further particularize corollary 7.5.3 to get a "Galois theorem without Galois assumption", in the context of chapter 5. We keep writing

$$Z_{\mathcal{C}}: \mathcal{C}^{\mathrm{op}} \longrightarrow \mathsf{CAT}, \quad C \mapsto \mathcal{C}/C,\ f \mapsto f^*$$

for the "pulling back" pseudo-functor on a category $\mathcal{C}$ with pullbacks.

Theorem 7.6.1 *Let us consider*

- *two categories $\mathcal{A}$, $\mathcal{P}$ with pullbacks,*
- *a functor $\mathcal{S}: \mathcal{A} \longrightarrow \mathcal{P}$,*
- *a pseudo-natural-transformation $\alpha: Z_{\mathcal{P}} \circ \mathcal{S} \Rightarrow Z_{\mathcal{A}}$ all of whose components*

$$\alpha_A: \mathcal{P}/\mathcal{S}(A) \longrightarrow \mathcal{A}/A$$

are full and faithful,

- *$\sigma: S \longrightarrow R$ a morphism of $\mathcal{A}$ which is of effective descent in the sense of definition 4.4.1,*
- *$\mathbb{G}_\sigma$ the kernel pair of σ, seen as an internal groupoid in $\mathcal{A}$ (see example 4.6.2).*

When $\mathcal{S}(\mathbb{G}_\sigma)$ remains an internal groupoid in $\mathcal{P}$, the following equivalence of categories holds:

$$\mathsf{Split}_\alpha(\sigma) \approx \mathcal{P}^{\mathcal{S}(\mathbb{G}_\sigma)}.$$

where the right hand side denotes the category of internal presheaves on $\mathbb{G}$ *in the usual sense of section 4.6.*

Proof In corollary 7.5.3, we put $H = S$, $K = Z_{\mathcal{P}}$ and $G = Z_{\mathcal{A}}$. Proposition 7.4.4 implies the existence of a precategorical decomposition

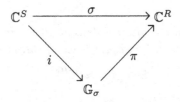

where $\mathbb{G}_\sigma$ *is the kernel pair of* σ. *Proposition 7.4.4 also asserts that* $\big(\sigma, (i, \mathbb{G}_\sigma, \pi)\big)$ *is an effective descent structure relatively to the pseudofunctor* $Z_{\mathcal{A}}$. *Corollary 7.5.3 then yields the equivalence*

$$\mathsf{Split}_\alpha(\sigma) \approx Z_{\mathcal{P}}^{\mathcal{S}(\mathbb{G}_\sigma)}.$$

The desired equivalence follows at once from corollary 7.1.3 and definition 7.2.6.   □

Proposition 7.6.2 *Let us consider*

- *a category* $\mathcal{A}$ *with pullbacks,*
- *a semi-left-exact reflection* $S \dashv i\colon \mathcal{P} \underset{\longrightarrow}{\longleftarrow} \mathcal{A}$,
- *an effective descent morphism* $\sigma\colon S \longrightarrow R$ *in* $\mathcal{A}$,
- *the kernel pair* $\mathbb{G}_\sigma$ *of* σ, *seen as a groupoid in* $\mathcal{A}$.

In these conditions, there is an equivalence of categories

$$\mathsf{Split}_R(\sigma) \approx Z_{\mathcal{P}}^{\mathcal{S}(\mathbb{G}_\sigma)}$$

where $\mathsf{Split}_R(\sigma) \subseteq \mathcal{A}/R$ *is the full subcategory of split objects in the sense of definition 5.1.7, with* $\overline{\mathcal{A}}$ *the class of all arrows in* $\mathcal{A}$ *and* $\overline{\mathcal{P}}$ *the class of all arrows in* $\mathcal{P}$. *If moreover* $\mathcal{S}(\mathbb{G}_\sigma)$ *is still a groupoid in* $\mathcal{P}$, *one obtains an equivalence of categories*

$$\mathsf{Split}_R(\sigma) \approx \mathcal{P}^{\mathcal{S}(\mathbb{G}_\sigma)}.$$

Proof By proposition 5.5.2, for every object $A \in \mathcal{A}$ we have a full and faithful functor

$$\alpha_A\colon \mathcal{P}/\mathcal{S}(A) \longrightarrow \mathcal{A}/A$$

with left adjoint $\mathcal{S}_A$. This can be rewritten

$$\alpha_A = i_A\colon (Z_{\mathcal{P}} \circ S)(A) \longrightarrow Z_{\mathcal{A}}(A)$$

and the pseudo-naturality of $\alpha\colon Z_{\mathcal{P}} \circ \mathcal{S} \Rightarrow Z_{\mathcal{A}}$ is obvious.

To conclude the proof by theorem 7.6.1, it remains to observe that the two notions of split object

$$\mathsf{Split}_{\alpha}(\sigma), \quad \mathsf{Split}_R(\sigma)$$

coincide, a fact which is attested by lemma 5.1.12. $\qquad\square$

At this stage it is certainly enlightening to compare more carefully the present proposition 7.6.2 and the Galois theorem of Janelidze (see 5.1.24).

Considering classes $\overline{\mathcal{A}}$ and $\overline{\mathcal{P}}$ in theorem 5.1.24 reduces, in theorem 7.5.2, to defining

$$F(A) = \overline{\mathcal{P}}/\mathcal{S}(A), \quad G(A) = \overline{\mathcal{A}}/A.$$

Lemma 5.1.4 still yields a pseudo-natural-transformation α with components

$$\alpha_A\colon \overline{\mathcal{P}}/\mathcal{S}(A) \longrightarrow \overline{\mathcal{A}}/A.$$

A first difference occurs in the fact that in theorem 5.1.24, one works under the minimal assumption

α_S *is full and faithful*

while in theorem 7.5.2, it is required that

each α_A is full and faithful.

This difference is rather unessential.

Indeed, in theorem 7.5.2, let us write

$$C_2 \ \substack{\xrightarrow{f_0} \\ \xrightarrow{m} \\ \xrightarrow{f_1}} \ C_1 \ \substack{\xrightarrow{d_0} \\ \xleftarrow{n} \\ \xrightarrow{d_1}} \ C_0$$

for the precategory $\mathbb{C}$ which is involved. Let us also write

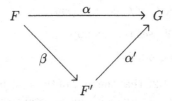

for the factorization of $\alpha\colon F \Rightarrow G$ in α', full and faithful in each component, and β, essentially surjective on the objects in each component. By theorem 7.5.2, we obtain an equivalence

$$\mathsf{Split}_{\alpha'}(\sigma) \approx F'^{\mathbb{C}}.$$

If we assume simply that α_{C_0} is full and faithful, then β_{C_0} is an equivalence, from which we get at once the equivalence

$$\mathsf{Split}_{\alpha'}(\sigma) \approx \mathsf{Split}_{\alpha}(\sigma),$$

via definition 7.5.1. If moreover we assume that α_{C_1} and α_{C_2} are also full and faithful, then β_{C_1} and β_{C_2} become equivalences as well and by definition 7.2.6, we get another equivalence

$$F'^{\mathbb{C}} \approx F^{\mathbb{C}}.$$

So the conclusion of theorem 7.5.2 still holds, namely the existence of an equivalence

$$\mathsf{Split}_{\alpha}(\sigma) \approx F^{\mathbb{C}}.$$

Now when the corresponding precategory $\mathbb{C}$ has a rather particular form, as in theorem 5.1.24 where it is the kernel pair of the morphism σ,

$$(S \times_R S) \times_S (S \times_R S) \overset{\overset{f_0}{\underset{m}{\longrightarrow}}}{\underset{\underset{f_1}{\longrightarrow}}{\longrightarrow}} S \times_R S \overset{\overset{d_0}{\underset{n}{\longleftarrow}}}{\underset{\underset{d_1}{\longrightarrow}}{\longrightarrow}} S,$$

a further weakening of the assumptions can occur (in the presence of the assumption of Galois descent) to arrive finally at the single assumption of α_S being full and faithful. We shall not insist on these variations on the minimal hypothesis.

It is more important to comment on the presence in theorem 5.1.24, and the absence in proposition 7.6.2, of a *Galois* descent assumption. In fact,

- theorem 5.1.24 uses the assumption

  $\sigma\colon S \longrightarrow R$ *is of Galois descent,*

- proposition 7.6.2 uses the assumption

  $\mathcal{S}(\mathbb{G}_\sigma)$ *is a groupoid.*

We know by lemma 5.1.22 that the first of these assumptions implies the second one. But let us observe with more details the argument in both

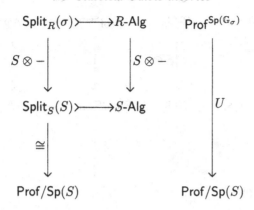

Diagram 7.15

proofs. For simplicity, we write the argument in the context of rings and algebras, as in section 4.7.

In theorem 4.7.15, a special case of theorem 5.1.24, one in fact considers the situation of diagram 7.15. Of course, several arguments were formally developed in the opposite categories of algebras, but for simplicity we avoid writing the exponent $(-)^{op}$ and thus allow some functors to be contravariant. The fact that σ is of Galois descent is used to prove that the functor $(S \otimes -)$ restricts to the categories of split algebras. Therefore $\mathsf{Split}_R(\sigma)$ becomes monadic over $\mathsf{Prof}/\mathsf{Sp}(S)$ because $(S \otimes -)$ is monadic by assumption, as a morphism of effective descent. The fact that σ is of Galois descent is used a second time to prove that $\mathsf{Sp}(\mathbb{G}_\sigma)$ is a groupoid, from which a classical result implies that $\mathsf{Prof}^{\mathsf{Sp}(\mathbb{G}_\sigma)}$ is monadic over $\mathsf{Prof}/\mathsf{Sp}(S)$. Since both monads involved are isomorphic, one gets the expected equivalence.

In proposition 7.6.2 and thus theorem 7.5.2, again expressed in the special context of theorem 4.7.15 one considers instead the squares of diagram 7.16, which become two dimensional pullbacks. Using the effective descent assumption on σ, one gets

$$R\text{-Alg} \cong R\text{-Alg}^{\mathbb{G}_\sigma} \quad \text{and thus} \quad \mathsf{Split}_R(\sigma) \cong \mathsf{Prof}^{\mathsf{Sp}(\mathbb{G}_\sigma)},$$

where the Galois groupoid $\mathbb{G}_\sigma$ is an internal groupoid in the opposite category of R-algebras (where it becomes discrete). The essential point in the approach of the present chapter has been to observe that the left hand square yields a Galois theorem as long as $R\text{-Alg} \cong R\text{-Alg}^{\mathbb{G}_\sigma}$ (condition of effective descent), just by describing $\mathsf{Prof}^{\mathsf{Sp}(\mathbb{G}_\sigma)}$ via the second square, which is again a two dimensional pullback.

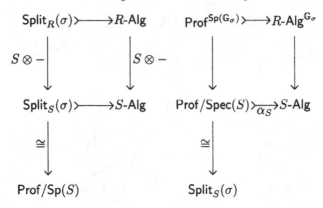

Diagram 7.16

Let us recall in particular (corollary 4.4.5) that every field extension is an effective descent morphism. Proposition 7.6.2 is thus a Galois theorem for every field extension. But this Galois theorem for a non-Galois extension of fields refers to a "Galois pregroup" and not a "Galois group". Indeed, if $K \subseteq L$ is the field extension, one has $\mathsf{Sp}(L) = \{*\}$ and the corresponding pregroup has the form

$$\mathsf{Sp}\big((L \otimes_K L) \otimes_L (L \otimes_K L)\big) \xrightarrow{\quad\quad\quad} \mathsf{Sp}(L \otimes_K L) \xleftarrow{\quad\quad\quad} \mathsf{Sp}(L) \cong \{*\}.$$

Let us conclude this section with a historical remark. The present section, allowing a Galois theorem for non-Galois extensions, is inspired by [48], but such a theorem can already be found in [41], under a slightly different form. However, the case of commutative rings was partly understood already by Magid (see [67]) in 1974 and, in some sense, even by Grothendieck a long time ago. In particular Grothendieck knew that the fundamental group of $\mathbb{Q}$ can be described as $\pi_0(\mathbb{C} \otimes_{\mathbb{Q}} \mathbb{C})$, although $\mathbb{Q} \subset \mathbb{C}$ is not a Galois extension.

7.7 Grothendieck toposes

The rest of this chapter intends to show that the Galois theorem for toposes, due to Joyal and Tierney (see [57]), is a special instance of the theory developed in section 7.5. Doing this requires of course a deep knowledge of topos theory and we want to avoid it, since this book is not a priori intended for specialists in topos theory. In fact Joyal and

Tierney develop their theory over an arbitrary base topos: we shall only handle the case of Grothendieck toposes over the base topos of sets.

To achieve our goal, we devote this section to an elementary and rather unusual course on Grothendieck toposes. We give short but nevertheless explicit proofs, omitting only the routine calculations which can be found in every textbook on this topic, for example [55], [65] or [8], volume 3.

We recall that given a small category $\mathcal{C}$, a presheaf on $\mathcal{C}$ is a functor $P\colon \mathcal{C}^{\mathrm{op}} \longrightarrow \mathsf{Set}$. We write $[\mathcal{C}^{\mathrm{op}}, \mathsf{Set}]$ for the category of presheaves on $\mathcal{C}$ and natural transformations between them.

Definition 7.7.1 A Grothendieck topos is a localization of a category of presheaves. More precisely, $\mathcal{E}$ is a Grothendieck topos when it can be presented as a full reflective subcategory

$$\mathcal{E} \underset{i}{\overset{a}{\underset{\longrightarrow}{\longleftarrow}}} [\mathcal{C}^{\mathrm{op}}, \mathsf{Set}], \quad a \dashv i$$

where $\mathcal{C}$ is a small category and the reflection a preserves finite limits.

Among Grothendieck toposes, we find the categories of sheaves on a locale (see 4.2.9), which are called the localic toposes. We can view a locale L as a category $\mathcal{L}$ whose objects are the elements of L and where a unique morphism $u \longrightarrow v$ between two elements exists precisely when $u \leq v$. The category of presheaves on the locale L is the usual category of contravariant presheaves on the corresponding category $\mathcal{L}$. When $u \leq v$ in L, that is, when there is a morphism $u \longrightarrow v$ in $\mathcal{L}$, the image of this morphism by a presheaf P is simply written as

$$P(u \leq v)\colon P(v) \longrightarrow P(u), \quad x \mapsto x|_u.$$

Definition 7.7.2 A sheaf on a locale L is a presheaf F on L which satisfies the following axiom:

$$\begin{aligned}
&\text{if} \quad u = \bigvee_{i \in I} u_i \quad \text{in} \quad L, \\
&\quad \forall i \in I \quad x_i \in F(u_i), \\
&\quad \forall i, j \in I \quad x_i|_{u_i \wedge u_j} = x_j|_{u_i \wedge u_j} \\
&\text{then} \quad \exists! x \in F(u) \quad \forall i \in I \quad x|_{u_i} = x_i.
\end{aligned}$$

A family $(x_i)_{i \in I}$ as above is called a compatible family in P along the covering $u = \bigvee_{i \in I} u_i$.

When in the axiom above, only the uniqueness of x holds, not necessarily the existence, F is called a separated presheaf.

Theorem 7.7.3 *The category* $\mathsf{Sh}(L)$ *of sheaves on a locale is a localization of the category of presheaves on L, thus is a Grothendieck topos.*

Definition 7.7.4 A *localic topos* is a category which is equivalent to the topos of sheaves on a locale.

Proof of theorem   Consider a presheaf P on L. Given $u \in L$, define $P'(u)$ as the set of all compatible families $(x_i)_{i \in I}$ in P (see 7.7.2) for all possible coverings $u = \bigvee_{i \in I} u_i$. Two compatible families $(x_i \in P(u_i))_{i \in I}$, $(y_j \in P(v_j))_{j \in J}$ are considered equivalent when

$$\forall i \in I \ \forall j \in J \ \ x_i|_{u_i \wedge v_j} = y_j|_{u_i \wedge v_j}.$$

The set $\widetilde{P}(u)$ is then the quotient of $P'(u)$ by the equivalence relation generated by these pairs.

The presheaf structure on P induces at once a presheaf structure on $\widetilde{P}$. From its definition, it follows at once that $\widetilde{P}$ is separated. You are certainly convinced that, from its definition, $\widetilde{P}$ must in fact trivially be a sheaf; for your peace of mind, spend an hour in unsuccessful efforts to write down the details of this evidence, and then five minutes to produce an easy counterexample. What is straightforward is the fact that when P is already separated, then $\widetilde{P}$ is a sheaf. Putting together these two observations, we conclude that $\widetilde{\widetilde{P}}$ is a sheaf. It is again routine to verify that it is the sheaf reflection of the presheaf P.

It remains to prove the left exactness of the sheaf reflection. The terminal sheaf $\mathbf{1}$ is such that $\mathbf{1}(u)$ is a singleton for each $u \in L$. A subpresheaf $R \rightarrowtail \mathbf{1}$ is thus entirely determined by those $u \in L$ for which $R(u)$ is a singleton; let us write $\mathsf{Supp}(R)$ for the family of those elements. The fact that R is a presheaf reduces to

$$\big(u \in \mathsf{Supp}(R) \ \text{ and } \ v \leq u\big) \Rightarrow \big(v \in \mathsf{Supp}(R)\big).$$

Given a compatible family $(x_i)_{i \in I}$ in a presheaf P along a covering $u = \bigvee_{i \in I} u_i$, the compatibility of the family implies that it extends to a compatible family on all $u \in L$ which are smaller than some u_i. Thus in the construction of $\widetilde{P}$, we can equivalently restrict our attention to those families $(u_i)_{i \in I}$ which are downward directed, that is, which correspond to subpresheaves of $\mathbf{1}$. Now given a covering $\mathsf{Supp}(R) = (u_i)_{i \in I}$ of u corresponding to a subpresheaf $R \rightarrowtail \mathbf{1}$, a compatible family of elements in P along this covering is just a natural transformation $R \Rightarrow P$, since $R(u_i)$ is a singleton for each $i \in I$ and is empty otherwise. Therefore

$\widetilde{P}(u)$ can be defined as the colimit

$$\widetilde{P}(u) = \operatorname*{colim}_{R} \operatorname{Nat}(R, P)$$

where R runs through those subobjects of $\mathbf{1}$ which correspond to coverings of u. If R, S correspond respectively to the coverings $(u_i)_{i \in I}$ and $(v_j)_{j \in J}$ of u, then $R \cap S$ corresponds to the family $(u_i \wedge v_j)_{(i,j) \in I \times J}$. Since L is a locale,

$$\bigvee_{i \in I,\, j \in J} u_i \wedge v_j = \bigvee_{i \in i} \left(u_i \wedge \bigvee_{j \in J} v_j \right) = \bigvee_{i \in I} (u_i \wedge u) = \bigvee_{i \in I} u_i = u.$$

Thus $R \cap S$ corresponds again to a covering of u, proving that the colimit defining $\widetilde{P}$ is filtered. Since filtered colimits of sets commute with finite limits, the functor $\widetilde{(-)}$ preserves finite limits, and thus also the associated sheaf functor $a = \widetilde{\widetilde{(-)}}$ does. $\qquad\square$

Proposition 7.7.5 *In the category of sheaves on a locale L, the representable presheaves are exactly all the subsheaves of the terminal sheaf $\mathbf{1}$. They constitute a locale isomorphic to L.*

Proof Consider a locale L viewed as a category $\mathcal{L}$. We have already considered the subpresheaves $R \rightarrowtail \mathbf{1}$ of the terminal sheaf in the proof of 7.7.3; each of them corresponds to a downward directed family $\mathsf{Supp}(R)$ of elements. Assume now that R is a sheaf. Since R takes only values "singleton" and "empty set", the family of all existing elements $* \in R(u)$ is trivially compatible, from which there exists a unique glueing of that family at the level $u = \bigvee \mathsf{Supp}(R)$, by the sheaf axiom. This means exactly that $R(u)$ is a singleton and u is the largest element of L with that property. But then R is just the representable functor $\mathcal{L}(-, u)$. And since all representable functors $\mathcal{L}(-, u)$ are obviously sheaves, we conclude that the subsheaves of $\mathbf{1}$ are exactly the representable functors, thus constitute a locale isomorphic to L. $\qquad\square$

The next proposition lists various interesting properties of limits and colimits in a topos.

Proposition 7.7.6 *A Grothendieck topos $\mathcal{E}$ is complete and cocomplete. Moreover,*

(i) *colimits are universal,*

(ii) *finite limits commute with filtered colimits,*

(iii) *each functor $A \times (-)\colon \mathcal{E} \longrightarrow \mathcal{E}$ preserves colimits,*

(iv) *coproducts are disjoint,*

(v) *a union of a downward directed family of subobjects is their colimit,*

(vi) *the subobjects of every object constitute a locale.*

Proof   With the notation of definition 7.7.1, the category of sets is complete and cocomplete, thus so is the category $[\mathcal{C}^{op}, \mathsf{Set}]$ of presheaves, where limits and colimits are computed pointwise. The Grothendieck topos $\mathcal{E}$ is thus complete as well, with limits computed as in $[\mathcal{C}^{op}, \mathsf{Set}]$; it is also cocomplete, the colimit of a diagram being obtained by applying the reflection a to the colimit in $[\mathcal{C}^{op}, \mathsf{Set}]$.

A colimit is universal when, pulling it back along any morphism, one gets again a colimit. A coproduct is disjoint when the canonical injections of the coproduct are monomorphisms and the intersection of two of them is the initial object.

We use again the notation of 7.7.1. Conditions (i) to (iv) express properties relating some colimits and some finite limits. These properties are valid in Set, thus they are valid in $[\mathcal{C}^{op}, \mathsf{Set}]$ where colimits and (finite) limits are computed pointwise. Since a preserves colimits, as a left adjoint, and finite limits, by assumption, all these properties transfer to $\mathcal{E}$.

Given $E \in \mathcal{E}$ and a monomorphism $f \colon P \rightarrowtail E$ in $[\mathcal{C}^{op}, \mathsf{Set}]$, we get at once a commutative triangle

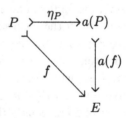

where $a(f)$ is a monomorphism by left exactness of a and i, and the unit η_P of the adjunction is a monomorphism since f is. The universal property of $a(P)$ indicates at once that it is the smallest subobject of E in $\mathcal{E}$ which contains P. But the union of a family of subobjects is the smallest subobject which contains them. Thus unions in $\mathcal{E}$ are obtained by first taking the set theoretical pointwise union in $[\mathcal{C}^{op}, \mathsf{Set}]$ and next applying the associated sheaf functor.

The previous remark implies conditions (v) and (vi), since again these conditions are valid in Set, thus are valid pointwise in $[\mathcal{C}^{op}, \mathsf{Set}]$, and finally are preserved by a. □

Next, we focus on properties of monomorphisms and epimorphisms. The reader familair with the notion of "regular category" will in particular recognize that proposition 7.7.7 below implies that a topos is a regular category. Let us mention that a topos is even an *exact* category in the sense of [4], but we shall not need this result.

Proposition 7.7.7 *In a Grothendieck topos $\mathcal{E}$:*

(i) *every monomorphism is regular,*

(ii) *a morphism which is both a monomorphism and an epimorphism is an isomorphism,*

(iii) *every epimorphism is regular,*

(iv) *epimorphisms are stable under change of base,*

(v) *the product of two epimorphisms is still an epimorphism,*

(vi) *every morphism factors as an epimorphism followed by a monomorphism.*

Proof   We use the notation of definition 7.7.1. In the category of sets, every monomorphism is the equalizer of its cokernel pair, that is, its pushout with itself. The same conclusion thus applies pointwise in $[\mathcal{C}^{op}, \mathsf{Set}]$. But a monomorphism $f\colon A \rightarrowtail B$ in $\mathcal{E}$ remains a monomorphism in $[\mathcal{C}^{op}, \mathsf{Set}]$, thus f is an equalizer $f = \mathsf{Ker}\,(u,v)$ in $[\mathcal{C}^{op}, \mathsf{Set}]$. Applying the functor a, we get $f = \mathsf{Ker}\,\big(a(u), a(v)\big)$.

In the same situation, if f is also an epimorphism in $\mathcal{E}$, then from $a(u) \circ f = a(v) \circ f$ we get $a(u) = a(v)$ and therefore $f = \mathsf{Ker}\,\big(a(u), a(v)\big)$ is an isomorphism.

Using again the notation of 7.7.1, we prove now the existence of images. Given a morphism $f\colon A \longrightarrow B$ in $\mathcal{E}$, we compute its kernel pair (u, v) in $\mathcal{E}$, that is, the pullback of f with itself, which is thus also its kernel pair in $[\mathcal{C}^{op}, \mathsf{Set}]$:

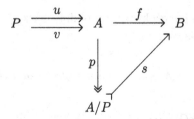

Next in $[\mathcal{C}^{op}, \mathsf{Set}]$ we consider the coequalizer p of (u, v) and the factorization s from this coequalizer, resulting from the equality $f \circ u = f \circ v$. In the category of sets, s would be injective since this is the standard image factorization of f. Thus the same conclusion holds pointwise in the category $[\mathcal{C}^{op}, \mathsf{Set}]$, and therefore p is an epimorphism and s a monomorphism in $[\mathcal{C}^{op}, \mathsf{Set}]$. Applying the reflection a, which preserves colimits as a left adjoint and finite limits by assumption, we obtain the following diagram in $\mathcal{E}$:

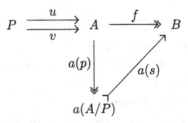

with $a(p)$ an epimorphism and $a(s)$ a monomorphism.

With the same notation, let now $f\colon A \twoheadrightarrow B$ be an epimorphism in $\mathcal{E}$. We must prove that f is a regular epimorphism, that is, a coequalizer. Going back to the last diagram, the monomorphism $a(s)$ is now an epimorphism as well, because so is f. By the first part of the proof, $a(s)$ is an isomorphism and f is isomorphic to $a(p)$, which is the coequalizer of (u, v).

Finally given two epimorphisms $f\colon A \twoheadrightarrow B$ and $g\colon C \twoheadrightarrow D$, the morphism $f \times g$ is the upper composite in the following diagram, where both parallelograms are pullbacks:

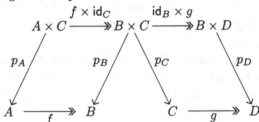

Since f and g are epimorphisms, by pulling back, $f \times \mathrm{id}_C$ and $\mathrm{id}_B \times g$ are epimorphisms as well. $\qquad\square$

Lemma 7.7.8 *Let $f\colon A \longrightarrow B$ be a morphism in a Grothendieck topos. The map*

$$f^{-1}\colon \mathsf{SubObj}(B) \longrightarrow \mathsf{SubObj}(A), \quad S \mapsto f^{-1}(S)$$

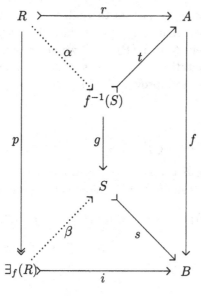

Diagram 7.17

between locales of subobjects has both a left adjoint written $\exists_f$ *and a right adjoint written* $\forall_f$.

Proof In the topos **Set** of sets, given a map $f\colon A \longrightarrow B$ and subobjects $R \rightarrowtail A$, $S \rightarrowtail B$, we define

$$\exists_f(R) \;=\; \{b \in B | \exists a \in A \;\; f(a) = b, \; a \in R\} = f(R),$$
$$\forall_f(R) \;=\; \{b \in B | \forall a \in A \;\; f(a) = b \Rightarrow a \in R\}.$$

It is immediate that

$$\exists_f(R) = f(R) \subseteq S \Leftrightarrow R \subseteq f^{-1}(S)$$

proving $\exists_f \dashv f^{-1}$. On the other hand

$$S \subseteq \forall_f(R) \Leftrightarrow \big(f(a) \in S \Rightarrow a \in R\big) \Leftrightarrow f^{-1}(S) \subseteq R$$

proves the adjunction $f^{-1} \dashv \forall_f$. This example justifies the notation and fixes the intuition.

In the case of a Grothendieck topos, given a subobject $r\colon R \rightarrowtail A$ and a morphism $f\colon A \longrightarrow B$, let us define $\exists_f(R)$ as the image factorization of the composite $f \circ r$ (see 7.7.7(vi)). Consider the diagram 7.17 where the right hand quadrilateral is thus a pullback. If $R \subseteq f^{-1}(S)$, using the set theoretical notation, we get a morphism α making the diagram commutative. Thus $f \circ r$ factors through the subobject S; since the image of $f \circ r$ is the smallest subobject through which $f \circ r$ factorizes, $\exists_f(R) \subseteq S$. Conversely if $\exists_f(R) \subseteq S$, we get a morphism β making the diagram commutative. This yields a factorization α through the pullback defining $f^{-1}(S)$, proving $R \subseteq f^{-1}(S)$. This proves already the adjunction $\exists_f \dashv f^{-1}$.

Next we define

$$\forall_f(R) = \bigcup \{S \,|\, S \subseteq B, \ f^{-1}(S) \subseteq R\}.$$

The family of those S is obviously downward directed, thus the corresponding union is a colimit and therefore is universal (see 7.7.6). Thus

$$
\begin{aligned}
f^{-1}\big(\forall_f(R)\big) &= f^{-1}\Big(\bigcup \{S \,|\, S \subseteq B, \ f^{-1}(S) \subseteq R\}\Big) \\
&= \bigcup \{f^{-1}(S) \,|\, S \subseteq B, \ f^{-1}(S) \subseteq R\} \\
&\subseteq R.
\end{aligned}
$$

It follows at once that

$$
\begin{aligned}
f^{-1}(S) \subseteq R &\Rightarrow S \subseteq \forall_f(R), \\
S \subseteq \forall_f(R) &\Rightarrow f^{-1}(S) \subseteq f^{-1}\big(\forall_f(R)\big) \subseteq R,
\end{aligned}
$$

which proves the adjunction $f^{-1} \dashv \forall_f$. $\square$

7.8 Geometric morphisms

We now turn our attention to the morphisms of toposes.

Definition 7.8.1 Given two categories $\mathcal{E}$ and $\mathcal{F}$ with finite limits, a geometric morphism $f \colon \mathcal{E} \longrightarrow \mathcal{F}$ between them is a pair of adjoint functors

$$f_* \colon \mathcal{E} \longrightarrow \mathcal{F}, \quad f^* \colon \mathcal{F} \longrightarrow \mathcal{E}, \quad f^* \dashv f_*$$

with the additional property that f^* preserves finite limits.

We shall use this definition of geometric morphism both when $\mathcal{E}$, $\mathcal{F}$ are toposes, and when they are locales, viewed as categories.

Given two categories $\mathcal{A}$, $\mathcal{B}$ with finite limits, we write $\mathsf{Geom}(\mathcal{A}, \mathcal{B})$ for the category whose objects are the geometric morphisms $f \colon \mathcal{A} \longrightarrow \mathcal{B}$ and whose morphisms $\alpha \colon f \Rightarrow g$ are the natural transformations $\alpha \colon f^* \Rightarrow g^*$. It is routine to observe that one gets an equivalent category by choosing $\alpha \colon g^* \Rightarrow f^*$, but this is unnecessary for our purposes.

Lemma 7.8.2 *Given two locales L, M, the categories $\mathsf{Geom}(\mathcal{L}, \mathcal{M})$ and $\mathsf{Geom}\big(\mathsf{Sh}(L), \mathsf{Sh}(M)\big)$ of geometric morphisms are equivalent.*

Proof Consider first a geometric morphism $f \colon \mathsf{Sh}(L) \longrightarrow \mathsf{Sh}(M)$ between the toposes of sheaves. The functor f_* preserves the subobjects of 1 because it has a left adjoint f^*, while f^* has the same property because it preserves finite limits. Therefore, applying proposition 7.7.5, f induces a geometric morphism of locales

$$\sigma(f) = (f_*, f^*) \colon \mathcal{L} \longrightarrow \mathcal{M}.$$

And trivially, given another geometric morphism $g \colon \mathsf{Sh}(L) \longrightarrow \mathsf{Sh}(M)$, every natural transformation $\alpha \colon f^* \Rightarrow g^*$ restricts to a corresponding natural transformation $\sigma(\alpha) \colon \sigma(f)^* \Rightarrow \sigma(g)^*$.

Consider now a geometric morphism of locales $f \colon \mathcal{L} \longrightarrow \mathcal{M}$. Given a sheaf F on L, the composite

$$\mathcal{M}^{\mathrm{op}} \xrightarrow{\;f^*\;} \mathcal{L}^{\mathrm{op}} \xrightarrow{\;F\;} \mathsf{Set}$$

is a sheaf on M. Indeed, a covering in M is just a colimit in the corresponding category $\mathcal{M}$. Thus f^* preserves coverings, because it has a right adjoint f_*. But given a covering $v = \bigvee_{j \in J} v_j$ in M, a compatible family in $F \circ f^*$ along that covering is also a compatible family in F along the covering $f^*(v) = \bigvee_{j \in J} f^*(v_j)$, thus has a unique glueing in $F\big(f^*(v)\big)$. Composing with f^* thus yields a functor

$$\mathsf{Sh}(f)_* \colon \mathsf{Sh}(L) \longrightarrow \mathsf{Sh}(M), \quad F \mapsto F \circ f^*.$$

We must prove the existence of a left adjoint $\mathsf{Sh}(f)^*$ to this functor $\mathsf{Sh}(f)_*$. It is well known that a presheaf on every category is a colimit of representable presheaves (see [8], volume 1). Since on a locale L the representable functors are already sheaves (see 7.7.5), every sheaf G on the locale M is thus a colimit of representable sheaves. More precisely,

$$G = \operatorname*{colim}_{v \in M,\ y \in G(v)} \mathcal{M}(-, v).$$

We define

$$\mathsf{Sh}(f)^*(G) = \operatorname*{colim}_{v \in M,\ y \in G(v)} \mathcal{L}\big(-, f^*(v)\big).$$

The functoriality of $\mathsf{Sh}(f)^*(G)$ is obvious and the expected adjunction is proved as follows, using the previous notation:

$$\mathsf{Nat}\big(\mathsf{Sh}(f)^*(G), F)\big) \;\cong\; \mathsf{Nat}\Big(\operatorname*{colim}_{v,y} \mathcal{L}\big(-, f^*(v)\big), F\Big)$$

$$\cong \ \lim_{v,y} \mathsf{Nat}\Big(\mathcal{L}\big(-,f^*(v)\big),F\Big)$$

$$\cong \ \lim_{v,y} F\big(f^*(v)\big)$$

$$\cong \ \lim_{v,y} \mathsf{Sh}(f)_*(F)(v)$$

$$\cong \ \lim_{v,y} \mathsf{Nat}\big(\mathcal{M}(-,v),\mathsf{Sh}_*(f)(F)\big)$$

$$\cong \ \mathsf{Nat}\big(\mathrm{colim}_{v,y}\,\mathcal{M}(-,v),\mathsf{Sh}_*(f)(F)\big)$$

$$\cong \ \mathsf{Nat}\big(G,\mathsf{Sh}_*(f)(F)\big).$$

Finally, to have a geometric morphism $\mathsf{Sh}(f)$, it remains to prove that $\mathsf{Sh}(f)^*$ preserves finite limits. We have the following situation, given an element $u \in L$:

$$\mathsf{Elts}(G) \xrightarrow{\ \phi_G\ } \mathcal{M} \xrightarrow{\ \mathcal{L}(u,f^*(-))\ } \mathsf{Set}$$

with $\mathcal{M} \xrightarrow{G} \mathsf{Set}$

with $\mathsf{Elts}(G)$ the category of elements of G, that is, the category of those pairs (v,y), with $y \in G(v)$ indexing the previous colimit. The functor ϕ_G is the forgetful functor $\phi_G(v,y) = v$. Given an element $u \in L$, let us define

$$\mathsf{Sh}(f)'(G)(u) = \mathrm{colim}_{v,y}\,\mathcal{L}\big(u,f^*(v)\big) \in \mathsf{Set}.$$

This set is the so-called tensor product of the contravariant presheaf G with the covariant presheaf $\mathcal{L}\big(u,f^*(-)\big)$; it is a classical result (see for example [8], volume 1) that this tensor product is commutative. More explicitly, consider now the situation

$$\mathsf{Elts}\Big(\mathcal{L}\big(u,f^*(-)\big)\Big) \xrightarrow{\ \phi\ } \mathcal{M} \xrightarrow{\ G\ } \mathsf{Set}$$

with $\mathcal{M} \xrightarrow{\mathcal{L}(u,f^*(-))} \mathsf{Set}$

where $\mathsf{Elts}\Big(\mathcal{L}\big(u,f^*(-)\big)\Big)$ is now the category of elements of $\mathcal{L}\big(u,f^*(-)\big)$, that is the category of pairs (v,z) with $z \in \mathcal{L}\big(u,f^*(v)\big)$. But $\mathcal{L}\big(u,f^*(v)\big)$ is a singleton when $u \le f^*(v)$ and the empty set otherwise. So our category of elements reduces to the category of those $v \in M$ such that $u \leq f^*(v)$. The commutativity of the tensor product of presheaves thus yields

$$\mathsf{Sh}(f)'(G)(u) = \operatorname*{colim}_{v,y} \mathcal{L}\big(u, f^*(v)\big) \in \mathsf{Set} = \operatorname*{colim}_{v,\ u \leq f^*(v)} G(v).$$

Since f^* preserves finite infima,

$$u \leq f^*(v) \text{ and } u \leq f^*(v') \Rightarrow u \leq f^*(v) \wedge f^*(v') = f^*(v \wedge v').$$

This proves that the second colimit is filtered, thus using it to define $\mathsf{Sh}(f)'$ implies at once that the functor

$$\mathsf{Sh}(f)' \colon \mathsf{Sh}(M) \longrightarrow [\mathcal{L}^{\mathrm{op}}, \mathsf{Set}]$$

preserves finite limits. Indeed, finite limits commute in $[\mathcal{L}^{\mathrm{op}}, \mathsf{Set}]$ with filtered colimits (see 7.7.6). Now the colimit defining $\mathsf{Sh}(f)^*(G)$ in $\mathsf{Sh}(L)$ is obtained by considering the first colimit defining $\mathsf{Sh}(f)'(G)$ and applying the associated sheaf functor to it. In other words, $\mathsf{Sh}(f)^*$ is the composite

$$\mathsf{Sh}(f)^* \colon \mathsf{Sh}(M) \xrightarrow{\mathsf{Sh}(f)'} [\mathcal{L}^{\mathrm{op}}, \mathsf{Set}] \xrightarrow{a} \mathsf{Sh}(L).$$

Since both $\mathsf{Sh}(f)'$ and a preserve finite limits, $\mathsf{Sh}(f)^*$ preserves finite limits as well and $\mathsf{Sh}(f)$ is a geometric morphism of locales.

It is obvious that given a second geometric morphism $g \colon \mathcal{L} \longrightarrow \mathcal{M}$, a natural transformation $\alpha \colon f^* \longrightarrow g^*$ extends uniquely, by unique factorization through the corresponding colimits, as a natural transformation $\mathsf{Sh}(\alpha) \colon \mathsf{Sh}(f)^* \Rightarrow \mathsf{Sh}(g)^*$.

It is also obvious that both constructions yield the expected equivalence. $\square$

Proposition 7.8.3 *The category* Loc *of locales and geometric morphisms has finite limits.*

Proof To prove the existence of finite limits, it suffices to prove the existence of a terminal object and pullbacks (see [8], volume 1).

The terminal topological space is the singleton, whose locale of open subsets is isomorphic to the lattice $\{0 < 1\}$. Let us prove that the locale $\{0 < 1\}$ is the terminal object of the category Loc of locales and geometric morphisms. Given a locale L, the following data obviously define a geometric morphism $h \colon L \longrightarrow \{0 < 1\}$:

$$h_*(u) = 1 \quad \Leftrightarrow \quad u = 1,$$

$$h^*(1) = 1, \qquad h^*(0) = 0.$$

Since h^* must preserve the initial and the terminal object, its definition is imposed, proving the uniqueness.

If X and Y are topological spaces, with locales $\mathcal{O}(X)$ and $\mathcal{O}(Y)$ of open subsets, the topological product $X \times Y$ is not at all such that $\mathcal{O}(X \times Y) = \mathcal{O}(X) \times \mathcal{O}(Y)$. In fact, an open subset in $X \times Y$ has the form

$$\bigcup_{i \in I} U_i \times V_i, \quad U_i \in \mathcal{O}(X), \ V_i \in \mathcal{O}(Y).$$

That form of an open subset is clearly not unique; for example one has formulæ like

$$\left(\bigcup_{i \in I} U_i \right) \times V = \bigcup_{i \in I} (U_i \times V)$$

and analogously in the second variable. Now if we consider a pullback of topological spaces

$$
\begin{array}{ccc}
P & \longrightarrow & X \\
\downarrow & & \downarrow{\scriptstyle f} \\
Y & \xrightarrow{\ g\ } & Z
\end{array}
$$

the topology of P is that induced by the topology of $X \times Y$. Given open subsets $U \in \mathcal{O}(X)$, $V \in \mathcal{O}(Y)$ and $W \in \mathcal{O}(Z)$, one also has

$$P \cap \Big(\big((U \cap f^{-1}(W)) \big) \times V \Big) = P \cap \Big(U \times (g^{-1}(W) \cap V) \Big).$$

Indeed, given $(x, y) \in P$, this means just

$$x \in U, \ f(x) \in W, \ y \in V \Leftrightarrow x \in U, \ g(y) \in W, \ y \in V$$

which is obvious since $f(x) = g(y)$. This suggests how to construct the corresponding pullback of the locales of open subsets and, more generally, a pullback of locales.

Let us thus consider three locales L, M, N and two geometric morphisms f, g as in the following diagram.

$$P \xrightarrow{\ k\ } L$$

$$h \downarrow \qquad\qquad \downarrow f$$

$$M \xrightarrow[\ g\ ]{} N$$

We want to construct the corresponding pullback P. With the previous intuition in mind and using a technique analogous to that for constructing the tensor product of modules, we consider first all the formal expressions

$$\bigvee_{i \in I} l_i \otimes m_i, \quad I \in \mathsf{Set}, \ l_i \in L, \ m_i \in M.$$

The locale P is then the quotient of the set of all these formal expressions by the congruence generated by

$$\left(\bigvee_{i \in I} l_i \right) \otimes m = \bigvee_{i \in I} (l_i \otimes m),$$

$$l \otimes \left(\bigvee_{i \in I} m_i \right) = \bigvee_{i \in I} (l \otimes m_i),$$

$$\left(l \wedge f^*(n) \right) \otimes m = l \otimes \left(g^*(n) \wedge m \right)$$

with $l, l_i \in L$; $m, m_i \in M$; $n \in N$. The morphisms h and k are then simply determined by

$$k^*(l) = l \otimes 1, \quad h^*(m) = 1 \otimes m$$

with 1 denoting the top element. The rest is routine verifications which can be found in [8], volume 3. □

Let us now observe a peculiar property of Grothendieck toposes with respect to geometric morphisms.

Lemma 7.8.4 *For every Grothendieck topos $\mathcal{E}$, there exists a unique (up to isomorphism) geometric morphism $g \colon \mathcal{E} \longrightarrow \mathsf{Set}$.*

Proof The functor $g^* \colon \mathsf{Set} \longrightarrow \mathcal{E}$ must preserve finite limits, but also colimits since it is a left adjoint. Therefore

$$g^*(X) \cong g^* \left(\coprod_X 1 \right) \cong \coprod_X g^*(1) \cong \coprod_X 1,$$

which proves the uniqueness.

For the existence, let us use the notation of 7.7.1. One has the following situation:

$$\mathcal{E} \underset{i}{\overset{a}{\rightleftarrows}} [\mathcal{C}^{op}, \mathsf{Set}] \underset{\overset{\text{colim}}{\longrightarrow}}{\overset{\Delta}{\underset{\text{lim}}{\longleftarrow}}} \mathsf{Set}$$

where

- $a \dashv i$ with a preserving finite limits,
- Δ maps the set X onto the constant functor on X,
- colim maps a presheaf onto its colimit,
- lim maps a presheaf onto its limit.

The adjunctions colim $\dashv \Delta \dashv$ lim hold just by definition of a colimit or a limit (see [8], volume 1); in particular Δ preserves limits. This shows that (i, a) and (lim, Δ) constitute geometric morphisms, from which we get a geometric morphism

$$(\text{lim} \circ i, a \circ \Delta) \colon \mathcal{E} \longrightarrow \mathsf{Set}. \qquad \square$$

It can be easily shown that if the category $\mathcal{C}$ considered in the proof of 7.8.4 is connected, then colim $\dashv \Delta \dashv$ lim considered there is a special case of $I \dashv H \dashv \Gamma$ as in the comments following 6.2.1.

Theorem 7.8.5 *Consider a Grothendieck topos $\mathcal{E}$. There exist a locale L and a geometric morphism*

$$h \colon \mathcal{E} \longrightarrow \mathsf{Sh}(L)$$

with the properties:

(i) *h^* is full and faithful,*

(ii) *up to an equivalence, h^* identifies $\mathsf{Sh}(L)$ with the full subcategory of $\mathcal{E}$ of those objects which are the unions of their atomic subobjects (an object is atomic when it is a subobject of $\mathbf{1}$).*

The locale L and the geometric morphism h are unique, up to isomorphisms, for the previous properties. The locale L is isomorphic to the locale of subobjects of $\mathbf{1}$ in $\mathcal{E}$. Via condition (ii), $\mathsf{Sh}(L)$ as a subcategory of $\mathcal{E}$ is saturated for subobjects.

Definition 7.8.6 A geometric morphism which satisfies the conditions of theorem 7.8.5 is called hyperconnected.

Proof of theorem   In every category, if $u\colon U \rightarrowtail 1$ is a subobject of the terminal object, two morphisms $\alpha, \beta\colon X \underset{\longrightarrow}{\longrightarrow} U$ are necessarily equal since $u \circ \alpha = u \circ \beta$, with u a monomorphism. Therefore every morphism $f\colon U \longrightarrow Y$ is a monomorphism.

Let us also recall an argument already used in the proof of lemma 7.8.2: every sheaf F on the locale L can be written as a colimit

$$F = \operatorname*{colim}_{u \in L,\ x \in F(u)} \mathcal{L}(-, u)$$

of representable sheaves.

Let us now turn to the proof of our theorem. If two localic toposes $\mathsf{Sh}(L)$ and $\mathsf{Sh}(M)$ are equivalent, then by lemma 7.8.2, the locales L and M are equivalent. But in a locale, viewed as a category, only the identities are isomorphisms. Thus equivalent locales L and M are in fact isomorphic. The uniqueness requirement in the statement thus follows at once from conditions (i) and (ii).

To prove the existence, let us choose for L the locale of subobjects of 1 in $\mathcal{E}$; as usual, we write $\mathcal{L}$ when we view L as a category. We first define the functor

$$h^*\colon \mathsf{Sh}(L) \longrightarrow \mathcal{E}.$$

Given a sheaf $F \in \mathsf{Sh}(L)$, we write it as a colimit of representable sheaves

$$F = \operatorname*{colim}_{U \in L,\ x \in F(U)} \mathcal{L}(-, U) \in \mathsf{Sh}(L)$$

and we define in $\mathcal{E}$

$$h^*(F) = \operatorname*{colim}_{U \in L,\ x \in F(U)} U.$$

This definition extends immediately to the morphisms $f\colon F \longrightarrow G$ in $\mathsf{Sh}(L)$: for each term of the colimit presenting F one has the composite

$$\mathcal{L}(-, U) \xrightarrow{\ s_{U,x}\ } F \xrightarrow{\ f\ } G$$

which is a term of the colimit presenting G; here $s_{U,x}$ denotes the natural transformation corresponding to $x \in F(U)$ by the Yoneda lemma. This yields a natural transformation from the diagram defining $h^*(F)$ to the diagram defining $h^*(G)$ and thus a corresponding factorization $h^*(f)\colon h^*(F) \longrightarrow h^*(G)$ between the colimits. The observant reader will have recognized the classical formula defining the Kan extension of $\mathcal{L} \rightarrowtail \mathcal{E}$ along $\mathcal{L} \rightarrowtail \mathsf{Sh}(L)$, when the locale L is identified respectively with the categories of subobjects of 1 in $\mathcal{E}$ and $\mathsf{Sh}(L)$.

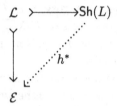

Observe at once that given $U \in L$, the colimit diagram expressing $\mathcal{L}(-, U)$ as colimit of representable sheaves admits precisely $\mathcal{L}(-, U)$ as the image of the terminal object. Thus the corresponding diagram defining $h^*(\mathcal{L}(-, U))$ admits U as the image of the terminal object, proving that $h^*(\mathcal{L}(-, U)) = U$. In other words, h^* induces an isomorphism between the categories of subobjects of $\mathbf{1}$ in $\mathsf{Sh}(L)$ and $\mathcal{E}$.

It is rather easy to guess what the possible right adjoint h_* of h^* can be. Given $U \in L$ and $E \in \mathcal{E}$, one must have, applying the Yoneda lemma, the expected adjunction $h^* \dashv h_*$ and the definition of h^*,

$$h_*(E)(U) \cong \mathsf{Nat}\big(\mathcal{L}(-, U), h_*(E)\big) \cong \mathcal{E}\big(h^*(\mathcal{L}(-, U)), E\big) \cong \mathcal{E}(U, E).$$

We thus define $h_* \colon \mathcal{E} \longrightarrow \mathsf{Sh}(L)$ by the formula

$$h_*(E) \colon \mathcal{L}^{\mathsf{op}} \longrightarrow \mathsf{Set}, \quad U \mapsto \mathcal{E}(U, E)$$

with obvious action on the morphisms. Observe that $h_*(E)$ is a sheaf. Indeed, using the preservation of colimits by representable functors (see [8], volume 1), for a downward directed family $(U_i)_{i \in I}$ in L, one has

$$h_*(E)\left(\bigvee_{i \in I} U_i\right) \cong \mathcal{E}\left(\bigvee_{i \in I} U_i, E\right)$$

$$\cong \mathcal{E}\left(\operatorname*{colim}_{i \in I} U_i, E\right)$$

$$\cong \lim_{i \in I} \mathcal{E}(U_i, E)$$

$$\cong \lim_{i \in I} h_*(E)(U_i).$$

So the elements of $h_*(E)(\bigvee_{i \in I} U_i)$ are in bijection with the elements of $\lim_{i \in I} h_*(E)(U_i)$, that is, with the compatible families in the sense of definition 7.7.2. This proves that $h^*(E)$ is a sheaf on L.

Proving the adjunction $h^* \dashv h_*$ is also easy. With previous notation and arguments, choosing $F \in \mathsf{Sh}(L)$ and $E \in \mathcal{E}$,

$$\mathcal{E}\big(h^*(F), E\big) \cong \mathcal{E}\left(\operatorname*{colim}_{U \in L, \; x \in F(U)} U, E\right)$$

$$\cong \lim_{U \in L, \ x \in F(U)} \mathcal{E}(U, E)$$

$$\cong \lim_{U \in L, \ x \in F(U)} h_*(E)(U)$$

$$\cong \lim_{U \in L, \ x \in F(U)} \mathsf{Nat}\big(\mathcal{L}(-, U), h_*(E)\big)$$

$$\cong \mathsf{Nat}\left(\operatorname*{colim}_{U \in L, \ x \in F(U)} \mathcal{L}(-, U), h_*(E)\right)$$

$$\cong \mathsf{Nat}\big(F, h_*(E)\big).$$

The unit of the adjunction $h^* \dashv h_*$ is thus given, for every sheaf $F \in \mathsf{Sh}(L)$ and every $V \in L$, by a map

$$\eta_{F,V} \colon F(V) \longrightarrow \mathcal{E}\big(V, h^*(F)\big).$$

Via the Yoneda lemma and the equality $V = h^*\big(\mathcal{L}(-, V)\big)$, this map is just

$$\mathsf{Nat}\big(\mathcal{L}(-, V), F\big) \longrightarrow \mathcal{E}\big(V, h^*(F)\big), \quad \alpha \mapsto h^*(\alpha).$$

We want now to prove that these maps are bijective. This will prove that the unit of the adjunction $h^* \dashv h_*$ is an isomorphism, and therefore that h^* is full and faithful (see [8], volume 1).

For this we construct an inverse

$$\mu_{F,V} \colon \mathcal{E}\big(V, h^*(F)\big) \longrightarrow \mathsf{Nat}\big(\mathcal{L}(-, V), F\big)$$

to $\eta_{F,V}$. We consider again the presentation of F as colimit of representable functors,

$$s_{U,x} \colon \mathcal{L}(-, U) \longrightarrow F, \quad U \in L, \ x \in F(U),$$

and the corresponding colimit defining $h^*(F)$,

$$\sigma_{U,x} \colon U \longrightarrow h^*(F), \quad U \in L, \ x \in F(U).$$

Given $\beta \colon V \longrightarrow h^*(F)$, we compute the pullbacks in $\mathcal{E}$

$$
\begin{array}{ccc}
W_{U,x} & \rightarrowtail & U \\
\downarrow & & \downarrow \scriptstyle{\sigma_{U,x}} \\
V & \underset{\beta}{\rightarrowtail} & h^*(F)
\end{array}
$$

where $W_{U,x} \subseteq V$ is thus a subobject of **1**. By universality of colimits (see 7.7.6),

$$V = \operatorname*{colim}_{U,x} W_{U,x} = \bigcup_{U,x} W_{U,x}$$

in $\mathcal{E}$, thus also in $\mathsf{Sh}(L)$, since h^* induces an isomorphism at the level of subobjects of **1**. It is also clear that

$$\mathcal{L}(-, W_{U,x}) \rightarrowtail \mathcal{L}(-, U)$$

is a compatible family of morphisms in $\mathsf{Sh}(L)$ and thus induces a factorization between the corresponding colimits, which we choose as

$$\mu_{F,V}(\beta)\colon \mathcal{L}(-, V) \longrightarrow F.$$

Proving that $\eta_{F,V}$ and $\mu_{F,V}$ are mutually inverse is just an easy routine. An arrow $\alpha\colon \mathcal{L}(-, V) \longrightarrow F$ corresponds to an element $a \in F(V)$ by the Yoneda lemma and thus $\alpha = s_{V,a}$. By construction of h^*, $h^*(\alpha) = h^*(s_{V,a}) = \sigma_{V,a}$. By construction of $\mu_{F,V}$, the left hand pullback in

$$
\begin{array}{ccc}
V \;=\!=\!=\; V & \qquad & \mathcal{L}(-,V) \;=\!=\!=\; \mathcal{L}(-,V) \\
\Big\| \qquad\quad \Big\downarrow{\scriptstyle \sigma_{V,a}} & & \Big\| \qquad\qquad \Big\downarrow{\scriptstyle s_{V,a}} \\
V \;\rightarrowtail\; h^*(F) & & \mathcal{L}(-,V) \;\rightarrowtail\; F \\
\quad {\scriptstyle \sigma_{V,a}} & & \qquad\quad {\scriptstyle \mu_{F,V}(\sigma_{V,a})}
\end{array}
$$

yields the commutativity of the right hand square. This proves

$$(\mu_{F,V} \circ \eta_{F,V})(\alpha) = \mu_{F,V}(\sigma_{V,a}) = s_{V,a} = \alpha.$$

Conversely, let us start with $\beta\colon V \longrightarrow h^*(F)$ in $\mathcal{E}$. We use the notation above for defining $\mu_{F,V}(\beta)$. By definition of h^*, every time the left hand diagram below is commutative –

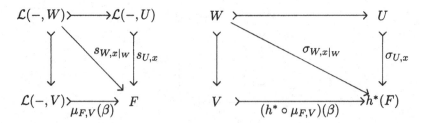

– the right hand diagram is commutative as well. But among those W we have all the $W_{U,x}$, by definition of $\mu_{F,V}$. Thus β and $(h^* \circ \mu_{F,V})(\beta)$ coincide on all $W_{U,x}$ whose union is V. Therefore $\beta = (h^* \circ \mu_{F,V})(\beta)$, which concludes the proof that $\eta_{F,V}$ and $\mu_{F,V}$ are inverses to each other.

Let us consider now the counit of the adjunction $h^* \dashv h_*$. Given $E \in \mathcal{E}$, $(h^* \circ h_*)(E)$ is obtained by considering the diagram of all

$$\mathcal{L}(-, U) \rightarrowtail \mathcal{E}(-, E), \quad \text{with } U \rightarrowtail 1.$$

By the Yoneda lemma, this reduces to considering all the elements of $\mathcal{E}(U, E)$ for all possible U, that is, all the atomic subobjects of E. The object $(h^* \circ h_*)(E)$ is then the colimit of all these atomic subobjects of 1. Since this family of atomic subobjects is trivially downward directed, the colimit is in fact a union. Thus $(h^* \circ h_*)(E)$ is the union of all atomic subobjects of E, which is of course a subobject of E. This proves that the counit of the adjunction is a monomorphism.

Identifying $\mathsf{Sh}(L)$ via h^* with a full subcategory of $\mathcal{E}$, an object $E \in \mathcal{E}$ is in $\mathsf{Sh}(L)$ precisely when at E, the counit of the adjunction is an isomorphism, that is, when E is the union of its atomic subobjects.

Now if $E = \bigcup_{i \in I} U_i \in \mathcal{E}$ expresses E as union of its atomic subobjects, for every subobject $S \rightarrowtail E$ in $\mathcal{E}$, one gets

$$S = S \cap E = S \cap \left(\bigcup_{i \in I} U_i \right) = \bigcup_{i \in I} S \cap U_i$$

in the locale of subobjects of E (see 7.7.6(vi)). But since each U_i is atomic, the same holds for $S \cap U_i$ and thus S is a union of atomic subobjects, thus a fortiori the union of all its atomic subobjects. This proves that $\mathsf{Sh}(L)$ is saturated in $\mathcal{E}$ on subobjects.

It remains to prove that h^* preserves finite limits, that is, that $\mathsf{Sh}(L)$, viewed as a full subcategory of $\mathcal{E}$, is closed under finite limits. We have already seen that h^* is an isomorphism at the level of subobjects of 1, thus the terminal object of $\mathcal{E}$ is also that of $\mathsf{Sh}(L)$. Next if $\alpha, \beta \colon F \rightrightarrows G$ are two morphisms in $\mathsf{Sh}(L)$, their equalizer $k \colon K \rightarrowtail A$ in $\mathcal{E}$ is in $\mathsf{Sh}(L)$, since this subcategory is saturated on subobjects.

Finally we must consider the case of binary products. First each functor $(- \times X)$ preserves colimits and therefore unions of downward directed families (see 7.7.6). Next, if U, V are subobjects of 1, then $U \times V$ is still a subobject of $1 \times 1 = 1$. Now choose $E = \bigcup_{i \in I} U_i$ and $F = \bigcup_{j \in J} V_j$ two objects of $\mathsf{Sh}(L)$, expressed in $\mathcal{E}$ as unions of their atomic subobjects. One has at once

$$E \times F = \left(\bigcup_{i \in I} U_i \right) \times \left(\bigcup_{j \in J} V_j \right) = \bigcup_{i \in I,\ j \in J} (U_i \times V_j),$$

proving that $E \times F$ is a union of atomic subobjects, thus the union of all its atomic subobjects. $\square$

Corollary 7.8.7 *Let $\mathcal{E}$ be a Grothendieck topos and $\sigma(\mathcal{E})$ its locale of subobjects of $\mathbf{1}$. For every locale M, there is an equivalence of categories*

$$\mathsf{Geom}\Big(\mathcal{E}, \mathsf{Sh}(M) \Big) \approx \mathsf{Geom}\Big(\mathsf{Sh}(\sigma(\mathcal{E})), \mathsf{Sh}(M) \Big).$$

Proof Given the Grothendieck topos $\mathcal{E}$, theorem 7.8.5 yields a geometric morphism $h\colon \mathcal{E} \longrightarrow \mathsf{Sh}(\sigma(\mathcal{E}))$. For simplicity of notation, let us keep writing L for the locale $\sigma(\mathcal{E})$.

If $f\colon \mathcal{E} \longrightarrow \mathsf{Sh}(M)$ is a geometric morphism, then both f^* and f_* map a subobject of $\mathbf{1}$ onto a subobject on $\mathbf{1}$: f^* by definition of a geometric morphism and f_* because it has a left adjoint f^*. Therefore f induces a geometric morphism of locales which for simplicity we write

$$\varphi = (\varphi_*, \varphi^*)\colon L = \sigma(\mathcal{E}) \longrightarrow M,$$

where φ_* and φ^* are thus the restrictions of f_* and f^*. By lemma 7.8.2, this defines a geometric morphism $\mathsf{Sh}(L) \longrightarrow \mathsf{Sh}(M)$, up to isomorphism.

Conversely, every geometric morphism $g\colon \mathsf{Sh}(L) \longrightarrow \mathsf{Sh}(M)$ yields, by composition with h, a geometric morphism $g \circ h\colon \mathcal{E} \longrightarrow \mathsf{Sh}(M)$.

Starting from $g\colon \mathsf{Sh}(L) \longrightarrow \mathsf{Sh}(M)$ above, restricting it to the subobjects of $\mathbf{1}$ and re-extending this restriction by colimits to the toposes of sheaves obviously yields a geometric morphism isomorphic to g.

Conversely, starting from $f\colon \mathcal{E} \longrightarrow \mathsf{Sh}(M)$ and the corresponding geometric morphism φ as above, we must prove that the triangle below is commutative up to an isomorphism:

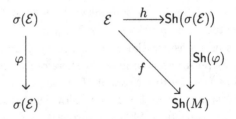

By the definitions of $\mathsf{Sh}(\varphi)^*$, given $w \in M$,

$$\mathsf{Sh}(\varphi)^*\big(\mathcal{M}(-,w)\big) = \operatorname*{colim}_{v \in M,\ y \in \mathcal{M}(v,w)} \mathcal{L}\big(-, f^*(v)\big)$$

$$= \operatorname*{colim}_{v \in M,\ v \leq w} \mathcal{L}\big(-, f^*(v)\big)$$

$$= \mathcal{L}\big(-, f^*(w)\big)$$

since $\mathcal{L}\big(-, f^*(w)\big)$ is the terminal object of the diagram defining the last colimit. Moreover

$$h^*\Big(\mathcal{L}\big(-, f^*(w)\big)\Big) = f^*(w).$$

This proves that $h^* \circ \varphi^*$ and f^* coincide on the subobjects of $\mathbf{1}$. Since every sheaf on M is a colimit of subobjects of $\mathbf{1}$ and all functors h^*, φ^*, f^* preserve colimits (they have right adjoints), the isomorphism $h^* \circ \varphi^* \cong f^*$ holds. The isomorphism $\varphi_* \circ h_* \cong f_*$ follows at once by adjunction.

It is now routine to extend these arguments to the case of natural transformations to get the expected equivalence. □

7.9 Two dimensional category theory

This section formalizes in particular the notions of pseudo-functor and two dimensional limit used previously in this chapter. The two dimensional limits of the previous sections were always bilimits in the sense of the present section, and often even pseudo-limits.

In this section we allow a category to have a class of objects and, between two objects, a class of morphisms.

The theorem of Joyal and Tierney presented in the following section requires a slight generalization of our Galois theory of section 7.5, due to the fact that the category **Groth** of Grothendieck toposes does not admit kernel pairs, but "bikernel pairs". The slogan for defining a bilimit could be

Replace all equalitites by isomorphisms.

More precisely, one has to work with categories in which it makes sense to speak of isomorphic arrows: the prototype is **CAT**, the category of categories and functors, where natural transformations can be defined between functors, allowing us to speak of isomorphic functors.

Now the generalization is not just a triviality. Indeed, for example, a diagram entirely constituted of identities is of course commutative.

But a diagram entirely constituted of isomorphisms has no reason to be commutative. Thus replacing identities by isomorphisms requires at the same time introducing compatibility axioms between the various isomorphisms which enter the problem. This is the essence of two dimensional category theory, whose details can be found in [59] or [8], volume 1. We just sketch here its essential ingredients.

Bearing in mind the example of CAT, we state the following definition; in the example of CAT, the objects are the categories and given two categories $\mathcal{A}$, $\mathcal{B}$, the category $\mathsf{CAT}(\mathcal{A}, \mathcal{B})$ is that of functors from $\mathcal{A}$ to $\mathcal{B}$ and natural transformations between them.

Definition 7.9.1 A 2-category $\mathcal{C}$ consists of

- a class of objects,
- for each pair (A, B) of objects, a category $\mathcal{C}(A, B)$,
- for each triple (A, B, C) of objects, a composition functor

$$\Gamma_{A,B,C} \colon \mathcal{C}(A, B) \times \mathcal{C}(B, C) \longrightarrow \mathcal{C}(A, C).$$

The following terminology and notation are classical:

- the objects of $\mathcal{A}(A, B)$ are called the arrows of $\mathcal{C}$ and are written as $f \colon A \longrightarrow B$;
- the arrows of $\mathcal{A}(A, B)$ are called the 2-cells of $\mathcal{C}$ and are written as $\alpha \colon f \Rightarrow g$
- the composition of 2-cells in the categories $\mathcal{C}(A, B)$ is called the vertical composition and written as $\alpha \circ \beta$ or just $\alpha\beta$;
- for the vertical composition, the identity 2-cell on an arrow f is written as 1_f;
- the composition of arrows given by the functors $\Gamma_{A,B,C}$ is written as $g \circ f$ or just gf –

$$\Gamma_{A,B,C}(f, g) = g \circ f;$$

- the composition of 2-cells via the functors $\Gamma_{A,B,C}$ is called the horizontal composition and as written $\gamma \star \delta$.

The axioms for a 2-category are just the usual associativity and identity axioms on the functors $\Gamma_{A,B,C}$, namely

(i) given four objects (A, B, C, D), the following square commutes –

$$\Gamma_{A,B,C} \times \text{id}$$
$$\mathcal{C}(A,B) \times \mathcal{C}(B,C) \times \mathcal{C}(C,D) \xrightarrow{\hspace{2cm}} \mathcal{C}(A,C) \times \mathcal{C}(C,D)$$

$$\text{id} \times \Gamma_{B,C,D} \downarrow \qquad\qquad \downarrow \Gamma_{A,C,D}$$

$$\mathcal{C}(A,B) \times \mathcal{C}(B,D) \xrightarrow[\Gamma_{A,B,D}]{\hspace{2cm}} \mathcal{C}(A,D)$$

(ii) for each object $A \in \mathcal{C}$, there exists an arrow $\text{id}_A \colon A \longrightarrow A$ which is an identity for the composition of arrows, while 1_{id_A} is an identity for the horizontal composition of 2-cells.

The examples of 2-categories we use in this book are just

(i) the 2-category CAT of categories, functors and natural transformations,

(ii) given a category $\mathcal{A}$, the 2-category $[\mathcal{A}^{\text{op}}, \text{CAT}]$ of functors, natural transformations and modifications,

(iii) the 2-categories Groth of Grothendieck toposes, LOC of localic toposes and Loc of locales, with the geometric morphisms as arrows and the natural transformations $\alpha \colon f^* \Rightarrow g^*$ as 2-cells,

(iv) the ordinary categories $\mathcal{C}$, viewed as 2-categories in which each category $\mathcal{C}(A,B)$ is the discrete category on the set $\mathcal{C}(A,B)$ of arrows.

There are corresponding obvious notions of 2-functor, 2-natural-transformation and 2-modification, but we shall not need them explicitly. What we need in fact is the possibility, due to the presence of 2-cells, of weakening the notions of functor and limit by replacing equalities by isomorphic 2-cells. Again we do not need this in full generality, so that we introduce these notions here only in the special case where they will be useful in this book.

Definition 7.9.2 Let $\mathcal{P}$ be an ordinary category and $\mathcal{C}$ a 2-category. A pseudo-functor $F \colon \mathcal{P} \longrightarrow \mathcal{C}$ consists in giving

- for each object $P \in \mathcal{P}$, an object $F(P) \in \mathcal{C}$,
- for each arrow $f \colon A \longrightarrow B$ in $\mathcal{P}$, an arrow $F(f) \colon F(A) \longrightarrow F(B)$ in $\mathcal{P}$,
- for each pair $A \xrightarrow{f} B \xrightarrow{g} C$ of composable arrows in $\mathcal{P}$, an isomorphic 2-cell $\gamma_{g,f} \colon F(g) \circ F(f) \Rightarrow F(g \circ f)$;
- for each object $A \in \mathcal{P}$ an isomorphic 2-cell $\iota_A \colon F(\text{id}_A) \Rightarrow \text{id}_{F(A)}$.

These data must satisfy the expected coherence axioms respectively to composition and identities, namely

(i) given composable arrows $A \xrightarrow{f} B \xrightarrow{g} C \xrightarrow{h} D$ in $\mathcal{P}$,

$$\gamma_{h,gf} \circ \left(1_{F(h)} \star \gamma_{g,f}\right) = \gamma_{hg,f} \circ \left(\gamma_{h,g} \star 1_{F(f)}\right),$$

(ii) given arrows $A \xrightarrow{f} B \xrightarrow{g} C$ in $\mathcal{P}$,

$$\iota_B \star 1_{F(f)} = \gamma_{\mathrm{id}_B, f}, \quad 1_{F(g)} \star \iota_B = \gamma_{g, \mathrm{id}_B}.$$

When all the 2-cells $\gamma_{g,f}$ and ι_A are identities, we thus recapture exactly the ordinary notion of functor.

Of course, this notion of pseudo-functor could be stated in a more general context, namely, when $\mathcal{P}$ is itself a 2-category.

It is now a little bit fastidious, but in any case straightforward, to write down the coherence axioms for a pseudo-natural-transformation between pseudo-functors. Ignoring the technical axioms in the two following definitions will certainly not prevent the reader from understanding the use we make of these notions.

Definition 7.9.3 Consider two pseudo-functors $F, F' \colon \mathcal{P} \rightrightarrows \mathcal{C}$ from a category $\mathcal{P}$ to a 2-category $\mathcal{C}$. A pseudo-natural-transformation $\alpha \colon F \Rightarrow F'$ consists in giving

- for each object $A \in \mathcal{P}$, an arrow $\alpha_A \colon F(A) \longrightarrow F'(A)$ in $\mathcal{C}$,
- for each arrow $f \colon A \longrightarrow B$ in $\mathcal{P}$, an isomorphic 2-cell $\tilde{\alpha}_f \colon \alpha_B \circ F(f) \Rightarrow F'(f) \circ \alpha_A$.

These data must satisfy the obvious coherent axioms, namely, given arrows $f \colon A \longrightarrow B$ and $g \colon B \longrightarrow C$ in $\mathcal{P}$:

(i) $\left(\gamma'_{g,f} \star 1_{\alpha_A}\right) \circ \left(1_{F'(g)} \star \tilde{\alpha}_f\right) \circ \left(\tilde{\alpha}_g \star 1_{F(f)}\right) = \tilde{\alpha}_{g \circ f} \circ \left(1_{\alpha_C} \star \gamma_{g,f}\right)$;

(ii) $\left(\iota'_A \star 1_{\alpha_A}\right) \circ \tilde{\alpha}_{\mathrm{id}_A} = 1_{\alpha_A} \star \iota_A$.

Finally, to be complete, it remains to write down the axioms for a modification.

Definition 7.9.4 Consider two pseudo-functors $F, F' \colon \mathcal{P} \rightrightarrows \mathcal{C}$ from a category $\mathcal{P}$ to a 2-category $\mathcal{C}$ and two pseudo-natural-transformations $\alpha, \alpha' \colon F \Rightarrow F'$. A modification $\theta \colon \alpha \rightsquigarrow \alpha'$ consists in giving

- for each object $a \in \mathcal{P}$, a 2-cell $\theta_A \colon \alpha_A \Rightarrow \alpha'_A$.

These data must satisfy the obvious coherence axiom, namely, given an arrow $f \colon A \longrightarrow B$ in $\mathcal{P}$

(i) $\tilde{\alpha}'_f \circ \left(\theta_B \star 1_{F(f)}\right) = \left(1_{F'(f)} \star \theta_A\right) \circ \tilde{\alpha}_f.$

To extend the analogy with the case of ordinary categories, observe that every object $A \in \mathcal{C}$ of a 2-category determines a contravariant functor

$$\mathcal{C}(-, A) \colon \mathcal{C}^{\mathrm{op}} \longrightarrow \mathsf{CAT}, \quad B \mapsto \mathcal{C}(B, A)$$

yielding the corresponding Yoneda embedding

$$Y_{\mathcal{C}} \colon \mathcal{C} \longrightarrow [\mathcal{C}^{\mathrm{op}}, \mathsf{CAT}], \quad A \mapsto \mathcal{C}(-, A),$$

which is full and faithful in the sense that there exists an isomorphism of categories

$$\mathcal{C}(A, B) \overset{\cong}{\longrightarrow} 2\text{-}\mathsf{Nat}\Big(\mathcal{C}(-, A), \mathcal{C}(-, B)\Big)$$

between the category $\mathcal{C}(A, B)$ and the category of 2-natural-transformations and modifications between $\mathcal{C}(-, A)$ and $\mathcal{C}(-, B)$. This should be put in parallel with the considerations in corollary 7.2.5.

The notion of bilimit transposes in the same spirit the ordinary notion of limit. Given an ordinary functor $F \colon \mathcal{P} \longrightarrow \mathcal{C}$ between ordinary categories, one writes $\Delta_C = \Delta(C) \colon \mathcal{P} \longrightarrow \mathcal{C}$ for the constant functor corresponding to object C. A cone with vertex C over F is exactly a natural transformation $\sigma \colon \Delta_C \Rightarrow F$. A limit natural transformation is then a cone $\pi \colon \Delta_L \Rightarrow F$ such that, for every object $C \in \mathcal{C}$, the composition with π

$$\mathcal{C}(C, L) \longrightarrow \mathsf{Nat}(\Delta_C, F), \quad h \mapsto \pi \circ \Delta_h$$

is a bijection, with Δ_h the constant natural transformation corresponding to h.

Definition 7.9.5 Consider a pseudo-functor $F \colon \mathcal{P} \longrightarrow \mathcal{C}$ from a category $\mathcal{P}$ to a 2-category $\mathcal{C}$. The bilimit of F, when it exists, is a pair (L, π) where L is an object of $\mathcal{C}$ and $\pi \colon \Delta_L \Rightarrow F$ is a pseudo-natural-transformation such that composition with π

$$\mathcal{C}(C, L) \longrightarrow \mathsf{PsNat}(\Delta_C, F), \quad h \mapsto \pi \circ \Delta_h$$

induces an equivalence between the category $\mathcal{C}(C, L)$ and the category of pseudo-natural-transformations and modifications from Δ_C to F.

In particular, since everything is now "pseudo", all the triangles which normally commute in an ordinary limit are replaced by triangles which commute "up to an isomorphic 2-cell". Moreover, the definition requires an equivalence of categories, not an isomorphism: this implies again that the factorization through the bilimit is no longer unique, but unique up to an isomorphic 2-cell. Of course, this last fact implies that the bilimit, when it exists, is no longer defined up to an isomorphism, but only up to an equivalence. That is, two bilimits L and L' are connected by arrows $L \underset{\longrightarrow}{\overset{\longleftarrow}{}} L'$ such that both composites are isomorphic to the identity.

Just for information, let us mention that when one requires an isomorphism of categories in the previous definition, one recaptures the notion of pseudo-limit.

It is a classical result that CAT admits all small bilimits. But we shall just need the following result.

Proposition 7.9.6 *The 2-category* CAT *of categories, functors and natural transformations admits bipullbacks, thus in particular bikernel pairs.*

Proof The bipullback of two functors F, G is easily seen to be given by

$$
\begin{array}{ccc}
\mathcal{P} & \overset{F'}{\longrightarrow} & \mathcal{B} \\
\Big\downarrow{\scriptstyle G'} & & \Big\downarrow{\scriptstyle G} \\
\mathcal{A} & \underset{F}{\longrightarrow} & \mathcal{C}
\end{array}
$$

where the objects of $\mathcal{P}$ are the triples

$$(A, \alpha, C, \beta, B)) \text{ with } \alpha\colon F(A) \overset{\cong}{\longrightarrow} C, \quad \beta\colon G(B) \overset{\cong}{\longrightarrow} C$$

where thus $A \in \mathcal{A}$, $B \in \mathcal{B}$, $C \in \mathcal{C}$ and α, β isomorphisms in $\mathcal{C}$. $\square$

The following proposition is a deep result in topos theory. We refer to [55] for a proof.

Proposition 7.9.7 *The 2-category* Groth *of Grothendieck toposes and geometric morphisms, with 2-cells as above, has bipullbacks, thus in particular bikernel pairs.* $\square$

Corollary 7.9.8 *The 2-category* LOC *of localic toposes and geometric morphisms is stable for finite bilimits in the 2-category* Groth *of Grothendieck toposes and geometric morphisms.*

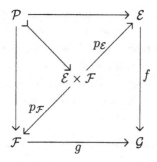

Diagram 7.18

Proof Corollary 7.8.7 can be rephrased as the fact that the 2-category LOC of localic toposes is "bireflective" in the 2-category Groth of Grothendieck toposes. Extending a classical argument for ordinary limits, this at once implies the result. □

Applying lemma 7.8.2, it is easy to observe that a finite bilimit of localic toposes is obtained as the topos of sheaves on the limit of the corresponding diagram of locales. The form of limits in the category of locales (see proposition 7.8.3) indicates at once that a bilimit of localic toposes, and thus a bilimit of Grothendieck toposes, is not at all computed like the usual limit of underlying categories.

The next result extends a little bit the stability property of corollary 7.9.8.

Corollary 7.9.9 *In the category* Groth *of Grothendieck toposes and geometric morphisms, the bipullback of two localic toposes over a Grothendieck topos is localic as well.*

Proof We refer to diagram 7.18, where the toposes $\mathcal{E}$ and $\mathcal{F}$ are localic; we must prove that the topos $\mathcal{P}$ is localic as well. The bipullback can be reconstructed via the biproduct $\mathcal{E} \times \mathcal{F}$ and the biequalizer $\mathsf{Ker}(f \circ p_{\mathcal{E}}, g \circ p_{\mathcal{F}})$. By corollary 7.9.8, $\mathcal{E} \times \mathcal{F}$ is localic. Applying again theorem 7.8.5 and its corollary 7.8.7, we get the diagram

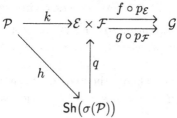

which commutes up to an isomorphism. So k is a "biequalizer" in **Groth**, which has bipullbacks. A classical argument for equalizers extends to prove that since k is a biequalizer, h is a biequalizer as well. But since the inverse image of h is full and faithful, it follows at once that h is a "biepimorphism" in **Groth**. Thus h is both an epimorphism and a biequalizer, and again a classical argument extends to prove that h is an equivalence. Thus $\mathcal{P}$ is localic. $\qquad\Box$

It remains to apply and generalize all this to the case of precategories, introduced in definition 7.2.1.

For this, consider again the category $\mathbb{P}$ defined in section 7.2:

$$P_2 \xrightarrow[\substack{f_1}]{\substack{f_0 \\ m}} P_1 \xleftarrow[d_1]{n} \xrightarrow{d_0} P_0$$

with the conditions

$$d_0 \circ f_1 = d_1 \circ f_0, \qquad d_1 \circ m = d_1 \circ f_1, \qquad d_0 \circ m = d_0 \circ f_0,$$
$$d_0 \circ n = \mathrm{id}_{C_0}, \qquad d_1 \circ n = \mathrm{id}_{C_0}.$$

Definition 7.9.10 An internal pseudo-precategory in a 2-category $\mathcal{C}$ is a pseudo-functor $\mathbb{C}\colon \mathbb{P} \longrightarrow \mathcal{C}$.

Definition 7.9.11 Let us consider

- a 2-category $\mathcal{C}$,
- an internal pseudo-precategory $\mathbb{C}\colon \mathbb{P} \longrightarrow \mathcal{C}$,
- a pseudo-functor $F\colon \mathcal{A}^{\mathrm{op}} \longrightarrow \mathsf{CAT}$.

The category $F^{\mathbb{C}}$ of covariant internal F-presheaves on $\mathcal{C}$ is the bilimit of the composite $F \circ \mathbb{C}$.

All the results of sections 7.1 to 7.5 admit expected straightforward translations to the present context of 2-categories. We shall freely use those results without further proofs.

7.10 The Joyal–Tierney theorem

Again we restrict our attention to the case of Grothendieck toposes and sketch the various proofs, referring to the works in the bibliography for a detailed account.

Let us first state the Galois theorem of Joyal and Tierney in the special case of Grothendieck toposes. This section is entirely devoted to "proving" this theorem, or at least to showing how it enters the context of our Galois theory of section 7.5.

Theorem 7.10.1 (Galois theorem) *For every Grothendieck topos* $\mathcal{E}$, *there exists an open localic groupoid* $\mathbb{G}$ *such that* $\mathcal{E}$ *is equivalent to the category of étale presheaves on* $\mathbb{G}$.

First of all, we have to explain the terminology used in the statement of this Galois theorem.

Definition 7.10.2 A geometric morphism $f\colon L \longrightarrow M$ of locales is open when

(i) f^* admits a left adjoint $f_!$,

(ii) $\forall u \in L \ \forall v \in M \ \ f_!\big(u \wedge f^*(v)\big) = f_!(u) \wedge v$.

Condition (ii) is called the Frœbenius condition.

Example 7.10.3 Consider a continuous map $f\colon X \longrightarrow Y$ of topological spaces. Write $\mathcal{O}(X)$ and $\mathcal{O}(Y)$ for the corresponding locales of open subsets. One gets at once a geometric morphism

$$(f^*, f_*)\colon \mathcal{O}(X) \longrightarrow \mathcal{O}(Y)$$

by defining, for $U \in \mathcal{O}(X)$ and $V \in \mathcal{O}(Y)$,

$$f^*(V) = f^{-1}(V),$$
$$f_*(U) = \bigcup\{V \,|\, f^{-1}(V) \subseteq U\}.$$

These definitions imply at once

$$f^*(V) \subseteq U \Leftrightarrow f^{-1}(V) \subseteq U \Leftrightarrow V \subseteq f_*(U).$$

Moreover the inverse image process preserves arbitrary unions and finite intersections of open subsets, since these are computed set theoretically. Now when f turns out to be an open map, that is, when f maps an open subset onto an open subset, we further define $f_!(U) = f(U)$ and obviously

$$f_!(U) \subseteq V \Leftrightarrow f(U) \subseteq V \Leftrightarrow U \subseteq f^{-1}(V).$$

This proves $f_! \dashv f^*$. Moreover

$$
\begin{aligned}
y \in f(U) \cap V \ &\Leftrightarrow\ \exists x \in U \ \ y = f(x) \in V \\
&\Leftrightarrow\ \exists x \in U \ \ x \in f^{-1}(V) \ \ y = f(x) \\
&\Leftrightarrow\ y \in f\big(U \cap f^{-1}(V)\big)
\end{aligned}
$$

which is the required Frœbenius condition.   □

Definition 7.10.4 A geometric morphism $f: L \longrightarrow M$ of locales is étale when

- f is an open morphism,
- there exists a covering $1 = \bigvee_{i \in I} u_i$ of the top element $1 \in L$ such that for each index $i \in I$, the restriction

$$h_! : \{u \in L \mid u \leq u_i\} \longrightarrow \{v \in M \mid v \leq h_!(u_i)\}$$

is an isomorphism of locales.

Of course, in the context of example 7.10.3, a local homeomorphism between topological spaces, which is automatically open, yields a corresponding étale morphism of locales.

Definition 7.10.5

(i) A localic groupoid is a groupoid in the category of locales.
(ii) A localic groupoid is open when the "domain" and "codomain" operations $d_0, d_1 : G_1 \overrightarrow{\longrightarrow} G_0$ are open.

It should be noticed that, in 7.10.5.(ii), the openness of d_0 is equivalent to the openness of d_1.

Definition 7.10.6 Let $\mathbb{G}$ be an open localic groupoid. With the notation of section 7.1, an internal presheaf (P_0, p_0, δ_1) on $\mathbb{G}$ is étale when the morphism $p_0 : P_0 \longrightarrow G_0$ is étale.

This finishes the explanation of the terminology used in 7.10.1. Our intention is now to deduce this Galois theorem from the Galois theory of section 7.5, via the considerations of section 7.9. To achieve this, we shall freely use the following "localic covering" theorem, whose proof occupies a big part of the work of Joyal and Tierney (see [57]). Again we state the theorem only for Grothendieck toposes.

Theorem 7.10.7 (Localic covering) *For every Grothendieck topos $\mathcal{E}$, there exist a locale L and a geometric morphism $\lambda: \mathsf{Sh}(L) \longrightarrow \mathcal{E}$ which is open and of effective descent.*   □

Of course we shall not prove this theorem here, but again we shall now explain the terminology used in this statement.

Definition 7.10.8 A geometric morphism $p\colon \mathcal{E} \longrightarrow \mathcal{F}$ between Grothendieck toposes is open when p^* commutes with the functors $\forall_f$ of lemma 7.7.8. More explicitly, given a morphism $f\colon A \longrightarrow B$ in $\mathcal{F}$ and a subobject $S \rightarrowtail A$, one has $p^*\big(\forall_f(S)\big) = \forall_{p^*(f)}\big(p^*(S)\big)$.

The relation with definition 7.10.2 is not clear a priori, but let us mention that a morphism of locales $f\colon L \longrightarrow M$ is open in the sense of definition 7.10.2 precisely when the corresponding geometric morphism $\mathsf{Sh}(f)\colon \mathsf{Sh}(L) \longrightarrow \mathsf{Sh}(M)$ is open in the sense of definition 7.10.8 (see [65]).

Definition 7.10.9 Let $\varphi\colon \mathcal{F} \longrightarrow \mathcal{E}$ be a geometric morphism between Grothendieck toposes. We consider the bikernel pair $\mathbb{G}_\varphi$ of φ in the 2-category **Groth** of Grothendieck toposes and view it as a pseudo-precategory, in the spirit of example 4.6.2,

$$
(\mathcal{F} \times_{\mathcal{E}} \mathcal{F}) \times_{\mathcal{F}} (\mathcal{F} \times_{\mathcal{E}} \mathcal{F}) \underset{\underset{f_1}{\xrightarrow{\hspace{1.2cm}}}}{\overset{\overset{f_0}{\xrightarrow{\hspace{1.2cm}}}}{\xrightarrow[m]{\hspace{1.2cm}}}} \mathcal{F} \times_{\mathcal{E}} \mathcal{F} \underset{\underset{d_1}{\xrightarrow{\hspace{1.2cm}}}}{\overset{\overset{d_0}{\xrightarrow{\hspace{1.2cm}}}}{\xleftarrow[\hspace{1.2cm}]{n}}} \mathcal{F}
$$

using the usual pullback notation to denote bipullbacks in **Groth**. A descent datum for φ is a pair (F, θ) where

   (i) F is an object of $\mathcal{F}$,
   (ii) $\theta\colon d_1^*(F) \longrightarrow d_0^*(F)$ is a morphism of $\mathcal{F} \times_{\mathcal{E}} \mathcal{F}$,
   (iii) $n^*(\theta) \cong \mathsf{id}_F$,
   (iv) $m^*(\theta) \cong f_0^*(\theta) \circ f_1^*(\theta)$,

where conditions (iii) and (iv) are convenient abbreviations for the commutativity of the diagrams 7.19 and 7.20, involving the isomorphisms appearing in the structure of pseudo-functors of n^*, m^*, f_0^* and f_1^*. A morphism of descent data $(F, \theta) \longrightarrow (F', \theta')$ is a morphism $f\colon F \longrightarrow F'$ in $\mathcal{F}$ which commutes with the descent data, that is, $\theta' \circ d_1^*(f) = d_0^*(f) \circ \theta$. We shall write $\mathsf{Desc}(\varphi)$ for this category of descent data.

Condition (iii) in 7.10.9 means the commutativity of diagram 7.19 while condition (iv) expresses the commutativity of the hexagon in diagram 7.2, which people familiar with classical descent theory will certainly recognize.

Definition 7.10.10 A geometric morphism $\varphi\colon \mathcal{F} \longrightarrow \mathcal{E}$ between Grothendieck toposes is of effective descent when the functor

$$
\mathcal{E} \longrightarrow \mathsf{Desc}(\varphi), \quad E \mapsto \big(\varphi^*(E), (d_1^* \circ \varphi^*)(E) \overset{\cong}{\longrightarrow} (d_0^* \circ \varphi^*)(E)\big)
$$

$$n^* d_1^*(F) \xrightarrow{\quad n^*(\theta) \quad} n^* d_0^*(F)$$

$$\cong \qquad\qquad \cong$$

$$F =\!=\!=\!=\!= F$$

Diagram 7.19

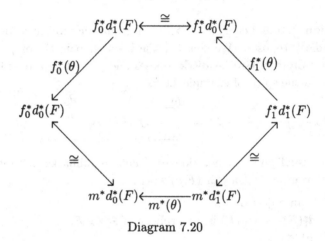

Diagram 7.20

is an equivalence of categories.

Lemma 7.10.11 *Let $\varphi\colon \mathcal{F}\longrightarrow \mathcal{E}$ be a geometric morphism of toposes which is of effective descent. The category $\mathsf{Desc}(\varphi)$ of descent data is equivalent to the bilimit in* CAT *of the diagram*

$$(\mathcal{F} \times_{\mathcal{E}} \mathcal{F}) \times_{\mathcal{F}} (\mathcal{F} \times_{\mathcal{E}} \mathcal{F}) \overset{\xleftarrow{\quad f_0^* \quad}}{\underset{\xleftarrow{\quad f_1^* \quad}}{\xleftarrow{\quad m^* \quad}}} \mathcal{F} \times_{\mathcal{E}} \mathcal{F} \overset{\xleftarrow{\quad d_0^* \quad}}{\underset{\xleftarrow{\quad d_1^* \quad}}{\xleftarrow{\quad n^* \quad}}} \mathcal{F}$$

where the pullback notation is used to denote bipullbacks in Groth.

Proof The equivalence in definition 7.10.10 shows that in the present case, for every descent datum (F, θ), the arrow θ is an isomorphism. The bilimit of the present statement has for objects those $F \in \mathcal{F}$ whose images along the various functors of the diagram constitute a compatible family up to compatible isomorphism. In particular we must have isomorphisms $\theta\colon d_1^*(F)\overset{\cong}{\longrightarrow}d_0^*(F)$ and the various required compatibilities reduce precisely to the definition of a descent datum. $\square$

We are now ready to exhibit the connection with the Galois theory of section 7.5. We shall apply corollary 7.5.3 – or more precisely, its straightforward generalization to the two dimensional context – to the following data, where we use freely the notation of 7.8.5 and 7.8.7:

- $\mathcal{A}$ = Groth is the 2-category of Grothendieck toposes, geometric morphisms and natural transformations between them (see 7.8.1);
- $\mathcal{X}$ = Loc is the 2-category of locales, geometric morphisms and natural transformations between them (see 7.8.1);
- $H\colon \mathcal{A}\longrightarrow\mathcal{X}$ is the functor $\sigma\colon$ Groth $\longrightarrow$ Loc, which maps a Grothendieck topos $\mathcal{F}$ onto its locale $\sigma(\mathcal{F})$ of subobjects of $\mathbf{1}$;
- $G\colon \mathcal{A}^{\mathrm{op}}\longrightarrow$ CAT is the forgetful functor

$$\text{Groth}\longrightarrow\text{CAT},\quad \mathcal{F}\mapsto\mathcal{F},\quad f\mapsto f^*$$

which is indeed contravariant;
- $K\colon \mathcal{X}^{\mathrm{op}}\longrightarrow$ CAT is the functor

$$\text{Loc}\longrightarrow\text{CAT},\quad L\mapsto\text{Sh}(L),\quad f\mapsto f^*$$

which is also indeed contravariant;
- $\alpha\colon K\circ H\Rightarrow G$ is given, for each Grothendieck topos $\mathcal{E}$, by

$$\alpha_{\mathcal{E}}\colon \text{Sh}\big(\sigma(\mathcal{E})\big)\longrightarrow\mathcal{E},\quad \alpha_{\mathcal{E}}=h^*$$

where h is the hyperconnected geometric morphism of 7.8.5;
- fixing a Grothendieck topos $\mathcal{E}$, $\sigma\colon S\longrightarrow R$ is the localic covering $\lambda\colon \text{Sh}(L)\longrightarrow\mathcal{E}$ of theorem 7.10.7;
- the required precategorical decomposition is replaced by a pseudo-precategorical decomposition; in the spirit of proposition 7.4.4 and via proposition 7.9.7, we use for this the bikernel pair $\mathbb{G}_\lambda$ of the localic covering $\lambda\colon \text{Sh}(L)\longrightarrow\mathcal{E}$, viewed as a pseudo-precategory; thus this yields the situation of diagram 7.21;
- let us finally recall that, in view of theorem 7.10.7 and definition 7.10.10, we have an equivalence of categories $\mathcal{E}\approx\text{Desc}(\lambda)$.

Lemma 7.10.12 *In the above situation, the two categories*

$$\text{Split}_\alpha(\lambda)=\mathcal{E}$$

are equal, where $\text{Split}_\alpha(\lambda)$ *is defined as after definition 7.5.1.*

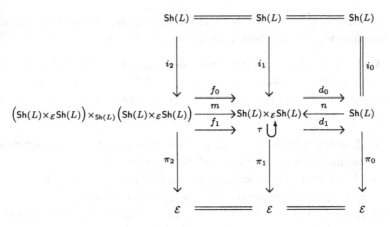

Diagram 7.21

Proof $\mathsf{Split}_\alpha(\lambda)$ is the full subcategory of $G(\mathcal{E}) = \mathcal{E}$ generated by the objects $E \in \mathcal{E}$ such that there exists $F \in KH\big(\mathsf{Sh}(L)\big) \cong \mathsf{Sh}(L)$ with $\lambda^*(E) \cong G(\lambda)(E) \cong \alpha_{\mathsf{Sh}(L)}(F)$. But $\alpha_{\mathsf{Sh}(L)}$ is just the identity on $\mathsf{Sh}(L)$, thus every object of $F \in \mathsf{Sh}(L)$ has the form $\alpha_{\mathsf{Sh}(L)}(F)$. This holds in particular for the objects $\lambda^*(E)$, proving that all objects of $\mathcal{E}$ are in $\mathsf{Split}_\alpha(\lambda)$. $\qquad\qquad\qquad\qquad\qquad\qquad\qquad\qquad\qquad\qquad\qquad\square$

Lemma 7.10.13 *In the above situation, the functor* $\sigma \colon \mathsf{Groth} \longrightarrow \mathsf{Loc}$ *transforms the pseudo-precategory* $\mathbb{G}_\lambda$, *bikernel pair of* λ, *into an open localic groupoid.*

Proof Applying corollary 7.9.9, we know that all toposes appearing in the definition of $\mathbb{G}_\lambda$ are localic. Lemma 7.8.2 then implies at once that $\sigma(\mathbb{G}_\lambda)$ is a "bigroupoid" in Loc. But since in a locale, only the identities are isomorphisms, two isomorphic geometric morphisms between locales must be equal. Therefore $\sigma(\mathbb{G}_\lambda)$ is an actual groupoid in Loc. $\qquad\square$

Lemma 7.10.14 *Let* $\mathbb{G}$ *be an open localic groupoid. The category of étale internal presheaves on* $\mathbb{G}$ *is equivalent to the bilimit, in* CAT, *of the diagram*

$$\mathsf{Sh}(G_1 \times_{G_0} G_1) \underset{\underset{f_1^*}{\longleftarrow}}{\overset{\overset{f_0^*}{\longleftarrow}}{\underset{m^*}{\longleftarrow}}} \mathsf{Sh}(G_1) \underset{\underset{d_1^*}{\longleftarrow}}{\overset{\overset{d_0^*}{\longleftarrow}}{\underset{n^*}{\longrightarrow}}} \mathsf{Sh}(G_0),$$

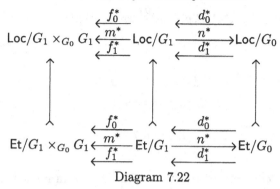

Diagram 7.22

that is, with the previous notation and that of definition 7.2.6, to the category $K^{\mathbb{G}}$.

Proof Étale morphism are stable under pulling back, from which we get the commutative diagram 7.22, where Et denotes the category of locales and étale geometric morphisms between them; it is a classical result that for a given locale L, the subcategory $\mathsf{Et}/L \subseteq \mathsf{Loc}/L$ is full. The limit of the top line is the category of internal presheaves on $\mathbb{G}$, while the limit of the bottom line is the category of étale presheaves on $\mathbb{G}$. But this bottom line is, up to equivalences, precisely that of the statement: indeed, it is another classical result that the topos of sheaves on a locale L is equivalent to the category Et/L. This concludes the sketch of the proof. We refer to [8], volume 3, for the details. $\qquad\square$

Lemma 7.10.15 *In the above situation,* $\big(\lambda, (i, \mathbb{G}_\lambda, \pi)\big)$ *is an effective descent structure with respect to the functor G, in the sense of definition 7.4.3.*

Proof By lemma 7.2.9, $G^{\mathbb{G}\varepsilon} = G(\mathcal{E}) = \mathcal{E}$. On the other hand the category $G^{\mathbb{G}\lambda}$ is the limit in CAT of the diagram (see 7.2.6)

$$(\mathcal{F} \times_{\mathcal{E}} \mathcal{F}) \times_{\mathcal{F}} (\mathcal{F} \times_{\mathcal{E}} \mathcal{F}) \underset{\underset{f_1^*}{\overset{m^*}{\longleftarrow}}}{\overset{f_0^*}{\longleftarrow}} \mathcal{F} \times_{\mathcal{E}} \mathcal{F} \underset{\underset{d_1^*}{\longleftarrow}}{\overset{d_0^*}{\underset{n^*}{\longrightarrow}}} \mathcal{F}$$

and we have observed in lemma 7.10.11 that this limit is equivalent to the category $\mathsf{Desc}(\lambda)$ of descent data, which is itself equivalent to $\mathcal{E}$ (see definition 7.10.10). $\qquad\square$

We are now able to conclude the proof of the expected Galois theorem 7.10.1.

$$n^* d_1^*(P,p) \xrightarrow{\ n^*(\theta)\ } n^* d_0^*(P,p)$$

$$\cong \uparrow \downarrow \qquad\qquad \cong \uparrow \downarrow$$

$$(P,p) =\!\!=\!\!=\!\!=\!\!= (P,p)$$

Diagram 7.23

Proof of the Galois theorem 7.10.1   This is now an immediate consequence of corollary 7.5.3 and lemmas 7.10.12 to 7.10.15:

$$\mathcal{E} \cong \mathsf{Split}_\alpha(\lambda) \cong K^{H \circ \mathbb{G}_\lambda}$$

and this last category is equivalent to that of étale presheaves on $H(\mathbb{G}_\lambda)$,
which is an open localic groupoid.   □

In fact, to conclude that our theorem 7.10.1 yields exactly the theorem
stated by Joyal and Tierney, one point remains to be checked. Indeed,
Joyal and Tierney, instead of the usual definition of internal presheaf of
section 7.1, use the following definition.

Definition 7.10.16 Consider a category $\mathcal{C}$ with pullbacks and an internal groupoid $\mathbb{G}$ in $\mathcal{C}$:

$$G_2 \ \substack{\xrightarrow{\ f_0\ } \\ \xrightarrow{\ m\ } \\ \xrightarrow{\ f_1\ }} \ G_1 \ \substack{\xrightarrow{\ d_0\ } \\ \xleftarrow{\ n\ } \\ \xrightarrow{\ d_1\ }} \ G_0.$$

A covariant internal presheaf on $\mathbb{G}$ in the sense of Joyal and Tierney is
a triple (P, p, θ) where

(i) $(P,p) \in \mathcal{C}/G_0$,
(ii) $\theta\colon d_1^*(P,p) \longrightarrow d_0^*(P,p)$ in $\mathcal{C}/G_1$,
(iii) $n^*(\theta) \cong \mathrm{id}$,
(iv) $m^*(\theta) \cong f_0^*(\theta) \circ f_1^*(\theta)$,

where conditions (iii) and (iv) are convenient abbreviations for the commutativity of diagrams 7.23 and 7.24, involving the isomorphisms appearing in the structure of pseudo-functors of n^*, m^*, f_0^* and f_1^*.

One should compare this with definition 7.10.9. Axiom (iii) in 7.10.16
means the commutativity of diagram 7.23 while axiom (iv) again yields
the classical hexagonal commutative diagram 7.24.

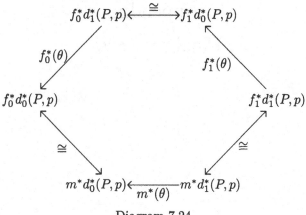

Diagram 7.24

Lemma 7.10.17 *In the situation of the Galois theorem 7.10.1, consider the open localic groupoid $H(\mathbb{G}_\lambda)$ of lemma 7.10.13. On this groupoid, the category of étale internal presheaves in the sense of Joyal and Tierney is equivalent to the category of étale internal presheaves of definition 7.10.6.*

Proof The internal presheaf (P, p, θ) of Joyal and Tierney is étale when p is étale. In both cases, an étale presheaf yields an étale morphism of locales

$$p\colon P \longrightarrow \sigma\big(\mathsf{Sh}(L)\big) = L.$$

But the category of étale morphisms over L is equivalent to the category of sheaves on L. Thus giving the pair (P, p) with p étale reduces to giving a sheaf F on L. The structure of an internal presheaf on (P, p) in the sense of Joyal and Tierney reduces to that of a descent datum on the sheaf F, in the sense of definition 7.10.9. The category of étale presheaves in the sense of Joyal and Tierney is thus equivalent to the bilimit of lemma 7.10.11. Via lemma 7.10.13, this bilimit is also that of lemma 7.10.14, where the localic groupoid is chosen to be $H(\mathbb{G}_\lambda)$. This lemma 7.10.14 says precisely that this bilimit is a category equivalent to that of étale internal presheaves in the sense of definition 7.10.6.   □

Appendix
Final remarks

A.1 Separable algebras

The notion of separable algebra is everywhere present in this book, even if it does not appear explicitly. For example, in the classical context of Galois theory, a K-algebra A on a field K is separable when it is split by some field extension $K \subseteq L$.

In order to throw more light on the historical development of the Galois theories presented in this book, we will take the notion of a separable algebra, and look at its evolution through various lines, from the classical theory to the most abstract level of general categories.

Most textbooks on algebra define this notion only in the special case of field extensions. They explain that for a finite dimensional (usually called just "finite") field extension the following conditions are equivalent.

(i) Every element of the extension is separable over the ground field; recall that an element is said to be separable if it is a simple root of its minimal polynonial, or equivalently, if that polynomial has a non-zero discriminant.

(ii) The extension is generated by its separable elements.

(iii) The separable degree of the extension coincides with its degree; recall that the separable degree is the number of algebra homomorphisms from the extension to the algebraic closure of the ground field, and the degree is the dimension of the extension as a vector space over that field.

And then they define the finite separable extensions as those which satisfy the equivalent conditions (i)–(iii). For the infinite (dimensional) extensions (iii) does not make sense, but (i) and (ii) are still equivalent and can be used as definitions (and in fact (i) is used in chapter 1 of this book). However sometimes it is also convenient to think of the transcendental extensions as being separable, and speak of algebraic separable and (general) separable extension; since non-algebraic separable extensions have no use in this book, we will not do that.

The first step of generalization is to replace the field extensions by the commutative algebras over fields. For a finite dimensional (commutative) algebra A over a field K, the following conditions are equivalent.

(iv) A is a finite product of separable field extensions of K.

(v) $L \otimes_K A$ has no nilpotent elements for any field extension L of K.

(vi) The L-algebra $L \otimes_K A$ is of the form $L \times \cdots \times L$ for some field extension L of K.

(vii) A is projective as an $(A \otimes_K A)$-module.

The equivalence (iv)⇔(v) is well known, and motivates (v), which is used for example by N. Bourbaki (see [9]), also in the case of infinite dimensional algebras, as the definition of a separable algebra; it is also used in the case of field extensions, in the definition of what we called above general (not necessarily algebraic) separable field extension. Each of the conditions (vi) and (vii) actually implies that A is finite dimensional, and N. Bourbaki uses (vi) as the definition of étale algebra, which agrees with A. Grothendieck's notion of étale covering of a scheme.

The second step of generalization is to replace the commutative algebras over fields by the commutative algebras over connected commutative rings. Recall that "connected" refers to the connectedness of the Zariski spectrum, which is equivalent to the absense of non-trivial idempotents. In other words a commutative ring is said to be connected if it is indecomposable into a non-trivial product of two rings (which also agrees with the categorical notion of connectedness). When K is such a ring, or even an arbitrary commutative ring, the condition (vii) above is used as the definition (also for non-commutative A, replacing then $A \otimes_K A$ by $A \otimes_K A^{op}$, where A^{op} is A with the opposite multiplication) of separable algebra. Note that such separable algebras have a section in S. Mac Lane's *Homology* (see [64]), and F. DeMeyer and E. Ingraham's book [24] is about them. However in order to develop their Galois theory, one has to require an additional condition, namely that A is projective also as K-module. At this point we should also mention that the notion of Galois extension of commutative rings was first defined (independently of A. Grothendieck) by M. Auslander and

O. Goldman (see [3]), and the Galois theory of those extensions was developed by S.U. Chase, D.K. Harrison, and A. Rosenberg (see [21]) and G.J. Janusz (see [54]), and many others (see the references in [24]). As suggested by the Galois theory of commutative rings, "the right notion" is what A.R. Magid (see [67]) later called a strongly separable algebra. It is a commutative algebra A over a connected commutative ring K satisfying the following equivalent conditions.

(viii) A is projective as a K-module and as an $(A \otimes_K A)$-module.

(ix) The L-algebra $L \otimes_K A$ is of the form $L \times \cdots \times L$ for some commutative K-algebra L which is (non-zero) finitely generated and projective as a K-module.

(x) The same condition with faithfully flat instead of finitely generated and projective.

(xi) The same condition with $K \longrightarrow L$ being a pure monomorphism of K-modules instead of L being finitely generated and projective.

Those algebras correspond to the finite dimensional separable algebras over fields, and their filtered colimits (="directed unions") which A.R. Magid calls locally strongly separable algebras correspond to the infinite dimensional ones. Note that the equivalences (viii)–(xi) actually involve A. Grothendieck's descent theory, and in particular (x)⇔(xi) even a more recent unpublished result of A. Joyal and M. Tierney ("effective descent=pure").

Before discussing the third step of generalization – to arbitrary commutative rings – let us come back to the old work on covering spaces, already commented on in the introduction.

If we begin again with the standard textbooks, in this case on algebraic topology, the notion we are looking for is the notion of a covering map of topological spaces as defined in chapter 6: a continuous map $f \colon A \longrightarrow B$ satisfying the condition:

(xii) Every point in B has an open neighbourhood U whose inverse image is a disjoint union of open subsets each of which is mapped homeomorphically onto U by f.

A covering space is the same thing as a covering map: $A = (A, f)$ is a covering space over B (or of B) when $f \colon A \longrightarrow B$ is a covering map.

One now understands better the analogy between the covering spaces and the separable algebras. If B is connected and locally connected, and has a universal (i.e. the "largest" connected) covering (E, p), then all connected coverings of B are quotients of (E, p) and there is a bijection between (the isomorphic classes of) them and the subgroups of the automorphism group $\mathsf{Aut}(E, p)$. That bijection is constructed precisely like the standard Galois correspondence for separable field extensions, but in the dual category $(\mathsf{Top}/B)^{\mathsf{op}}$ of bundles over B.

We recall that this result appears in most books on algebraic topology only in the special case when the Chevalley fundamental group $\mathsf{Aut}(E, p)$ is isomorphic to the usual Poincaré fundamental group of B. The general case was first studied by C. Chevalley (see [23], where however the Galois correspondence is not explicitly mentioned) who actually called $\mathsf{Aut}(E, p)$ the Poincaré group.

The work of A. Grothendieck (see [35] and other "SGA") in abstract algebraic geometry is most amazing by the number of important notions he discovered. One of them is the notion of étale covering of a scheme which is a "combination" of a separable algebra and a covering space. And what is called Grothendieck's Galois theory is the result of his idea that one should describe not just the above-mentioned Galois correspondence, but the whole category of coverings of a given space/scheme/field. Omitting the precise definition of étale covering, let us just observe that it generalizes to the context of Grothendieck toposes, and then to elementary toposes. The generalization makes the definition very simple: a morphism $f\colon A \longrightarrow B$ (in a topos with coproducts) with connected B is a covering morphism if

(xiii) (A, f) is locally a coproduct of terminal objects, i.e. there exists an epimorphism $p\colon E \longrightarrow\!\!\!\!\rightarrow B$ such that the pullback functor p^* sends (A, f) to a coproduct of objects of the form $(E, 1_E)$ (although there is a better definition equivalent to this one for "good" toposes).

Speaking of a theory of covering morphisms in a topos, one would refer first of all refer to the locally connected ($=$"molecular") and "locally simply connected" cases described by M. Barr and R. Diaconescu in [7], and then in order to get a reasonably complete list of references for more complicated cases (also for ordinary topological spaces), to the more recent paper [60] of J. Kennison. M. Barr's work (see [5] and [6]) is also to be mentioned here.

Another thing is that since the Grothendieck toposes first of all generalize the categories of sheaves over topological spaces, we have to mention that the following conditions for a sheaf F over a "good" topological space B are equivalent.

(xiv) $F \longrightarrow 1$ is a covering in the topos of sheaves over B.

(xv) The étale space over B corresponding to F is a covering of B.

(xvi) F is a locally constant sheaf.

An important area of investigation in homotopy theory is to study the category of simplicial sets and other categories admitting the Quillen homotopy structure (see [70]) and/or its various modifications. Accordingly the notion of a covering has been introduced for some of them (see [12] and [33]). Most importantly there are coverings of simplicial sets and of groupoids; the definitions are recalled in section A.3 (it is far in the book, but almost no previous material is required to understand those definitions).

As already mentioned in the introduction, the central extensions of groups are not usually considered as a part of Galois theory. However, not to speak of their deep relationship with the covering spaces, we recall that they turn out to be precisely the covering morphisms in a certain "non-Grothendieck" special case of categorical Galois theory, as explained in section 5.2. Moreover, the same is true for the central extensions of Ω-groups as defined by A.S.-T. Lue in [63] (see also A. Fröhlich [30] for the earlier definition for algebras over rings, and [31] for a complete version of the theory).

Now we return to commutative rings. A.R. Magid in [67] develops Grothendieck's Galois theory of commutative rings in full generality, and according to his work the "right notion" now becomes what he calls the componentially locally strongly separable (algebras), i.e. the K-algebras A such that

(xvii) The Pierce representation of A (see section 4.2) is a sheaf of locally strongly separable algebras.

Note that A.R. Magid uses what he calls Galois groupoids instead of Galois groups, and those groupoids have a profinite topology which makes them not equivalent (as topological groupoids) to any kind of a topological family of groups – unlike the ordinary groupoids (although R. Brown has good reasons to say that even ordinary groupoids should never be replaced by families of groups!). Note also that the groupoids appear in Galois theory of commutative rings first in the papers of O.E. Villamayor and D. Zelinsky (see [73] and [74].

Finally, the categorical Galois theory defines a covering morphism in an abstract category with respect to a pair of adjoint functors between that category and another one. Although all the notions considered above are special cases of it, the most important observation which actually was the starting point of the categorical approach (see [36]) is

> the covering morphisms (= "objects split over extensions" –
> see chapters 5 and 6 for the details) with respect to the Pierce
> spectrum functor

$$\mathsf{Sp}\colon \mathsf{Ring}^{\mathsf{op}} \longrightarrow \mathsf{Prof}$$

> and its right adjoint are precisely the componentially locally
> strongly separable algebras in the sense of A.R. Magid.

We would like to conclude that the last generalization (from commutative rings to general categories) simply indicates that the Galois theory has moved from classical algebra to category theory – just as a long time ago it moved from numbers to abstract fields. We know that those who believe that the fundamental theorem of Galois theory is a triviality and the "real" Galois theory begins with the class field theory will not agree with us. And yet, in this book we have tried to support the conclusion above by showing that the fundamental theorem of Galois theory has a purely categorical formulation and a purely categorical proof (as originally shown in various versions in [36], [37], [39], [41]).

An additional remark for category-theorists: three special cases of "Galois theory in variable categories" are mentioned in [48]:

(i) What we called above the categorical Galois theory.
(ii) R.H. Street's general theory of torsors (see [72]). Unfortunately neither the present book nor [48] or [72] describes the long list of investigations on various special types of torsors and their connections with Galois theories (another book is to be written about that!). Let us just say that various interesting articles of M. Bunge, J. Funk, J. Kennison, A. Kock, I. Moerdijk and others are devoted to the torsor approach to the fundamental groups in toposes, and that S.U. Chase and M.E. Sweedler's book (see [22]) devoted to so-called Galois objects (a modified notion of a torsor) had opened a new area of research in Galois theory of rings – on so-called Hopf Galois extensions.
(iii) The A. Joyal–M. Tierney theory in [57] (see section 7.10), which plays an important role in topos theory.

These three cases by no means cover each other, and only the first one really gives the fundamental theorem of Galois theory as an immediate corollary. Note also that [49], whose level of generality is strictly between

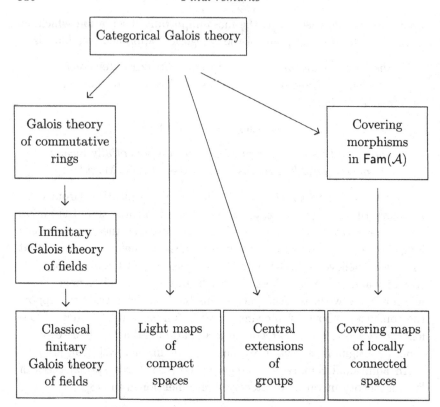

Scheme 2: Levels of generality in Galois theory

[48] and the categorical Galois theory, has nothing to do with [62] (which is related to [22]), but provides an approach to the Tannaka duality different from, for example, P. Deligne [26]. On the other hand [26] itself is in some sense similar to [57].

A.2 Back to the classical Galois theory

Excluding the "non-galoisian" situations studied in chapter 7, scheme 2 on this page presents roughly the levels of generality in Galois theory considered in this book so far.

In this section we will explain that the classical Galois theory is a special case of the theory of covering morphisms in FinFam($\mathcal{A}$), which is the obvious finite version of Fam($\mathcal{A}$). The important conclusion is that the classical theories of separable algebras and covering maps are much more similar to each other than one can conclude from the scheme on page 310 – and, not going into details, let us just mention that we are in fact speaking of a well-known similarity which Grothendieck's theory of étale coverings of schemes is based on.

Let K be a field and C_K the opposite category of finite dimensional commutative K-algebras; that is, a morphism $\alpha\colon A\longrightarrow B$ in C_K is a K-algebra homomorphism $\alpha\colon B\longrightarrow A$, and we assume $\alpha(1) = 1$ as in the previous chapters.

If e is an idempotent in some $A \in C_K$, then

$$Ae = \{ae \mid a \in A\} \subseteq A$$

on the one hand is the ideal in A generated by e, and on the other hand can be considered as an object in C_K with $1 = e$. Note that the map $A\longrightarrow Ae$ defined by $a \mapsto ae$ is a K-algebra homomorphism, although the inclusion $Ae\longrightarrow A$ is not. Also note that

$$A \cong Ae \times A(1-e)$$

as K-algebras, and more generally, given idempotents $e_1,\dots,e_n$ in A with $e_1 + \cdots + e_n = 1$ and $e_i e_j = 0$ for $i \neq j$, we have

$$A \cong \prod_{i=1}^{n} A_{e_i},$$

which becomes

$$A \cong \coprod_{i=1}^{n} Ae_i$$

in C_K. Conversely, any decomposition into a coproduct in C_K is up to isomorphism of this form: indeed, given $A \cong \coprod_{i\in I}^{n} A_i$ we take $e_1,\dots,e_n$ to be the idempotents in A corresponding to $(1,0,\dots,0),\dots,(0,\dots,0,1)$ in $\coprod_{i\in I}^{n} A_i$ which in fact is the cartesian product of the K-algebras $A_1,\dots,A_n$. Of course, all possible non-trivial coproducts in C_K ($=$ products of K-algebras) must be finite since we are dealing here with the finite dimensional algebras.

Since for the decompositions above we have $n \leq \dim [A : K]$, and every non-trivial idempotent in any Ae_i would give a further decomposition, we obtain

Lemma A.2.1 *Every $A \in C_K$ has a finite system $e_1,\dots,e_n$ of idempotents such that*

(i) *each Ae_i has no non-trivial idempotents,*

(ii) $e_1 + \cdots + e_n = 1$,

(iii) $i \neq j \Rightarrow e_i e_j = 0$.   □

Readers familiar with chapter 4 will immediately conclude (using lemma 4.2.5, or directly), that $\{e_1, \ldots, e_n\}$ in lemma A.2.1 is precisely the set of all minimal (non-zero) idempotents in A, and that therefore the boolean algebra of all idempotents in A is the unique (up to isomorphism) boolean algebra with 2^n elements – and that $\{e_1, \ldots, e_n\}$ can be identified with its Stone space. However, our intention in this section is not to examine the finite version of the results of "profinite" arguments of chapter 4, but to show a simple direct way of applying the finite version of the results of section 6.6 to obtain the fundamental theorem of classical Galois theory.

Lemma A.2.2 *If $C \in \mathcal{C}_K$ is a non-zero ring with no non-trivial idempotents, then the functor* $\mathsf{Hom}(C, -) \colon \mathcal{C}_K \longrightarrow \mathsf{Set}$ *preserves finite coproducts.*

Proof Since there is no K-algebra homomorphism from the zero K-algebra to C, we know that $\mathsf{Hom}(C, -)$ preserves the empty coproduct. In order to prove that it preserves the binary coproducts, we have to show that every K-algebra homomorphism $A_1 \times A_2 \longrightarrow C$ (where $\times$ is the cartesian product for K-algebras) factors through one of the projections $A_1 \times A_2 \longrightarrow A_i$, $i = 1, 2$. Since C has no non-trivial idempotents and

- $(1, 0)$ and $(0, 1)$ are idempotents in $A_1 \times A_2$,
- $(1, 0)(0, 1) = (0, 0)$ which is the 0 in $A_1 \times A_2$,
- $(1, 0) + (0, 1) = (1, 1)$ which is the 1 in $A_1 \times A_2$

– we conclude that under the homomorphism above we have either $(1, 0) \mapsto 1$ and $(0, 1) \mapsto 0$, or $(1, 0) \mapsto 0$ and $(0, 1) \mapsto 1$. That is, that homomorphism factors through one of the projections.   □

From lemmas A.2.1 and A.2.2, and the finite version of proposition 6.1.5, we obtain

Theorem A.2.3 *The category $\mathcal{C}_K$ is equivalent to $\mathsf{FinFam}(\mathcal{A}_K)$, where $\mathcal{A}_K$ is the full subcategory in $\mathcal{C}_K$ with objects all $A \in \mathcal{C}_K$ satisfying the following equivalent conditions:*

(i) *A is connected in $\mathcal{C}_K$;*

(ii) *A has no non-trivial idempotents;*

(iii) *A is indecomposable into a non-trivial product as a K-algebra (or, equivalently, as a ring).*   □

This theorem of course suggests identifying $\mathcal{C}_K$ with $\mathsf{FinFam}(\mathcal{A}_K)$ and examining the Galois theory in $\mathcal{C}_K$ corresponding to the adjunction

$$\mathcal{C}_K = \mathsf{FinFam}(\mathcal{A}_K) \xrightleftharpoons{\hspace{1.5em}} \mathsf{FinSet}$$

which is constructed similarly to $\mathsf{Fam}(\mathcal{A}) \xrightleftharpoons{\hspace{1.5em}} \mathsf{Set}$ used in sections 6.2–6.7, but with $\mathcal{A}_K$ instead of the abstract $\mathcal{A}$, and the finite families and finite sets instead of all families and all sets.

Bearing in mind that we intend to apply the Galois theorem only to the field extensions, the first step of the examination is

Observations A.2.4

(i) A morphism $\alpha\colon A \longrightarrow B$ in $\mathcal{C}_K$ with connected B is a trivial covering morphism if and only if (A, α) is a coproduct of a finite number of copies of $(B, 1_B)$ in $\mathcal{C}_K$, and therefore if and only if

$$A \cong B \times \cdots \times B = B^n \quad \text{(for some finite } n)$$

as B-algebras.

(ii) Let $p\colon E \longrightarrow B$ and $\alpha\colon A \longrightarrow B$ be morphisms in $\mathcal{C}_K$ with connected E. Then by (i), and since the pullbacks in $\mathcal{C}_K$ are tensor products of algebras, (A, α) is split by $p\colon E \longrightarrow B$ if and only if

$$E \otimes_B A \cong E \times \cdots \times E$$

as E-algebras.

(iii) Let $p\colon E \longrightarrow B$ be a morphism in $\mathcal{C}_K$, in which E and B are fields. Then by (ii), and since

- p is an effective descent morphism in $\mathcal{C}_K$, which easily follows from corollary 4.4.5 (or can be proved directly using the same arguments),
- therefore p is a morphism of Galois descent if and only if $(E, p) \in \mathsf{Split}_B(p)$ – see the remarks between the proof of proposition 6.6.6 and theorem 6.6.7 –

we conclude that p is a morphism of Galois descent if and only if

$$E \otimes_B E \cong E \times \cdots \times E$$

as E-algebras.

(iv) Let $\alpha\colon A \longrightarrow B$ be a morphism in $\mathcal{C}_K$, in which B is a field. Then, as easily follows from 6.6.3(i), the following conditions are equivalent:

    (a) $\alpha\colon A \longrightarrow B$ is a covering morphism;

    (b) $\alpha\colon A \longrightarrow B$ is split by a field extension $B \subseteq E$ (considered as a morphism $E \longrightarrow B$ in $\mathcal{C}_K$);

    (c) $\alpha\colon A \longrightarrow B$ is split by a Galois field extension.   □

The second step is to translate these algebraic conditions further into the classical language, which only needs picking up the appropriate simple results from chapter 2:

Observations A.2.5

(i) Theorem 2.3.3 together with A.2.4(ii) tells us that if $p\colon E \longrightarrow B$ and $\alpha\colon A \longrightarrow B$ are morphisms in $\mathcal{C}_K$ in which E and B are fields, then (A, α) is split by $p\colon E \longrightarrow B$ if and only if the field extension $B \subseteq E$ splits the B-algebra A in the sense of definition 2.3.1 (here $B \subseteq E$ up to isomorphism of course).

(ii) From (i) and A.2.4(iii) we conclude that if $p\colon E \longrightarrow B$ is a morphism in $\mathcal{C}_K$ correspondidng to a field extension $B \subseteq E$, then the following conditions are equivalent:

p is a morphism of Galois descent;

$(E, p) \in \mathsf{Split}_B(p)$;

$B \subseteq E$ is a (finite dimensional) Galois extension.   □

The third step is to give the following description of algebras split by a Galois extension (which could have been done already in chapter 2):

Proposition A.2.6 *Let $B \subseteq E$ be a finite dimensional Galois field extension, and A a finite dimensional B-algebra. Then the following conditions are equivalent:*

(i) *$B \subseteq E$ splits A;*

(ii) *there exist intermediate field extensions*

$$B \subseteq E_i \subseteq E, \quad i = 1, \ldots, m$$

such that

$$A \cong E_1 \times \cdots \times E_m$$

as B-algebras (in particular $B \in \mathcal{C}_K$ implies $A \in \mathcal{C}_K$).

Proof   It is easy to see that if $A \cong A_1 \times A_2$ then $B \subseteq E$ splits A if and only if it splits A_1 and A_2. This makes (ii) $\Rightarrow$ (i) obvious (just use the definition of Galois extension) and reduces (i) $\Rightarrow$ (ii) to the case of connected A. Since (i) implies that A is a B-subalgebra of $E \otimes_B A \cong E \times \cdots \times E$ (by A.2.5(i) and A.2.4(ii)), the connectedness easily implies $B \subseteq A \subseteq E$ as desired. $\square$

Now we are ready to make the final conclusion:

Conclusion A.2.7 The abstract Galois theorem 5.1.24 applied to the adjunction $\mathsf{FinFam}(\mathcal{A}_K) \overset{\longrightarrow}{\underset{\longleftarrow}{}} \mathsf{FinSet}$ gives the Galois theorem 2.4.3. Note that the isomorphism between the Galois groups used in 5.1.24 (the groupoid from 5.1.24 is of course a group in this case) and 2.4.3 can be deduced from theorem 5.1.24 just as it is done in theorem 6.7.4 for a universal covering. Also note that the explicit construction of the equivalence given in 2.4.3 is in fact "automatic": any category equivalence of the form $\mathcal{D}^{\mathsf{op}} \approx \mathsf{FinSet}^G$, where G is a finite group, can be described as $D \mapsto \mathsf{Hom}(D, L)$, where L is any fixed object in $\mathcal{D}^{\mathsf{op}}$ corresponding to G (acting on itself by the multiplication) under the equivalence. $\square$

Remark A.2.8

(i) In addition to the conclusion above, A.2.4(iv) tells us that the covering morphisms $\alpha \colon A \longrightarrow B$ in $\mathcal{C}_K$ with B a field are precisely the separable B-algebras, i.e. the B-algebras of the form $A_1 \times \cdots \times A_m$, where each A_i is a finite dimensional separable field extension of B. In fact this can also be extended for an arbitrary B, but replacing "separable" by "strongly separable" defined as follows: a B-algebra A is strongly separable if it is projective as B-module and as $(A \otimes_B A)$-module (recall again that all algebras are commutative and with unit).

(ii) The similarity with the covering maps of locally connected topological spaces suggests asking if there are universal covering morphisms in $\mathcal{C}_K$. The answer, obvious for readers familiar with field theory, is that this is not the case unless K is obtained from a separably closed field $\overline{K}$ by taking all elements in $\overline{K}$ fixed under a finite group of automorphisms of $\overline{K}$ (for example as for $K = \mathbb{R}$ and $\overline{K} = \mathbb{C}$). Since the "obvious good candidate" for the universal covering of an arbitrary field K is the separable closure $\overline{K}$ of K, which is always the "union" of all finite dimensional (and also of all) Galois extensions of K, it is reasonable to "add" it to $\mathcal{C}_K$. However, the tensor product $\overline{K} \otimes_K \overline{K}$ will (in general) have an infinite number of idempotents, which will force us to replace FinSet by the category of profinite topological spaces, and then there is no good reason not to work with the category of all commutative rings instead of $\mathcal{C}_K$. This conclusion is also supported by the existence of a good notion of strongly separable algebra (see above) over an arbitrary commutative ring. We will thus arrive at the Galois theory with respect to the adjunction between commutative rings and profinite spaces, i.e. at the situation considered in section 4.5. As shown in [37] (first briefly in [36], see also [19]) this brings us to A.R. Magid's Galois theory [67] with the so-called componentially locally strongly separable algebras as the covering morphisms. In that setting, every object (i.e. every commutative ring) has a "minimal" universal covering (connected when the object is connected), which is the separable closure in the sense of [67].

A.3 Exhibiting some links

No proofs are given in this section; we briefly describe various important links between concrete Galois/covering theories provided by a well-known abstract categorical theorem (see theorem A.3.1 below), and make some remarks on the corresponding viewpoints on the fundamental group.

Let us introduce some notation essentially already used in chapter 7: Given a functor $T\colon \mathcal{S} \longrightarrow \mathcal{A}$ from a small category $\mathcal{S}$ to a category $\mathcal{A}$ with (small Hom-sets and small) colimits, we construct the diagram

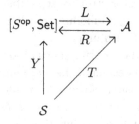

which we will call the fundamental diagram of the functor T, as follows:

- $[\mathcal{S}^{\mathrm{op}}, \mathsf{Set}]$ denotes the category of all functors $\mathcal{S}^{\mathrm{op}} \longrightarrow \mathsf{Set}$, and the functor $Y\colon \mathcal{S} \longrightarrow [\mathcal{S}^{\mathrm{op}}, \mathsf{Set}]$ is the Yoneda embedding, i.e. $Y(s) = \mathrm{Hom}(-, s)$ for each object s in $\mathcal{S}$;

- L is the functor defined by

$$L(F) = \mathsf{colim}\left(\mathsf{Elts}(F) \xrightarrow{\phi_H} S \xrightarrow{T} A\right),$$

where $\mathsf{Elts}(F)$ is the category of elements of F and ϕ_F the forgetful functor (see section 7.8);
- R is the functor defined by

$$R(A) = \mathsf{Hom}\left(T(-), A\right).$$

Theorem A.3.1 *The fundamental diagram above has the following properties:*

(i) $L \dashv R$, *i.e. L is left adjoint of R;*
(ii) $LY \cong T$ *and L is the unique (up to isomorphism) colimit preserving functor with this property.* $\square$

Many special cases of this theorem were already used in this book; let us list (most of) them:

Example A.3.2

(i) For an arbitrary functor $f \colon S \longrightarrow S'$ between small categories take T to be the composite of f and the Yoneda embedding $S' \longrightarrow [S'^{\mathsf{op}}, \mathsf{Set}]$. Then for an object A in $\mathcal{A} = [S'^{\mathsf{op}}, \mathsf{Set}]$ we have

$$R(A)(s) = \mathsf{Nat}\left(\mathsf{Hom}\left(-, f(s)\right), A\right) \cong Af(s),$$

and so the adjunction $L \dashv R$ can be identified with

$$[S^{\mathsf{op}}, \mathsf{Set}] \quad \begin{array}{c} \text{Left Kan extension along } f \\ \xrightarrow{\hspace{4cm}} \\ \xleftarrow{\hspace{4cm}} \\ \text{Composition with } f \end{array} \quad [S'^{\mathsf{op}}, \mathsf{Set}]$$

where we identify f with the dual functor f^{op}, as we already did many times above.
(ii) In the special case $S' = S$, $f = 1_S$, independently of (i) the property A.3.1(ii) tells us that L must be isomorphic to the identity functor, and then the formula defining L shows once again that every functor $S^{\mathsf{op}} \longrightarrow \mathsf{Set}$ is a colimit of representable functors. Moreover, A.3.1(i) then tells us that R must also be isomorphic to the identity functor, which is nothing but the Yoneda lemma.
(iii) The adjunction $L \dashv R$ from (i) was in fact used in the proof of lemma 7.8.2 in the case where f was a geometric morphism of locales.

(iv) If we take $S' = \mathbf{1}$ in (i), then $L \dashv R$ becomes $\mathsf{colim} \dashv \Delta$ from the proof of lemma 7.8.4.

(v) In (iv) we had $T(s) = 1 \in \mathsf{Set}$ for every object s in S. If we take another constant functor $S \longrightarrow \mathsf{Set}$, namely with $T(s) = \pi_0(S)$ ($=$ the set of connected components of S) instead of 1, then $L \dashv R$ becomes $I \dashv H$ from section 6.2 in the case $\mathsf{Fam}(\mathcal{A}) = [S^{\mathrm{op}}, \mathsf{Set}]$.

(vi) If we return to an arbitrary $\mathcal{A}$, but take $S = \mathbf{1}$, then the formulæ for L and R become

$$L(X) = X \cdot T(*), \quad R(A) = \mathsf{Hom}\big(T(*), A\big),$$

where $T(*)$ is the image in $\mathcal{A}$ of the unique object of $\mathbf{1}$ under T, and $X \cdot T(*)$ is the coproduct in $\mathcal{A}$ of the X-indexed family of $T(*)$ (where X is an arbitrary set). When $T(*) = 1$, this gives the adjunction considered in 6.2.2(iv), and therefore generalizes $H \dashv \Gamma$ of section 6.2 (in this case we do not need all colimits in $\mathcal{A}$ of course – just copowers).

(vii) For a moment let us write L_T and R_T instead of L and R respectively; since any $T \colon S \longrightarrow \mathcal{A}$ determines the dual functor $T^{\mathrm{op}} \colon S^{\mathrm{op}} \longrightarrow \mathcal{A}^{\mathrm{op}}$, the whole story above can be dualized, and the fundamental diagram of T^{op} displays as

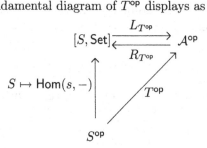

with

$$
\begin{aligned}
L_{T^{\mathrm{op}}}(F) &= \mathsf{lim}\big(\mathsf{Elts}(F) \longrightarrow S \xrightarrow{\ T\ } \mathcal{A}\big), \\
R_{T^{\mathrm{op}}}(A) &= \mathsf{Hom}\big(A, T(-)\big)
\end{aligned}
$$

(with the appropriate definition of $\mathsf{Elts}(F)$ and the forgetful functor from it to S). In particular the finite version of this adjunction (in a very special situation) turns out to be a category equivalence in the Galois theorem 2.4.3 – also see the last remark in A.2.7. $\square$

Example A.3.3 Let B be a topological space, S the locale $\mathcal{O}(B)$ of open subsets in B, considered as a category, and

$$T \colon S \longrightarrow \mathsf{Top}/B$$

the functor defined by $T(U) = (U, i_U)$, where $i_U \colon U \longrightarrow B$ is the inclusion map.

(i) $[S^{\mathrm{op}}, \mathsf{Set}]$ is the (ordinary) category of presheaves over the space B.

(ii) For $F \in [S^{\mathrm{op}}, \mathsf{Set}]$ and $L(F) = (A, \alpha)$, the space A can be identified with the space of germs $[U, b, x]$ of F with $\alpha \colon [U, b, x] \mapsto b$. Recall that such a germ is the equivalence class of a triple (U, b, x) in which U is an open subset of B, $b \in U$, and $x \in F(U)$, where (U, b, x) and (U', b', x') are said to be equivalent when $b = b'$ and there exists an open $V \subseteq U \cap U'$ with $b \in V$ and $x|_V = x'|_V$ – denoting by $x|_V$ the image of x under the map $F(U) \longrightarrow F(V)$. Briefly

$$A = \operatorname*{colim}_{(U,b,x)} U,$$

the (filtered!) colimit of U over all (U, b, x), and this also determines the topology in A as the colimit topology; readers of course noticed that the triples (U, b, x) are nothing but the objects of $\mathsf{Elts}(F)$ and that $U = \phi_F(U, b, x)$.

(iii) For $(A, \alpha) \in \mathsf{Top}/B$, $R(A, \alpha)$ is what is called the sheaf of local sections (or cross-sections) of (A, α) since

$$R(A, \alpha)(U) = \mathsf{Hom}\big(T(U), (A, \alpha)\big)$$
$$= \mathsf{Hom}\big((U, i_U), (A, \alpha)\big)$$
$$= \big\{u \colon U \longrightarrow A \text{ in } \mathsf{Top} \mid \alpha(b) = b \text{ for each } b \in U\big\}.$$

(iv) The adjunction $L \dashv R$ induces precisely the equivalence between the category $\mathsf{Sh}(B)$ of sheaves over B and the category $\mathsf{Et}(B)$, which is the full subcategory in Top/B with objects all (A, α) with étale α. This can be displayed as

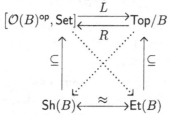

and means that a presheaf F is a sheaf if and only if the canonical morphism $F \longrightarrow RL(F)$ is an isomorphism, and a continuous map $\alpha \colon A \longrightarrow B$ is étale if and only if the canonical map $LR(A, \alpha) \longrightarrow (A, \alpha)$ is an isomorphism.

(v) By (iv), the two inclusion functors $\mathsf{Sh}(B) \longrightarrow \left[\mathcal{O}(B)^{\mathrm{op}}, \mathsf{Set}\right]$ and $\mathsf{Et}(B) \longrightarrow \mathsf{Top}/B$ have a left and a right adjoint respectively. The left adjoint of $\mathsf{Sh}(B) \longrightarrow \left[\mathcal{O}(B)^{\mathrm{op}}, \mathsf{Set}\right]$ is the associated sheaf functor, and it is of course isomorphic to the one constructed in the proof of theorem 7.7.3 when the locale involved is $\mathcal{O}(B)$. $\quad\square$

Here is another "famous" example playing a central role in homotopy theory, just as the previous one does in sheaf theory:

Example A.3.4 First we observe that since a finite dimensional vector space $\mathbb{R}^n$ can be considered as the coproduct of n copies of $\mathbb{R}$, there is a well-defined functor $\mathbb{R}^{(-)}$ from the category of finite sets to the category of real vector spaces. The set

$$\Delta = \left\{ (x_1, \dots, x_n) \in \mathbb{R}^{n+1} \middle| x_i \geq 0 \text{ for all } i, \text{ and } x_1 + \cdots + x_{n+1} = 1 \right\}$$

is called the standard n dimensional simplex; it is an n dimensional simplex which is "chosen nicely": in particular for every map

$$f \colon \{1, \dots, n+1\} \longrightarrow \{1, \dots, m+1\}$$

the map

$$\mathbb{R}^{n+1} = \mathbb{R}^{\{1, \dots, n+1\}} \xrightarrow{\ \mathbb{R}^f\ } \mathbb{R}^{\{1, \dots, m+1\}} = \mathbb{R}^{m+1}$$

induces a map $\Delta_f \colon \Delta_n \longrightarrow \Delta_m$. Moreover, since $\mathbb{R}$ is a topological field, its topology makes each Δ_n a topological space and each Δ_f a continuous map.

Now we take S to be the simplicial category Δ whose objects are all (non empty) sets of the form $[n] = \{1, \dots, n+1\}$ and morphisms all maps f between them satisfying

$$i \leq j \Rightarrow f(i) \leq f(j),$$

and $T \colon \Delta \longrightarrow \mathsf{Top}$ to be the functor defined by $T([n]) = \Delta_n$, $T(f) = \Delta_f$ in the notation above. The corresponding adjunction $L \dashv R$ according to the standard terminology has

- $[\Delta^{\mathrm{op}}, \mathsf{Set}] = \mathsf{SimplSet}$, the category of simplicial sets,
- $L = |-| \colon \mathsf{SimplSet} \longrightarrow \mathsf{Top}$, the geometric realization functor,
- $R \colon \mathsf{Top} \longrightarrow \mathsf{SimplSet}$ the "singular complex" or "nerve" functor.

Note that the adjunction $L \dashv R$ here does not induce an equivalence between the "images" of L and R as it did in example A.3.3. However

- it induces an equivalence between certain categories built up from Top and SimplSet, and the resulting category is the main object of investigation in homotopy theory,
- the image of L is still important: it produces the notion of CW-complex, which is a "homotopy-theoretic candidate" for the notion of a "good" space. □

Replacing spaces by small categories we obtain

Example A.3.5 Regarding the objects of Δ as small categories (each $[n]$ being an ordered set is a category!) and the morphisms as functors, we take $T: \Delta \longrightarrow \mathsf{Cat}$ to be the inclusion of Δ into the category Cat of all small categories. The corresponding adjunction $L \dashv R$ has R full and faithful, and so induces an equivalence between Cat and a full subcategory in SimplSet (which can be described as the "image" of R). The functor R is again called the nerve. Note that for a (small) category A we have

- $R(A)([0])$ is the set of objects in A,
- $R(A)([1])$ is the set of morphisms in A,
- $R(A)([2])$ is the set of composable pairs, and, more generally, $R(A)([n])$, $n \geq 2$ is the set of composable sequences

$$\bullet \xrightarrow{a_1} \bullet \xrightarrow{a_2} \bullet \cdots \bullet \xrightarrow{a_{n-1}} \bullet \xrightarrow{a_n} \bullet$$

of morphisms in A. □

Omitting many similar examples, let us consider only groupoids:

Example A.3.6 Consider the commutative (up to isomorphism) diagram A.1: in which

- Grpd is the category of small groupoids,
- the fraction groupoid functor is the left adjoint of the inclusion functor $\mathsf{Grpd} \longrightarrow \mathsf{Cat}$ – it is the universal construction making all morphisms of a given category invertible –
- the functor "codiscrete" sends a set X to the codiscrete groupoid on X, i.e. to the groupoid whose objects are all the elements of X, with $\mathrm{Hom}(x, y)$ having exactly one element for every x and every y in X.

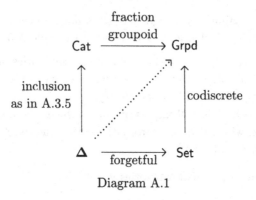

Diagram A.1

We take $T: \mathbf{\Delta} \longrightarrow \mathsf{Grpd}$ to be the resulting dotted arrow, i.e. any of the two isomorphic composites above. The corresponding adjunction $L \dashv R$ again has R full and faithful, and it is again called the nerve functor; for L the standard notation is

$$L(X) = \pi_1(X), \text{ or } \Pi_1(X),$$

and that groupoid is called the fundamental groupoid of the simplicial set X.   □

Among the links between Galois/covering theories provided by the examples above are

A.3.7 Let us first apply the results of sections 6.6 and 6.7 to the case $\mathcal{C} = \mathsf{Set}^G$, the category of G-sets, where G is a group; and let us consider only the case $B = 1$. It is easy to prove the following.

(i) A morphism $A \longrightarrow 1$ is a trivial covering morphism if and only if the action of G on A is trivial, i.e. $ga = a$ for all $g \in G$ and $a \in A$.

(ii) Every morphism $A \longrightarrow 1$ is a covering split by $G \longrightarrow 1$ (where G is considered as an object in Set^G with the G-action via the multiplication in G – as already used in the proof of theorem 6.7.4). In particular $G \longrightarrow 1$ is the unique (up to isomorphism) connected universal covering of 1.

(iii) Applying theorems 6.6.7 and 6.7.4 to $G \longrightarrow 1$ we obtain the trivial equivalence $\mathsf{Set}^G \approx \mathsf{Set}^G$.

Now let us compare this with the same theorems for an abstract $\mathcal{C}$ via the adjunction

$$\mathsf{Set}^G = [G, \mathsf{Set}] \underset{R}{\overset{L}{\underset{\longleftarrow}{\longrightarrow}}} \left((\mathcal{C}/B)^{\mathsf{op}}\right)^{\mathsf{op}} = \mathcal{C}/B$$

with $G = \mathsf{Aut}(p)^{\mathsf{op}}$ $\left(\approx \mathsf{Aut}(p)\right)$ obtained from the inclusion functor $\mathsf{Aut}(p)^{\mathsf{op}} \longrightarrow (\mathcal{C}/B)^{\mathsf{op}}$ as in A.3.2(vii); note that L and R here play the roles of $L_{T^{\mathsf{op}}}$ and $R_{T^{\mathsf{op}}}$ of A.3.2(vii). We observe the following.

- It is easy to deduce from the results of sections 6.6 and 6.7 that R induces an equivalence between the category of coverings of B in $\mathcal{C}$ and the category of coverings of 1 in Set^G itself (the finite version of this again agrees with the Galois theorem 2.4.3!).

- An independent investigation of the adjunction $L \dashv R$ could be called the "external Galois theory" in contrast with the "internal" one developed in chapter 5. Readers can judge now to what extent these two approaches agree. □

A.3.8 The equivalence $\mathsf{Sh}(B) \approx \mathsf{Et}(B)$ described in example A.3.3 tells us that $\mathsf{Et}(B)$ is a topos for every topological space B, and in particular so is LoCo/B for a locally connected B – which we could use to replace various topological arguments by the topos-theoretic ones; for example the fact that every epimorphism is an effective descent morphism holds in any topos. We could also use sheaf theory itself, expressing all constructions in the language of sheaves instead of étale maps; in particular the trivial covering maps correspond to the constant sheaves, and the covering maps to the locally constant sheaves. A good example where the competition between the two languages really occurs is the construction of the (connected) universal covering of a "good" space, say a CW-complex. Such a universal covering $p \colon E \longrightarrow B$ has E the set of equivalence classes of all paths f in B with $f(0) = b_0$ for a fixed $b_0 \in B$, and p is defined by $p([f]) = f(1)$ – but for the topology on E there is a choice between the following two definitions.

(i) The set of all paths f in B with $f(0) = b_0$ being a subset of all paths in B has the compact-open topology, and then we take the quotient topology on E.

(ii) Let $\mathcal{U}$ be the set of all open subsets U in B in which every path f with $f(0) = f(1)$ is equivalent in B to a constant path (if B were a CW-complex, or more generally locally simply connected and locally path-connected and connected, then we could simply take the set of all open simply connected subsets; however, the algebraic topologists prefer a still more general situation of semi-locally simply connected instead of locally simply connected). For $U \in \mathcal{U}$ we take $F(U)$ to be the set of all equivalence classes $[f]$ of paths f in B with $f(1) \in U$. It can be shown that the collection of all $F(U)$ produces a sheaf on B whose étale space is the E above.

Textbooks in algebraic topology usually give sufficient information to make the proof of equivalence of the two definitions an easy exercise; they also manage to avoid the words "sheaf" and "étale" in the second one.   □

A.3.9 As shown in Appendix One of the book [33] of P. Gabriel and M. Zisman (which we certainly recommend our readers to look at!) the functors L and R from example A.3.4 as well as those from example A.3.6 send covering morphisms to covering morphisms. Moreover, it is shown there that for every simplicial set B the induced adjunctions,

$$\mathsf{SimplSet}/B \rightleftarrows \mathsf{Grpd}/\Pi_1(B),$$

$$\mathsf{SimplSet}/B \rightleftarrows \mathsf{Top}/|B|$$

themselves induce equivalences between the categories of coverings: of B and of its fundamental groupoid $\Pi_1(B)$ in the first case, and of B and its geometric realization $|B|$ in the second case. The resulting equivalence between the coverings of $|B|$ in Top and the coverings of $\Pi_1(B)$ in Grpd easily gives an independent proof of the classification theorem of coverings of $|B|$ in terms of the Poincaré fundamental group (in the connected case, to which the general case easily reduces since $|B|$ is localy connected – but also in the general case with the Poincaré groupoid equivalent to $\Pi_1(B)$). In fact a more general classification theorem is proved in [33]. For those who prefer more geometrical language we would recommend R. Brown's book [12].

Let us also observe:

(i) A slightly different construction of what we call T in example A.3.4 is used in [33] (and also in [65]).

(ii) According to [33] a covering morphism $\alpha \colon A \longrightarrow B$ in Grpd can be defined as a functor α from the groupoid A to the groupoid B satisfying the following condition:

for every morphism $f: b \longrightarrow b'$ in B and every object
a in A with $\alpha(a) = b$ there exists a unique morphism
$g: a' \longrightarrow a$ in A with $\alpha(g) = f$.

This definition on the one hand clearly imitates the unique path-lifting property for topological spaces, and on the other hand tells us that α determines a functor

$$F_\alpha: B^{\mathrm{op}} \longrightarrow \mathsf{Set}$$

with $F_\alpha(b) = \alpha^{-1}(b)$. In fact $(-)^{\mathrm{op}}$ is irrelevant here since $B^{\mathrm{op}} \approx B$ and $\alpha: A \longrightarrow B$ satisfies the condition above if and only if $\alpha^{\mathrm{op}}: A^{\mathrm{op}} \longrightarrow B^{\mathrm{op}}$ does so (in the modern language, a functor between groupoids is a discrete fibration if and only if it is a discrete opfibration). This passage $\alpha \mapsto F_\alpha$ suggests that the fundamental groupoid ($=$ the "largest" Galois groupoid classifying all coverings) of B must be B itself, or equivalent to it. This can be shown independently, but we also need to prove that the definition above of a covering morphism of groupoids does agree with the general one. The proof is similar to the same proof for simplicial sets sketched in (iii) below, but uses in addition the description of effective descent morphisms in Grpd: they turn out to be the functors surjective on composable triples of morphisms. For the simplicial sets, on the one hand that description is automatic since they form a topos, and on the other hand it helps to obtain the result in Grpd.

(iii) Again imitating the geometrical unique lifting property, P. Gabriel and M. Zisman define the covering morphisms of simplicial sets via the liftings for the representable simplicial sets, i.e. those of the form

$$\mathsf{Hom}(-, [n]): \Delta \longrightarrow \mathsf{Set},$$

but how does this definition agree with definition 6.5.9? The answer is beautiful:

(a) Let X be the set of all morphisms from all $\mathsf{Hom}(-, [n])$ to B; by the Yoneda lemma it can be identified with the set of all elements of B (i.e. of all $B([n])$) and there is a canonical epimorphism

$$p: X \cdot \mathsf{Hom}(-, [n]) \longrightarrow\!\!\!\!\!\rightarrow B,$$

which is an effective descent morphism since $\mathsf{SimplSet}$ is a topos.

(b) It is easy to see (and it is "almost" mentioned in [33] – see the first proposition in [33], Appendix 1, 2.2) that a morphism $\alpha\colon A \longrightarrow B$ satisfies the "lifting definition" in [33] if and only if it is split by the morphism p from (a).

(c) The splitting by p above is equivalent to being a covering, as follows from 6.6.3(i) and the fact that $X \cdot \mathsf{Hom}\big(-,[n]\big)$ is a projective object in SimplSet. ☐

Remark A.3.10 The reformuation of the definition of a covering in terms of "unique lifting" is also useful: it says that $\alpha\colon A \longrightarrow B$ is a covering morphism (in the sense of definition 6.5.9, when A and B are connected) if and only if there exists an effective descent morphism $p\colon E \longrightarrow B$ such that for every commutative diagram

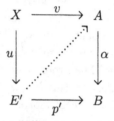

of solid arrows, where $p'\colon E' \longrightarrow B$ is a connected component of the morphism $p\colon E \longrightarrow B$, there exists a unique dotted arrow as above making the whole diagram commute.

The reader can once again return to finite dimensional separable field extensions and see that this lifting property (in the FinFam context) gives one of the classical definitions.

The geometrical unique lifting property used in section 6.8 is "less categorical": it also can be reformulated as a condition on a certain diagram being a pullback, but just in Set – not in Top. ☐

It is time now to ask J. Kennison's question (see [60]): *What is the fundamental group?*

The most general answer in the context of section 5.1 could be that the "fundamental groupoid" of an object B in $\mathcal{C}$ is the internal "progroupoid" in $\mathcal{X}$ formed by the Galois descent morphisms. However, this definition would only make good sense if there were enough Galois descent morphisms to split "all coverings", or otherwise we would be forced to replace groupoids by certain precategories as in chapter 7. Yet, under certain additional conditions on $\mathcal{X}$ (which hold when $\mathcal{X} = $ Set), the

precategories can be themselves replaced by their fundamental groupoids
defined now as for the simplicial sets in example A.3.6. And having any
notion of a fundamental groupoid, one can define the fundamental group
$\pi_1(B, b)$ for b being an object in the fundamental groupoid of B (i.e. a
morphism from 1 to the object of objects in the internal context), as the
internal automorphism group of b.

Omitting various intermediate levels of generality with many interesting examples, some of which are mentioned in [60], let us consider the
case where $\mathcal{C} = \mathsf{Fam}(\mathcal{A})$ and B is connected and has a universal covering
as in section 6.7. In this case we have the following.

- The fundamental groupoid of B is to be defined as the Galois
  groupoid of any universal covering morphism $p \colon E \longrightarrow B$, and
  since $\mathsf{Set}^{\mathsf{Gal}\,[p]} \approx \mathsf{Cov}(B)$ for each such p, the fundamental groupoid
  is then uniquely determined up to equivalence.

- The fundamental group of B is to be defined either as the automorphism group of any object of the fundamental groupoid, or as
  $\mathsf{Gal}\,[p]$ for a connected universal covering (E, p) of B. It is then
  determined uniquely up to isomorphism.

- As shown in section 6.7, the group $\mathsf{Gal}\,[p]$, for a connected universal
  covering (E, p) of B (which itself is determined uniquely up to
  isomorphism), is isomorphic to $\mathsf{Aut}(p)$ which is what we call the
  Chevalley fundamental group. However, depending on the way
  $p \colon E \longrightarrow B$ was constructed, it might be even easier to calculate
  $\mathsf{Gal}\,[p]$ using just its definition, i.e. as

$$
I(E \times_B E \times_B E) \xrightarrow{\hspace{2cm}} I(E \times_B E) \xleftarrow{\hspace{1cm}} I(E),
$$

  which since $\mathsf{Gal}\,[p]$ is a group now is convenient to write simply as
  $I(E \times_B E)$ (topologists would write $\pi_0(E \times_B E)$, or even $\pi_0(\tilde{B} \times_B \tilde{B})$ since $\tilde{B} \longrightarrow B$ seems to be a standard way to denote (the)
  universal covering of B). For instance, using the construction of
  $p \colon E \longrightarrow B$ via A.3.8(ii) (when $\mathcal{C} = \mathsf{LoCo}$ and B is a 'good' space)
  one arrives at the Poincaré fundamental group as $I(E \times_B E)$, and
  the same can be done via A.3.9.

- Another approach, which actually also works for 'less good' spaces,
  is to use proposition 6.4.2, which tells us that the surjective étale
  maps $\coprod_{\lambda \in \Lambda} U_\lambda \longrightarrow B$, obtained from the families $(U_\lambda)_{\lambda \in \Lambda}$ of open
  $U_\lambda \subseteq B$ whose union is B, are enough to split all coverings of B in
  LoCo. Using as above mentioned the corresponding precategories

and their fundamental groupoids, it can be shown that when each U_λ has no non-trivial coverings, the (Chevalley) fundamental group of B is the fundamental group(oid) (in the sense of A.3.6) of the simplicial set S defined by

$$S([n]) = I\left(\underbrace{\left(\coprod_{\lambda \in \Lambda} U\lambda\right) \times_B \cdots \times_B \left(\coprod_{\lambda \in \Lambda} U\lambda\right)}_{n + 1 \text{ times}}\right).$$

There are also similar constructions without the existence of universal coverings involving progroups which occur in M. Artin and B. Masur's book (see [2]) and later in a topos-theoretic context (see again [60] and references there) – not to mention A. Grothendieck's original ideas on descent and coverings.

A.4 A short summary of further results and developments

We list now some works closely related to the present book. Most of them could have constituted an additional chapter. We invite the reader to consult them.

A.4.1 As shown in A.2, (the Grothendieck form of) the fundamental theorem of classical Galois theory not only can be considered as a special case of the categorical one, but can actually be deduced from it. For Magid's Galois theory of commutative rings (see [67]) the same is done in [37]; also see [16] and [17] for the further categorical simplifications in two directions. Moreover, as shown in [19], the categorical approach helps to improve the description of Galois correspondence for commutative ring extensions with profinite Galois groupoids.

A.4.2 Reference [38] answers one of the questions kindly suggested by S. Mac Lane. And it turns out that not the Picard–Vessiot extensions themselves, but their subalgebras generated by a fundamental system of solutions of a given linear differential equation and the inverse of the corresponding Wronski determinant, occur in the Galois theory of the adjunction

$$\left(\begin{array}{c} \text{Differential} \\ \text{commutative rings} \end{array}\right)^{op} \xrightarrow[\supseteq]{\text{constants}} \left(\begin{array}{c} \text{Commutative} \\ \text{rings} \end{array}\right)^{op}.$$

A.4.3 As the title of [40] shows, it answers a question of R. Brown. The double central extensions are defined as the covering morphisms with respect to a Galois structure (= relatively admissible adjunction in the sense of section 5.1) involving

$$
\left(\begin{array}{c} \text{Group} \\ \text{extensions} \end{array}\right) \begin{array}{c} \xrightarrow{\text{``centralization''}} \\ \xleftarrow{\quad\supseteq\quad} \end{array} \left(\begin{array}{c} \text{Central group} \\ \text{extensions} \end{array}\right).
$$

This further extends to higher dimensions, and using the Brown–Ellis–Hopf formula from [13] presents the homology groups of groups as certain fundamental groups (see [43]).

A.4.4 The level of generality in the Barr–Diaconescu covering theory (see [7]) is stricly between the level considered in chapter 6 above and the level of topological spaces; this follows from the results of [42].

A.4.5 The theory of central extensions of universal algebras (and more generally, of objects in a Barr exact category) which on the one hand is a special case of the Galois theory, and on the other hand contains the case of Ω-groups studied by A. Fröhlich's school (the original definition is due to A.S.-T.Lue in [63]; also see J. Furtado-Coelho in [31] and references there), and in particular the ordinary central extensions of groups, is developed in [44]. A further comparison with the universal-algebraic notion of a centre is carried out in [46].

A.4.6 As already mentioned earlier, chapter 5 above should provide a good help for readers not familiar with category theory to understand [18], where the study of the relationship between the Galois theory and factorization systems (on the general categorical level) begins. Further results for the cases of "less well-behaved" coverings are obtained in [45], [51], [52], [53]. For example, as follows from the results of [45], the (purely inseparable, separable)-factorization for the finite dimensional field extensions extends to the category of commutative rings (although the class of homomorphisms imitating the purely inseparable extensions is yet to be investigated).

A.4.7 A categorical version of a so-called tautological proof of the van Kampen theorem is described in [14].

A.4.8 A theory of coverings with all Galois groupoids being (internal) equivalence relations, containing the case of light maps of compact spaces, the case of ring homomorphisms with semisimple kernels, and some others, is developed in [47].

A.4.9 The Galois theory of the adjunction

$$\left(\begin{array}{c} \text{Simplicial sets,} \\ \text{Kan fibrations} \end{array} \right) \xrightarrow[\text{nerve}]{\pi_1} \left(\begin{array}{c} \text{Groupoids,} \\ \text{fibrations} \end{array} \right),$$

which produces a new notion of second order covering map of simplicial sets, is studied in [15].

A.4.10 The approach to Galois theory in various "boolean" cases developed by Y. Diers (see [25]) can be deduced from the categorical one (described in chapter 5 above), as follows from results of [17].

A.4.11 The Galois theory in symmetric monoidal categories (see [49]) has a level of generality strictly between [48] and [41]; its main prupose is to provide a simplified categorical framework for the so-called Tannaka duality.

A.4.12 We would like to mention two Ph.D. theses: of B. Mesablishvili on Galois theory of commutative rings in toposes (see [68]), where in particular the relationship of the results of S.U. Chase and M.E. Sweedler (see [22]), of M. Barr (see [5], [6]) and of Th. Ligon (see [62]) was investigated; and of M. Gran (see [34]) who described the central extensions with respect to various adjunctions involving internal categories in so-called Mal'tsev categories.

Bibliography

[1] **J. Adamek and J. Rosicky**, *Locally presentable and accessible categories*, London Math. Soc. Lect. Notes **189**, Cambridge University Press, 1994

[2] **M. Artin and B. Mazur**, *Etale homotopy*, Springer Lect. Notes in Math. **100**, 1969

[3] **M. Auslander and O. Goldman**, The Brauer group of a commutative ring, *Trans. Amer. Math. Soc.* **97**, 1960, 367–409

[4] **M. Barr**, *Exact categories*, Springer Lect. Notes in Math. **236**, 1971, 1–120

[5] **M. Barr**, Abstract Galois theory, *J. of Pure and Applied Algebra* **19**, 1980, 21–42

[6] **M. Barr**, Abstract Galois theory II, *J of Pure and Applied Algebra* **25**, 1982, 227–247

[7] **M. Barr and R. Diaconescu**, On locally simply connected toposes and their fundamental groups, *Cahiers de Topologie et Géométrie Différentielle Catégorique* **XXII-3**, 1981, 301–314

[8] **F. Borceux**, *Handbook of categorical algebra*, 3 volumes, Cambridge University Press, 1994

[9] **N. Bourbaki**, *Algèbre*, chapitre IV, Masson, Paris, 1981

[10] **N. Bourbaki**, *Topologie générale*, Ch. 2, Hermann, Paris, 1971

[11] **D. Bourn**, The shift functor and the comprehensive factorization, *Cahiers de Topologie et Géométrie Différentielle Catégorique* **XXVIII-3**, 1987, 197–226,

[12] **R. Brown**, *Topology: a geometric account of general topology, homotopy types, and the fundamental groupoid*, Wiley, 1988

[13] **R. Brown and G. Ellis**, Hopf formulæ for the higher homology of a group, *Bull. London Math. Soc.* **20**, 1988, 124–128

[14] **R. Brown and G. Janelidze**, Van Kampen theorems for categories of covering morphisms in lextensive categories, *J. of Pure and Applied Algebra* **119**, 1997, 255–263

[15] **R. Brown and G. Janelidze**, Galois theory of second order covering maps of simplicial sets, *J. of Pure and Applied Algebra* **135**, 1999, 23–31

[16] **A. Carboni and G. Janelidze**, Decidable (=separable) objects and morphisms in lextensive categories, *J. of Pure and Applied Algebra* **110**, 1996, 219–240

[17] **A. Carboni and G. Janelidze**, Boolean Galois theories, in preparation

[18] **A. Carboni, G. Janelidze, G.M. Kelly, and R. Paré**, On localization and stabilization of factorization systems, *Applied Categorical Structures* **5**, 1997, 1–58

[19] **A. Carboni, G. Janelidze, and A.R. Magid**, A note on Galois correspondence for commutative rings, *J. of Algebra* **183**, 1996, 266–272

[20] **C. Cassidi, M. Hébert, and G.M. Kelly**, Reflective subcategories, localizations, and factorization systems, *J. of Australian Math. Soc.*, Ser. A **38**, 1985, 287–329

[21] **S.U. Chase, D.K. Harrison, and A. Rosenberg**, *Galois theory and cohomology of commutative rings*, Mem. Amer. Math. Soc. **52**, 1965

[22] **S.U. Chase and M.E. Sweedler**, *Hopf algebras and Galois theory*, Springer Lect. Notes in Math. **97**, 1969

[23] **C. Chevalley**, *Theory of Lie groups*, Princeton University Press, 1946

[24] **F. DeMeyer and E. Ingraham**, *Separable algebras over commutative rings*, Springer Lect. Notes in Math. **181**, 1971

[25] **Y. Diers**, *Categories of boolean sheaves of simple algebras*, Springer Lect. Notes in Math. **1187**, 1986

[26] **P. Deligne**, *Catégories tannakiennes*, "Grothendieck Festchrift", 2, Birkhäuser 1990, 111–195

[27] **R. et A. Douady**, *Algèbre et théories galoisiennes*, Fernand Nathan, 1977

[28] **S. Eilenberg**, Sur les transformations continues d'espaces métriques compacts, *Fund. Math.* **22**, 1934, 292–296

[29] **P. Freyd**, *Abelian categories*, Harper and Row, 1964

[30] **A. Fröhlich**, Baer-invariants of algebras, *Trans. Amer. Math. Soc.* **109**, 1963, 221–244.

[31] J. Furtado-Coelho, *Varieties of Ω-groups and associated functors,* Ph.D. Thesis, University of London, 1972

[32] P. Gabriel und F. Ulmer, *Lokal präsentierbare Kategorien,* Spinger Lect. Notes in Math. **221,** 1971

[33] P. Gabriel and M. Zisman, *Calculus of fractions and homotopy theory,* Springer, 1967

[34] M. Gran, *Central extensions for internal groupoids in Maltsev categories,* Ph.D. Thesis, Université catholique de Louvain, 1999

[35] A. Grothendieck, *Revêtements étales et groupe fondamental,* SGA1, exposé V, Springer Lect. Notes in Math. **224,** 1971

[36] G. Janelidze, Magid's theorem in categories, *Bull. Georgian Acad. Sci.* **114,** 3, 1984, 497–500 (in Russian)

[37] G. Janelidze, The fundamental theorem of Galois theory, *Math. USSR Sbornik* **64 (2),** 1989, 359–374

[38] G.Janelidze, *Galois theory in categories: the new example of differential fields,* Proc. Conf. Categorical Topology in Prague 1988, World Scientific, 1989, 369–380

[39] G. Janelidze, Pure Galois theory in categories, *J. of Algebra* **132,** 1990, 270–286

[40] G. Janelidze, What is a double central extension? (the question was asked by Ronald Brown), *Cahiers de Topologie et Géometrie Différentielle Catégorique* **XXXII-3,** 1991, 191–202

[41] G. Janelidze, *Precategories and Galois theory,* Springer Lect. Notes in Math. **1488,** 1991, 157–173

[42] G. Janelidze, A note on Barr–Diaconescu covering theory, *Contemp. Math.* **131,** 3, 1992, 121–124

[43] G. Janelidze, *Higher dimensional central extensions and the Brown–Ellis–Hopf formula,* International Meeting in Category Theory, Halifax (New Scotia) 1995, unpublished

[44] G. Janelidze and G.M. Kelly, Galois theory and a general notion of central extension, *J. of Pure and Applied Algebra* **97,** 1994, 135–161

[45] G. Janelidze and G.M. Kelly, The reflectiveness of covering morphisms in algebra and geometry, *Theory and Applications of Categories* **3,** 1997, 132–159

[46] G. Janelidze and G.M. Kelly, Central extensions in universal algebra: a unification of three notions, *Algebra Universalis,* to appear

[47] G. Janelidze, L. Marki, and W. Tholen, Locally semisimple coverings, *J. of Pure and Applied Algebra* **128,** 1998, 281–289

[48] **G. Janelidze, D. Schumacher, R. Street**, Galois theory in variable categories, *Applied Categorical Structures* 1, 1993, 103–110

[49] **G. Janelidze and R.H. Street**, Galois theory in symmetric monoidal categories, *J. of Algebra 220*, 1999, 174-187

[50] **G. Janelidze and W. Tholen**, Facets of descent, I, *Applied Categorical Structures* 2, 1994, 245–281

[51] **G. Janelidze and W. Tholen**, Extended Galois theory and dissonant morphisms, *J. of Pure and Applied Algebra*, to appear

[52] **G. Janelidze and W. Tholen**, Functorial factorization, well-pointedness and separability, *J. of Pure and Applied Algebra*, to appear

[53] **G. Janelidze and W. Tholen**, Strongly separable morphisms, in preparation

[54] **G.J. Janusz**, Separable algebras over commutative rings, *Trans. Amer. Math. Soc.* **122**, 1966, 461–479

[55] **P. Johnstone**, *Topos theory*, Academic Press, 1977

[56] **P. Johnstone**, Factorization theorems for geometric morphisms, I, *Cahiers de Topologie et Géométrie Différentielle*, **XVII-1**, 1981, 3–17

[57] **A. Joyal and M. Tierney**, *An extension of the Galois theory of Grothendieck*, Mem. Amer. Math. Soc. **51**, 309, 1984

[58] **Kelley**, *General topology*, Springer, 1971 (from Van Nostrand, 1955)

[59] **G.M. Kelly, R. Street**, *Review of the elements of 2-categories*, Springer Lect. Notes in Math. **420**, 1974, 75–103

[60] **J. Kennison**, What is the fundamental group? *J. of Pure and Applied Algebra* **59**, 1989, 187–200

[61] **A.G. Kuroš**, *Theory of groups*, 2 volumes, Chelsea Publishing Co. 1956

[62] **Th.S. Ligon**, *Galois-Theorie in monoidalen Kategorien*, Algebra Berichten, 1978

[63] **A.S.-T.Lue**, Baer-invariants and extensions relative to a variety, *Proc. Cambridge Philos. Soc.* **63**, 1967, 569–578

[64] **S. Mac Lane**, *Homology*, Springer, 1963

[65] **S. Mac Lane, I. Moerdijk**, *Sheaves in geometry and logic*, Universitext, Springer, 1992

[66] **S. Mac Lane**, *Categories for the working mathematician*, Second Ed., Springer, 1998

[67] **A.R. Magid**, *The separable Galois theory of commutative rings*, Marcel Dekker, 1974

[68] **B. Mesablishvili**, *Galois theory of connected commutative rings in topoi*, Ph.D. Thesis, University of Tbilisi, 1998

[69] **J. Milnor**, *Introduction to algebraic K-theory*, Princeton University Press, 1971

[70] **D. Quillen**, *Homotopical algebra*, Springer Lect. Notes in Math. **43**, 1967

[71] **I. Stewart**, *Galois theory*, Chapman and Hall, 1973

[72] **R.H. Street**, One-dimensional non-abelian cohomology, Preprint, *Macquarie Math. Reports* **81-0024**, 1981

[73] **O.E. Villamayor and D. Zelinsky**, Galois theory for rings with finitely many idempotents, *Nagoya Math. J.* **27**, 1966, 721–731

[74] **O.E. Villamayor and D. Zelinsky**, Galois theory with infinitely many idempotents, *Nagoya Math. J.* **35**, 1969, 83–98

[75] **G.T. Whyburn** *Open mappings on locally compact spaces*, Mem. Amer. Math. Soc. **1**, 1950

Index of symbols

336

General index

admissible
 – class, 117
 relatively – adjunction, 117
algebra, 15
 étale –, 23
 split –, 98
 strongly separable –, 315
 universal –, 329
algebraic
 – algebra, 17
 – element, 1, 17
 – extension, 1
atomic subobject, 280

bikernel pair, 297
bilimit, 291

category
 – of families, 186
 2-–, 288
 extensive –, 194
 finitely well-complete –, 149
 internal –, 102, 225
 internal discrete –, 246
 Mal'tsev –, 330
 simplicial –, 320
central
 – extension, 134
 double – extension, 329
Chevalley fundamental group, 213
Chinese lemma, 19
clopen, 51, 69
codiscrete groupoid, 321
cogroupoid, 110
colimit
 universal –, 269
commutator, 128
 elementary –, 128
compact-open topology, 323
compatible family, 267

complex
 CW-–, 321
 singular –, 320
conjugate elements, 3
connected
 – component, 168, 193
 – groupoid, 213
 – object, 187, 188
 – space, 168, 186
 – subset, 168
 locally – space, 197
 locally path – space, 217
 locally simply – space, 323
 open – component, 186
 path – space, 217
 semi-locally simply – space, 324
 simply – space, 217
coproduct
 disjoint –, 270
 free – completion, 189
covering
 – map, 205
 – morphism, 207
 localic –, 295
 trivial – map, 205
 universal –, 211
CW-complex, 321

decomposition
 precategorical –, 251
derivative, 4
descent
 – datum, 297
 effective –, 91
 effective – structure, 254
 Galois –, 98
 geometric morphism of effective –, 297
 morphism of Galois –, 162, 211
 relative – morphism, 118

338